ELECTRIC SAFETY

Practice and Standards

ELECTRIC SAFETY

Practice and Standards

Mohamed El-Sharkawi
University of Washington

CRC Press
Taylor & Francis Group
Boca Raton London New York

CRC Press is an imprint of the
Taylor & Francis Group, an **informa** business

CRC Press
Taylor & Francis Group
6000 Broken Sound Parkway NW, Suite 300
Boca Raton, FL 33487-2742

First issued in paperback 2017

© 2014 by Taylor & Francis Group, LLC
CRC Press is an imprint of Taylor & Francis Group, an Informa business

No claim to original U.S. Government works
Version Date: 20130617

ISBN 13: 978-1-138-07399-9 (pbk)
ISBN 13: 978-1-4665-7149-5 (hbk)

Visit the Taylor & Francis Web site at
http://www.taylorandfrancis.com

and the CRC Press Web site at
http://www.crcpress.com

This textbook is dedicated to my wife, Fatma, and my sons, Adam and Tamer. The book is also dedicated to linemen heroes who undertake dangerous jobs to keep our light shining.

Contents

Preface

Electricity is one of the best forms of energy known to man; it is clean, readily available, quiet, and highly reliable. Equipment powered by electricity is pollution free and is more compact than those powered by other forms of energy such as gas or oil. Because of the overwhelming advantages of electricity, it has now become widely available by just the flip of a switch.

Since the early days of the electrical revolution, electricity was recognized as hazardous to humans and animals. Today, in spite of the safety products available in the market, more than 1000 people are killed every year in the United States alone due to electric shocks and several thousands more are disabled or injured.

Performing maintenance work on an electric grid is a daring job that demands a high degree of skill and attention. It is therefore handled by highly trained linemen who deserve great appreciation for their dedication to maintain our light even under harsh and hazardous conditions.

To protect the public and workers from the hazards of electricity, several regulations and standards have been developed to address every electric safety issue known to man so far. On December 29, 1970, President Nixon signed into law the Occupational Safety and Health Act. The purpose of the legislation was to "assure so far as possible every working man and woman in the Nation safe and healthful working conditions." Before this legislation was implemented, about 15,000 job-related fatalities (not all related to electricity) were reported every year in addition to 2.5 million disabilities and 300,000 cases of illnesses.

To administer the act, the secretary of labor created a new division in the Labor Department called the Occupational Safety and Health Administration (OSHA). Its mission was to assure the safety and health of America's workers by developing and enforcing standards, providing safety training, and improving safety and health in the workplace. On May 29, 1971, the first group of standards was adopted to provide baseline for safety and health protection in American workplaces.

Today, OSHA and its state partners have over 2000 inspectors and more than 200 offices throughout the United States. In addition, it uses the service of engineers, scientists, standards writers, attorneys, and other technical and support personnel.

OSHA's Code of Federal Regulation (CFR) is a set of standards that deals with several job-related issues such as electric, mechanical, and chemical safety. The electric safety code can be found in OSHA's standard 29 CFR 1910 in sections such as 1910-137, 1910-269, and 1910-302.

The Institute of Electrical and Electronic Engineers (IEEE) has several working groups and standard committees devoted to electric safety.

This organization has an international outlook, and its work is the collective judgment of its members. It is therefore considered the source of all standards worldwide. As a matter of fact, OSHA rules, in most cases, are direct extractions from IEEE standards.

Based on my experience as a safety consultant and expert witness, I find that electric safety at worksites is often left to field workers whose experience is transferred from one generation to the next. Furthermore, the safety procedures are often based on having workers memorize a set of rules. Logical explanation of these rules may not be provided to field workers or be fully understood by them. When a rule is forgotten or incorrectly extrapolated, hazardous conditions occur at the worksite. To address this problem, electrical engineers must acquire more knowledge in the area of electric safety and must be able to enhance the training modality of field workers and to communicate instructions clearly to them. This cognitive learning will certainly improve the safety at worksites. However, unfortunately, most power engineers are not familiar with the electric safety practices governing site works.

The goal of this book is to provide electrical power engineers with the knowledge and analysis they need to be well versed in electric safety and to effectively transfer their skills to the relevant workers in their organizations. This book is very practical and can be used as a textbook worldwide, since electric safety codes are very similar everywhere.

Chapter 1 covers the fundamentals of electric circuits as well as electric and magnetic fields. Chapter 2 describes the main components of substations and transmission lines. Chapter 3 deals with the biological impact of electric current. Chapter 4 covers the ground resistance calculation and measurement. In Chapter 5, the hazards of electricity are discussed. Chapter 6 covers the induced voltage on metallic objects due to electric fields. It describes a computation method to evaluate the severity of the problem. Chapter 7 is dedicated to the impact of magnetic field on the induced voltage. De-energized line work is discussed in Chapter 8. Chapter 9 covers the energized line work, while Chapter 10 deals with arc flash. The impact of atmospheric discharge on site safety is discussed in Chapter 11. Chapter 12 is dedicated to the illusive stray voltage problem. Chapter 13 discusses the electric field profile in the right-of-way and vertical clearance of towers. Finally, Chapter 14 discusses the induced voltage in metal pipes, rail tracks, and communication cables.

Besides the engineering analysis in all the aforementioned areas, the book includes several real-life case studies. In addition, it has a number of examples to show that variations in implementing electric safety procedures can create sites with various safety levels. Each chapter has a set of exercises that reinforces the knowledge assimilated. A solution manual for the exercises is also available to instructors on request through the publisher.

Author

Mohamed A. El-Sharkawi is a fellow of IEEE. He received his PhD in electrical engineering from the University of British Columbia in 1980 and joined the University of Washington as a faculty member the same year. He is presently a professor of electrical engineering in the energy area and has previously served as the associate chair and the chairman of graduate studies and research.

Professor El-Sharkawi has also served as vice president, subcommittee chair, conference chair, and task force and working group chair of several IEEE technical societies. He is a member of the IEEE Standards Committee, a member of the editorial boards of various publications, and is the associate editor of several engineering journals. He has published over 250 papers and book chapters in his research areas and has authored two textbooks: *Fundamentals of Electric Drives* and *Electric Energy: An Introduction.* He has also authored and coauthored five research books in the area of power systems and applications.

Professor El-Sharkawi has organized and taught several international tutorials on electric safety, power systems, renewable energy, induction voltage, and intelligent systems. He holds five licensed patents in the area of renewable energy, VAR management, and minimum arc sequential circuit breaker switching. He is an expert witness and trainer in the area of electric safety.

For more information, please visit El-Sharkawi's website at http://cialab. ee.washington.edu or http://cialab.org.

List of Acronyms

AAAC	All aluminum alloy conductors
AAC	All aluminum conductors
AACSR	Aluminum alloy conductor steel reinforced
AC or ac	Alternating current
ACAR	Aluminum conductor alloy reinforced
ACGIH	American Conference of Governmental Industrial Hygienists
ACSR	Aluminum conductors steel reinforced
AS	Automatic sectionalizer
AV	Artificial ventilation
CB	Circuit breaker
CFR	Code of Federal Regulations
CGS	Centimeter–gram–second
CPR	Cardiopulmonary resuscitation
CT	Current transformer
CVT	Capacitor voltage transformer
DC or dc	Direct current
DoD	Department of Defence
DS	Disconnecting switch
EC	Electrical component
EF	Electric field
EFSP	Electric field strength profile
EGC	Equipment grounding conductor
EMF	Electromagnetic field
EPA	Environmental Protection Agency
EPS	Equipotential surface
FR	Flame resistant
GFCI	Ground fault circuit interrupter
GMR	Geometric mean radius
GPR	Ground potential rise
HV	High voltage
IAD	Insulating aerial device
ICNIRP	International Commission on Non-Ionizing Radiation Protection
IEEE	Institute of Electrical and Electronics Engineers
IPE	Isolating protective equipment
INT	Isolated neutral transformer
LAB	Limited approach boundary
LIM	Line isolation monitor
LRT	Light rail traction
MAD	Minimum approach distance
MCC	Motor Control Center

MF	Magnetic field
MNC	Maximum nonfibrillation current
MOV	Metal oxide varistor
MVC	Minimum vertical clearance
NASA	National Aeronautics and Space Administration
NEC	National Electrical Code
NERC	North American Electric Reliability Council
NESC	National Electrical Safety Code
NEV	Neutral to earth voltage
NFPA	National Fire Protection Association
NFRS	Negative floating return system
NOx	Nitrogen oxide
OCPL	Overhead contact power line
OHGW	Overhead ground wire
OSHA	Occupational Safety and Health
PAB	Prohibited approach boundary
PCR	Pipelines, railroads, and communication
PE	Potential energy
PG	Protective ground or temporary ground
PPE	Personal protective equipment
PSC	Public Service Commission
PT	Potential transformer
R	Recloser
RAB	Restricted approach boundary
RH	Reference height
RMS or rms	Root mean squares
ROW	Right of Way
SF6	Sulfur hexafluoride
STATCOM	Static compensator
TASER	Thomas A Swifts Electronic Rifle
TCR	Thyristor-controlled reactor
TES	Traction electrification system
TL	Transmission line
TPS	Traction power substation
TSC	Thyristor-switched capacitor
TW	Trapezoidal wire
USDA	US Department of Agriculture
VD	Voltage divider
VF	Ventricular fibrillation
WHO	World Health Organization
xfm	Transformer

Disclaimer

The contents of this book are based on the technical opinion of the author. Approximations and rule of thumb are often used in this book. The author makes no claim, promise, or guarantee about the accuracy, completeness, or adequacy of the contents of this book and expressly disclaims liability for errors and omissions in the contents of this book. Mohamed A. El-Sharkawi and CRC Press do not and cannot know all the facts of any particular electrical situation, and, as such, the information provided in this textbook is not intended to create any express or imply warranty to the reader. The content is for informational purposes only, and the reader's adoption and/or application is performed strictly at the reader's own risk. Users should conduct an independent investigation of the facts for their particular situations and exercise their own judgment as to the appropriate solution based upon the results thereof.

1

Fundamentals of Electricity

Atoms are made up of positively charged protons and negatively charged electrons. In its equilibrium form, the negative and positive charges are equal. The electric charge is measured in coulombs and is known as the elementary charge q

$$q = 1.602 \times 10^{-19} \text{ C} \tag{1.1}$$

where q is positive for protons and negative for electrons.

The total charge of any object Q is the multiple of the elementary charge q attained by the object. The electric charge can be static or dynamic. Static charge is acquired by the object and stayed in the object without any motion. The dynamic charge moves inside the object due to external forces.

The electric current i is defined as the amount of dynamic charges Q passing a given cross section of a conductor during a given time. The current is measured in coulombs/second which is *amperes*:

$$i = -\frac{dQ}{dt} \tag{1.2}$$

The negative sign in Equation 1.2 indicates that the direction of current is opposite to the direction of the flow of electrons. This is a long-established norm in electrical engineering.

The charges acquired by an object elevate the potential of the object. Ohm's law relates this potential to the current inside the object in a linear relationship

$$v = iz \tag{1.3}$$

where
  i is the current passing through the object; its unit is ampere or A
  z is the impedance of the object; its unit is ohm or Ω
  v is the potential difference across the object; its unit is *volt* or V

The voltage and current produce electromagnetic field (EMF) near the energized line. This EMF can reach a nearby metallic object and acts as an invisible link between the energized line and the object. This property is the basis for several applications including transformers, generators, motors, radio transmission, cellular phones, and microwave communications. On the other hand, the EMFs surrounding an energized power line can elevate the voltage in nearby metallic objects to unsafe level.

The EMF is composed of two fields: electric field (EF) and magnetic field (MF). The voltage of the energized object produces EF and the current produces MF. The magnitudes of these fields are directly related to the levels of voltage and current, respectively.

1.1 Electric Fields

Coulomb's law of force states that dissimilar charges create forces that attract them, and similar charges create forces that repel them, as shown in Figure 1.1. These forces can be expressed by

$$F = \frac{Q_1 Q_2}{4\pi\varepsilon_0 d^2} \tag{1.4}$$

where
  F is the attraction or repulsion force in Newton (N)
  d is the distance between the charges in meter (m)
  Q_1 and Q_2 are the charges in coulombs (C)
  ε_0 is a constant known as absolute permittivity (8.85×10^{-12} F/m)

The force represents the impact of one electric charge on another. The direction of this force forms the line of force.

The EF is defined as the force per unit charge. The direction of the field is the direction of the force. Figure 1.2 shows the lines of forces, which are also the lines of the EF. They radially leave the positive charge and radially fall on the negative charge, as shown in Figure 1.3. This EF is also referred to as the *electric field strength E*.

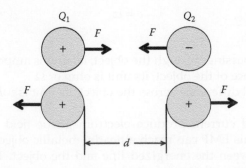

FIGURE 1.1
Coulomb's law of forces.

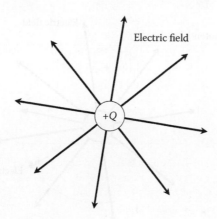

FIGURE 1.2
Electric field lines of a charge in space.

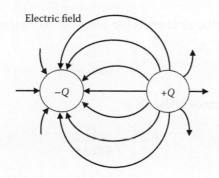

FIGURE 1.3
Electric field contours between two charges.

$$E = \frac{F}{Q} \tag{1.5}$$

where the unit of E is V/m.

The EF can also be computed by assuming an electron is placed in the line of force of a charge Q. In this case, Equation 1.4 can be written as

$$F = \frac{Qq}{4\pi\varepsilon_0 d^2} \tag{1.6}$$

where
  q is the charge of the electron
  d is the distance between the charge Q and the electron

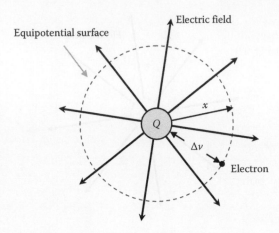

FIGURE 1.4
Equipotential surface.

The EF exerted on the electron is the force per its charge:

$$E = \frac{F}{q} = \frac{Q}{4\pi\varepsilon_0\, d^2} \tag{1.7}$$

At any arbitrary distance x, the EF is

$$E = \frac{F}{q} = \frac{Q}{4\pi\varepsilon_0\, x^2} \tag{1.8}$$

Figure 1.4 shows a charge Q and its EF. If we move around the charge while maintaining the distance x, the EF is constant.

The difference in potential between the charge Q and any arbitrary point x is defined as the cumulative effect of the EF lines from the surface of the charge to point x. In mathematical term, the potential difference is the integral of the EF strength

$$\Delta v = \int_r^x E\, dx \tag{1.9}$$

where
  r is the radius of the charge
  Δv is the potential difference between the surface of the charge and point x

The potential difference can then be computed using Equations 1.8 and 1.9.

$$\Delta v = \int_r^x \frac{Q}{4\pi\varepsilon_0\, x^2}\, dx = \left.\frac{-Q}{4\pi\varepsilon_0\, x}\right|_r^x = \frac{Q}{4\pi\varepsilon_0}\left(\frac{1}{r} - \frac{1}{x}\right) \tag{1.10}$$

Note that Δv is constant at any point on the surface of any concentric spheres as shown in Figure 1.4. Since the potential on the concentric sphere is constant, the surface of the spherical surface is called *equipotential surface*.

Example 1.1

A spherical object of 10 cm radius is suspended in air and carries a charge of 1.0 nC. Compute the potential difference between the object and a nearby point located 1 m away from the center of the charged object.

Solution:

Equation 1.10 computes the voltage between a charge and any nearby object at a distance x:

$$\Delta v = \frac{Q}{4\pi\varepsilon_0}\left(\frac{1}{r}-\frac{1}{x}\right) = \frac{10^{-9}}{4\pi\times 8.85\times 10^{-12}}\left(\frac{1}{0.1}-\frac{1}{1}\right) = 80 \text{ V}$$

Equation 1.10 is very useful and can help us compute the voltage anywhere near the charge. We can even compute the potential of the charge itself with respect to remote earth by setting $x = \infty$; hence

$$v = \frac{Q}{4\pi\varepsilon_0 r} \tag{1.11}$$

Example 1.2

A spherical object of 10 cm radius is suspended in air and carries a charge of 1.0 nC. Compute the potential of the object.

Solution:

In Equation 1.10, if we set $x = \infty$, we can compute the potential of the object as

$$\Delta v = \frac{Q}{4\pi\varepsilon_0}\left(\frac{1}{r}-\frac{1}{x}\right)$$

$$\Delta v = \frac{Q}{4\pi\varepsilon_0}\left(\frac{1}{r}\right) = \frac{10^{-9}}{4\pi\times 8.85\times 10^{-12}}\left(\frac{1}{0.1}\right) = 90 \text{ V}$$

Example 1.3

A person touches the charged object in the previous example. The object is fully discharged into the person in 1 μs. Estimate the current through the person.

Solution:

Equation 1.2 is the general form for computing the current due to discharging an object. The equation can be approximated as

$$i = -\frac{dQ}{dt} \approx -\frac{\Delta Q}{\Delta t} = -\frac{Q_1 - Q_2}{t_1 - t_2}$$

where index 1 refers to the initial conditions just before discharge and index 2 refers to the charge once it is fully dissipated. If we set $t_1 = 0$, then

$$i = -\frac{Q_1 - Q_2}{t_1 - t_2} = -\frac{10^{-9} - 0}{0 - 10^{-6}} = 1\,\text{mA}$$

1.2 Magnetic Fields

The rate of flow of charges inside a conductor is defined as the electric current, as given in Equation 1.2. The flow of current produces MF that surrounds the conductor, as shown in Figure 1.5. On the left side of the figure, the conductor carries current that is going into the page causing the MF to flow in a clockwise direction. On the right side of the figure, the current

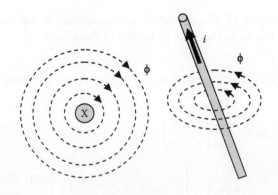

FIGURE 1.5
Magnetic field.

flowing in the conductor causes the MF to rotate counterclockwise. In both cases, the direction of the current with respect to the direction of the MF is the same.

If the conductor is surrounded by air, the relationship between the electric current and the magnetic flux is constant. This constant is known as the *inductance* of the conductor

$$L = \frac{\phi}{i} \tag{1.12}$$

where

ϕ is the MF (flux) surrounding the conductor; its unit is weber or volt second

i is the current inside the conductor

L is the inductance; its unit is henry or ohm second

The flux density B is defined as the flux that falls on a given area:

$$B = \frac{d\phi}{dA} \tag{1.13}$$

where the unit of B is Vs/m.

The flux density in a given area is the amount of flux falling perpendicularly on the area divided by the area itself. Consider the case in Figure 1.6 where the conductor carries current adjacent to a small area dA, at a distance x from the energized conductor. The flux density B_x of this area is

$$B_x = \frac{d\phi_x}{dA_x} = \frac{d\phi_x}{dl\, dx} \tag{1.14}$$

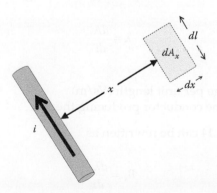

FIGURE 1.6
Flux density.

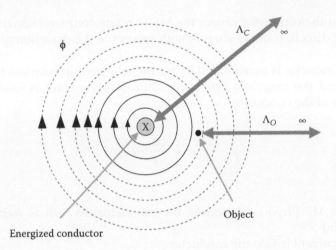

FIGURE 1.7
Flux linkage.

where
  ϕ_x is the flux falling on the area dA_x
  dl is the length of the area
  dx is the width of the area

An important variable often used in MF analysis is the *flux linkage* Λ. The term is used to describe the flux ϕ that surrounds an object. Examine Figure 1.7. The figure shows an energized conductor and an object nearby. The flux linkage of the energized conductor Λ_C is all the flux from the surface of the conductor to an infinite distance. The flux linkage of the nearby object Λ_O is all the flux that surrounds the object, which is the flux from the surface of the object to infinite distance (dashed circles).

The flux linkage Λ is often computed per unit length of the conductor

$$\lambda = \frac{d\Lambda}{dl} \tag{1.15}$$

where
  λ is the flux linkage per unit length (Vs/m)
  l is the length of the conductor producing the flux

Hence, Equation 1.14 can be rewritten as

$$B_x = \frac{d\lambda_x}{dx} \tag{1.16}$$

The MF strength H is defined as

$$H_x = \frac{B_x}{\mu_r \mu_0} \quad (1.17)$$

where

μ_r is the relative permeability; its value depends on the material carrying the flux (for air, it is equal to 1)

μ_0 is a constant known as the absolute permeability $\mu_0 = 4\pi \times 10^{-7}$; its unit is henry/m or ohm s/m

The unit of H is A/m

The relationship between the electric current and the MF strength is described by Ampere's law:

$$i = \oint H \, ds \quad (1.18)$$

Ampere's law states that the integral of the EF strength around a closed path is equal to the current enclosed by the path. The path in the Equation 1.18 is ds. At a distance x from the center of the conductor, the path is the circumference of a circle:

$$H_x(2\pi x) = i \quad (1.19)$$

Using Equations 1.16, 1.17, and 1.19, we can compute the flux linkage between any two arbitrary points a and b in air:

$$\lambda_{ab} = \int_a^b \frac{\mu_0 i}{2\pi x} \, dx = 2 \times 10^{-7} i \ln\left(\frac{b}{a}\right) \quad (1.20)$$

The voltage between the two arbitrary points can be computed using Equation 1.20:

$$v_{ab} = \frac{d\lambda_{ab}}{dt} = 2 \times 10^{-7} \ln\left(\frac{b}{a}\right) \frac{di}{dt} \quad (1.21)$$

where the unit of v_{ab} is volt/length of conductor (V/m).

Example 1.4

A conductor has a current surge of 100 kA/s. A parallel nearby de-energized conductor is located at the same height as the energized conductor. Its horizontal distance is 1 m away from the energized conductor and is located 2 m above ground. Compute the voltage on the de-energized conductor with respect to earth.

Solution:

First, we need to compute the two points where the voltage is to be computed by Equation 1.21. Point a is the location of the de-energized conductor, which is 1 m horizontally. Point b is the point on the ground below the de-energized conductor, which is

$$b = \sqrt{1^2 + 2^2} = 2.236 \text{ m}$$

In Equation 1.21, set $a = 1$ m and $b = 2.236$ m:

$$v_{ab} = 2 \times 10^{-7} \ln\left(\frac{b}{a}\right)\frac{di}{dt} = 2 \times 10^{-7} \ln\left(\frac{2.236}{1}\right)10^5 = 16.1 \text{ mV/m}$$

1.3 Alternating Current

The waveform of ac voltage is shown in Figure 1.8 and can be expressed mathematically by

$$v = V_{max} \sin \omega t \tag{1.22}$$

where
v is the instantaneous voltage
V_{max} is the peak (maximum) value of the voltage
ω is the angular frequency in radian/second (rad/s)
t is the time in s

ω is expressed by

$$\omega = 2\pi f \tag{1.23}$$

where f is the frequency of the ac waveform in cycles/s or hertz (Hz).

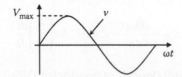

FIGURE 1.8
Sinusoidal waveform of ac voltage.

1.3.1 Root Mean Square

The waveform in Figure 1.8 changes with time and quantifying it is a challenge. The maximum value is a single-point measurement that is susceptible to noise. Using the average value is ineffective as it is always zero because of the symmetry of the waveform around the time axis. Therefore, the *root mean square* (rms) was selected by the end of the nineteenth century to quantify the ac waveforms. The concept of rms is shown in Figure 1.9 where the waveform is squared to eliminate the negative part. The average of the new waveform is then obtained, and finally the square root of this average is computed.

For the waveform in Equation 1.22, the rms value is

$$V_{rms} = \sqrt{\frac{1}{2\pi}\int_0^{2\pi} V_{max}^2 (\sin^2 \omega t)\, d\omega t} = \sqrt{\frac{1}{2\pi}\int_0^{2\pi} \frac{V_{max}^2}{2}(1-\cos 2\omega t)\, d\omega t} = \frac{V_{max}}{\sqrt{2}} \qquad (1.24)$$

Normally, the subscript rms is not used as just V implies an rms value.

Example 1.5

What is the maximum value of a sinusoidal voltage of 120 V (rms)?

Solution:

By direct substitution in Equation 1.24, we can compute the peak voltage

$$V_{rms} = \frac{V_{max}}{\sqrt{2}}$$

$$V_{max} = \sqrt{2}\, V_{rms} = \sqrt{2} \times 120 = 169.7 \text{ V}$$

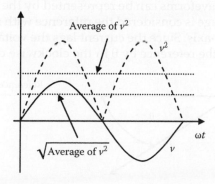

FIGURE 1.9
Concept of rms.

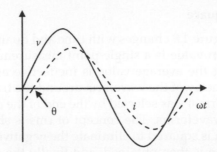

FIGURE 1.10
Lagging current.

1.3.2 Phase Shift

If the load in ac circuits is composed of elements such as resistances, capacitances, and inductances, the phase shift angle of the current θ is between −90° and +90°. Figure 1.10 shows the current and voltage waveforms of an inductive load that is composed of a resistance and inductance.

The current waveform in Figure 1.10 can be expressed mathematically by

$$i = I_{max} \sin(\omega t - \theta) \qquad (1.25)$$

where θ is the phase shift of the current with respect to voltage.

1.3.3 Concept of Phasors

Phasors are graphical representations of variables in ac circuits. They give quick information on the magnitude and phase shift of any waveform. The length of the phasor is proportional to the rms value of the waveform, and its phase angle represents the phase shift of the waveform from a reference waveform. For the waveforms in Figure 1.10, the current lags the voltage by the angle θ. These waveforms can be represented by the phasor diagram in Figure 1.11. The voltage is considered the reference with no phase shift, so it is aligned with the *x*-axis. Since the current lags the voltage by θ, the phasor for the current lags the reference by θ in the clockwise direction.

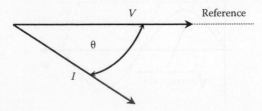

FIGURE 1.11
Phasor representation of the current and voltage in Figure 1.10.

1.3.4 Complex Numbers

Alternating current variables are analyzed mathematically by using complex number computations where any phasor is represented by a magnitude and an angle. For example, the voltage and current in Figure 1.10 or 1.11 can be represented by the following complex variable equations

$$\bar{V} = V \angle 0° \tag{1.26}$$

$$\bar{I} = I \angle -\theta° \tag{1.27}$$

where

$\bar{V}$ is the phasor representation of the voltage; it has a bar on the top indicating that the variable is a complex number

V is the magnitude of the voltage in rms

The symbol $\angle$ is followed by the angle of the phasor, which is zero for voltage and $-\theta$ for the lagging current.

The complex forms in Equations 1.26 and 1.27 are called *polar* forms (also known as *trigonometric* forms).

Example 1.6

The voltage applied across a load impedance is 120 V. The current passing through the load is 10 A and lags the voltage by 30°. Compute the load impedance.

Solution:

Using Ohm's law, the impedance Z is the voltage divided by the current. The computation must be performed using complex number mathematics. We can take the voltage as our reference; hence

$$\bar{Z} = \frac{\bar{V}}{\bar{I}} = \frac{120 \angle 0°}{10 \angle -30°} = 12 \angle 30° \, \Omega$$

1.4 Three-Phase Systems

High-power equipment such as generators, transformers, and transmission lines are built as three-phase equipments. The three-phase system had been selected as a standard for power system equipment mainly because it

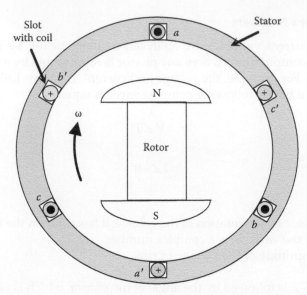

FIGURE 1.12
Three-phase generator.

produces rotating MF inside ac motors, therefore causing motors to rotate without the need for extra controls. Since electric motors constitute the majority of the electric energy consumed worldwide, having a rotating MF is a very important advantage of three-phase systems.

A simple schematic of the generator is shown in Figure 1.12. The generator consists of an outer frame called *stator* and a rotating magnet called *rotor*. At the inner perimeter of the stator, coils are placed inside slots. Each of the two slots separated by 180° houses a single coil (*a–a′*, *b–b′*, or *c–c′*). The coil is built by placing a wire inside a slot in one direction (e.g., *a*) and winding it back inside the opposite slot (*a′*). In Figure 1.12, we have three coils; each is separated by 120° from the other coils.

When the magnet spins inside the machine by an external prime mover, its MF cuts all coils and, therefore, induces a voltage across each of them. If each coil is connected to load impedance, a current flows into the load. The dot inside the coil indicates a current direction toward the reader, and the cross indicates a current in the opposite direction.

Figure 1.13 shows the produced voltage waveforms of the three coils. These are known as balanced three-phase. They are equal in magnitude but shifted by 120° from each other. To express the continuous waveforms of a balanced three-phase system mathematically, we need to select one of the waveforms as a reference and express all other waveforms in relation to it.

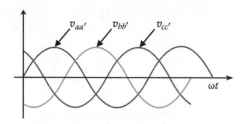

FIGURE 1.13
Three-phase waveforms.

Assuming that $v_{aa'}$ is our reference voltage, the waveforms in Figure 1.13 can be expressed by the following equations

$$v_{aa'} = V_{max} \sin(\omega t)$$

$$v_{bb'} = V_{max} \sin(\omega t - 120°) \tag{1.28}$$

$$v_{cc'} = V_{max} \sin(\omega t - 240°) = V_{max} \sin(\omega t + 120°)$$

where $v_{aa'}$, $v_{bb'}$, and $v_{cc'}$ represent the instantaneous voltages of the three phases *a*, *b*, and *c*, respectively. These voltages are known as *phase voltages*. Note that $v_{bb'}$ in Figure 1.13 lags $v_{aa'}$ by 120° and that $v_{cc'}$ leads $v_{aa'}$ by 120°. These waveforms can be written in the polar form as

$$\bar{V}_{aa'} = \frac{V_{max}}{\sqrt{2}} \angle 0° = V \angle 0°$$

$$\bar{V}_{bb'} = \frac{V_{max}}{\sqrt{2}} \angle -120° = V \angle -120° \tag{1.29}$$

$$\bar{V}_{cc'} = \frac{V_{max}}{\sqrt{2}} \angle 120° = V \angle 120°$$

As shown in Figure 1.12, the three-phase generator has three independent coils, and each coil has two terminals. To transmit the generated power from the power plant, six wires seem to be needed. Instead, the three-phase systems are connected in *wye* or *delta* configuration where only three wires are needed. These three wires are considered one circuit.

1.4.1 Wye-Connected Balanced Circuit

The wye connection is also referred to as "Y" or "star." For the waveforms in Figure 1.13, each phase voltage is generated by an independent coil inside a three-phase generator in Figure 1.12. To connect the coils in wye configuration, terminals *a'*, *b'*, and *c'* of the three coils are bonded into a common point called neutral, or *n*, as shown in Figure 1.14.

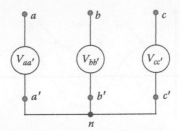

FIGURE 1.14
The connection of the three coils in wye.

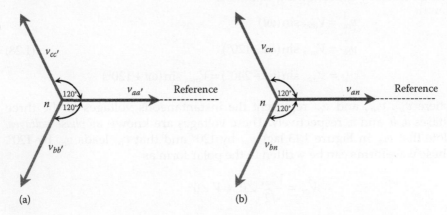

FIGURE 1.15
Phasor diagram of the three phases: (a) coil voltage and (b) coil voltage to neutral.

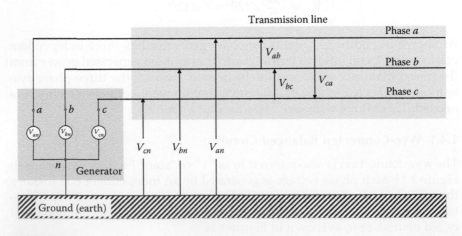

FIGURE 1.16
Three-phase wye generator connected to three-phase transmission line.

The phasor diagram of the three-phase voltages in wye configuration can be obtained by grouping the three phasors, as shown in Figure 1.15a. Since the common point is n, the phasors can be labeled $\overline{V}_{an}$, $\overline{V}_{bn}$, and $\overline{V}_{cn}$, as shown in Figure 1.15b.

The three terminals of the wye connection (a, b, and c) in Figure 1.14 are connected to a three-wire transmission line, as shown in Figure 1.16. The neutral point is often connected to the ground. The phase voltage of this system is the voltage between any line and the ground (or neutral). Hence, for a balanced system

$$V_{ph} = V_{an} = V_{bn} = V_{cn} \tag{1.30}$$

where V_{ph} is the magnitude of the phase voltage. Equation 1.29 can then be rewritten as

$$\overline{V}_{an} = V_{ph} \angle 0°$$

$$\overline{V}_{bn} = V_{ph} \angle -120° \tag{1.31}$$

$$\overline{V}_{cn} = V_{ph} \angle 120°$$

The voltage between any two lines is known as the *line-to-line* voltage. The line-to-line voltage between a and b is $\overline{V}_{ab}$; it is the potential of phase a minus the potential of phase b:

$$\overline{V}_{ab} = \overline{V}_{an} - \overline{V}_{bn} = V_{ph} \angle 0° - V_{ph} \angle -120° = \sqrt{3} V_{ph} \angle 30°$$

$$\overline{V}_{bc} = \overline{V}_{bn} - \overline{V}_{cn} = V_{ph} \angle -120° - V_{ph} \angle 120° = \sqrt{3} V_{ph} \angle -90° \tag{1.32}$$

$$\overline{V}_{ca} = \overline{V}_{cn} - \overline{V}_{an} = V_{ph} \angle 120° - V_{ph} \angle 0° = \sqrt{3} V_{ph} \angle 150°$$

The line-to-line voltages $\overline{V}_{ab}$, $\overline{V}_{bc}$, and $\overline{V}_{ca}$ of a balanced system are equal in magnitude and separated by 120° from each other. The magnitude of the line-to-line voltage is greater than the phase voltage by $\sqrt{3}$.

Example 1.7

The voltage of a Y-connected three-phase transmission line is 550 kV. Compute its phase voltage.

Solution:

The voltage of any three-phase circuit is given as line-to-line quantity V_{ll}. The phase voltage, which is also known as line-to-ground or phase-to-ground, is

$$V_{ph} = \frac{V_{ll}}{\sqrt{3}} = \frac{550}{\sqrt{3}} = 317.5 \text{ kV}$$

1.4.2 Delta-Connected Balanced Circuit

The delta (Δ) configuration can be made by cascading the coils in Figure 1.13; terminal a' is connected to b, b' is connected to c, and c' is connected to a, as shown in Figure 1.17. Since point a' is the same as b, b' is the same as c, and c' is the same as a;

$$\bar{V}_{aa'} = \bar{V}_{ab}$$

$$\bar{V}_{bb'} = \bar{V}_{bc} \qquad\qquad (1.33)$$

$$\bar{V}_{cc'} = \bar{V}_{ca}$$

The wiring connection and the phasor diagram of the delta configuration are shown in Figure 1.18. In the wiring connection, a', b', and c' are removed as they are the same points as b, c, and a respectively.

A delta-connected generator is attached to a three-phase transmission line, as shown in Figure 1.19. Notice that the delta connection has only three wires with no neutral terminal.

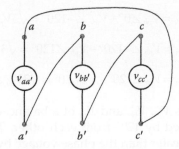

FIGURE 1.17
Delta connection.

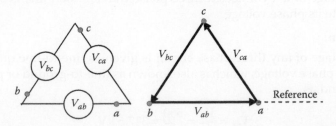

FIGURE 1.18
Delta connection and its phasor diagram.

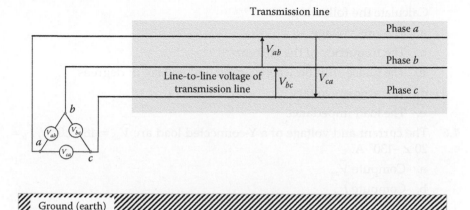

FIGURE 1.19
A three-phase generator connected to a three-phase transmission line.

Exercises

1.1 A spherical object of 10 cm radius is suspended in air and carries a charge of 1.0 nC. The center of the object is 1 m above ground. Compute the potential of the object with respect to the ground.

1.2 A spherical object of 20 cm radius is suspended in air and carries a charge of 10.0 nC. A grounded metal rod touches the object and fully discharges it in 10 ns. Estimate the current passing through the object.

1.3 A conductor carries a 60 Hz, 10 kA current. A parallel nearby small de-energized conductor is 1 m away from the energized conductor and is located 2 m above ground. Compute the voltage on the de-energized conductor with respect to earth.

1.4 The voltage and current of an electric load are

$$v = V_{max} \sin \omega t$$

$$i = I_{max} \cos(\omega t - 1.4)$$

Compute the phase shift of the current in degrees.

1.5 The voltage and current of an electric load can be expressed by

$$v = 208 \sin(377t + 0.5236)$$

$$i = 10 \sin(3778t + 0.87266)$$

Calculate the following:

a.   The rms voltage
b.   The frequency of the current
c.   The phase shift between current and voltage in degrees
d.   The average voltage
e.   The load impedance

1.6   The current and voltage of a Y-connected load are $\bar{V}_{ab} = 480 \angle 0°$ V, $\bar{I}_b =$ $20 \angle -130°$ A.

a.   Compute $\bar{V}_{an}$.
b.   Compute $\bar{I}_a$.

2

Basic Components of an Electric Grid

An electric grid is made up of a complex network of power plants, transformers, transmission and distribution lines, sensing systems, protective equipment, and control devices. The transmission lines are the links between high-voltage substations. The distribution lines are the links between high- and medium-voltage substations. They are also the links between medium- and low-voltage circuits. At power plants, the transformers are used to step up the voltage of the transmission lines to high levels (220–1200 kV) to reduce the current passing through the transmission lines, thus reducing the cross section of the transmission conductors and consequently reducing the size of the towers and increasing their spans. Near load centers, the voltage of the transmission lines is stepped down to medium levels (15–25 kV) for the distribution of power within city limits without the need for large towers. In residential areas, the voltage is further stepped down to a phase voltage of 100–240 V for household use worldwide.

The power system is extensively monitored and controlled. It has several layers of protections to minimize the effect of any damaged equipment on the system's ability to provide electricity to customers. The key devices and equipment that are related to electric safety are described in this chapter. These include power lines and substations.

Power lines include conductors, towers, transformers, fuses, sectionalizers, surge protectors, etc.

Substations include transformers, busbars, measurement equipment, circuit breakers, reclosers, bypasses, sectionalizers, isolators, load breakers, surge protectors, grounding systems, power factor corrections, voltage regulators, current limiters, auxiliary power, etc.

2.1 Power Lines

The bulk power is transmitted to distant locations within the power grid by high-voltage overhead conductors called *transmission lines*. The power lines that distribute the bulk power near and within city limits, are often called *medium-voltage distribution lines, subtransmission lines,* or *high-voltage distribution lines*. The lines that energize customers' loads are called *distribution lines*.

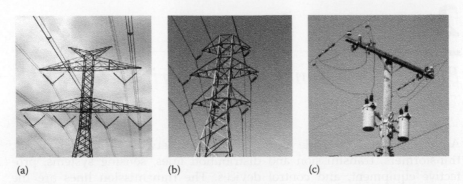

(a)   (b)   (c)

FIGURE 2.1
Transmission tower: (a) high-voltage tower, (b) medium-voltage tower, and (c) low-voltage pole.

Power line conductors are mounted on towers or poles depending on their voltage levels. For high-voltage transmission lines, the towers are normally made out of galvanized steel trusses that allow the tower to be flexible in windy conditions and in case of an earthquake. Medium-voltage transmission towers are often made of steel, concrete, or concrete-filled steel tubes. Low-voltage distribution lines are made of wood, concrete, or concrete-filled steel tubes. Some of the common towers are shown in Figure 2.1.

2.1.1 Conductors

Overhead conductors are made of aluminum strands. Some include steel reinforcement strands in the middle, as shown in Figure 2.2. The strands make the conductor flexible and the reinforcement makes it strong and rugged. The layers of a stranded conductor are spiraled in opposite directions to prevent the strands from unraveling. The most common types of conductors are given here:

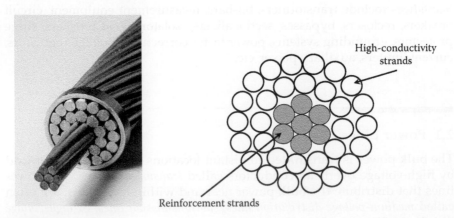

High-conductivity
strands

Reinforcement strands

FIGURE 2.2
Transmission line conductor. (Photo courtesy of Dave Bryant through Wikipedia.)

Aluminum conductor steel reinforced (ACSR): This has been used for about a century. It is made of an aluminum conductor stranded around a steel core. It can have different amount of steel reinforcement depending on the applications. For example, in areas with high ice loads, it is made with more steel reinforcement than in mild weather areas. The variations in the amount of aluminum and steel change the conductivity of the conductor.

All-aluminum alloy conductor (AAAC): This conductor is made of aluminum, magnesium, and silicon alloy. There is no need for reinforcement steel strands as the alloy is quite strong. Actually, for the same diameter, the AAAC is stronger, lighter, and more conductive than the ACSR. The AAAC, however, is more expensive, has a higher expansion coefficient and requires dampers to limit vibration in windy conditions.

Aluminum conductor alloy reinforced (ACAR): This type of conductor has concentrically stranded an aluminum around aluminum alloy core. It is a good choice for applications that demand high current-carrying capacity and low conductor weight.

Aluminum alloy conductor steel reinforced (AACSR): This is made of aluminum alloy concentrically stranded around a steel core. This is one of the toughest types of conductors.

All aluminum conductor (AAC): This is made of aluminum strands without any reinforcements. It is used in applications when little mechanical stress is exerted on the conductor and when the temperature is mild.

The cross section of the strands in most conductors is round. However, compacted conductors have strands that are shaped as *trapezoidal wires* (TW) so as not to leave any voids within the conductor's cross section. *ACAR/TW* means aluminum conductor alloy reinforced with trapezoidal-shaped strands.

Table 2.1 gives typical data for AAC. Note that the resistance/kilometer is inversely proportional to the diameter of the conductor. The allowable ampacity is based on conductor temperature of 75°C, ambient temperature of 25°C, and wind speed of 2 ft/s.

2.1.1.1 Cables

Cables are more complex than bare overhead conductors. This is mainly because the conductors in cables are very close to each other and are also close to ground potential. Therefore, the dielectric strength of the material separating conductors from ground as well as each other must be very high, much higher than air. There are four main types of cables:

1. Solid dielectric
2. Self-contained liquid-filled
3. High-pressure liquid-filled pipe
4. Submarine

TABLE 2.1

Typical Data for AAC

Code Word	Size AWG/kcmil	Stranding		Diameter		Cross-Sectional Area (in²)	Cross-Sectional Area (cm²)	Net Weight (lb/ft)	Net Weight (kg/m)	Rated Strength (lb)	Rated Strength (N)	Resistance Ω/1000ft		Resistance Ω/km		Allowable Ampacity (A)
		Number Wires	Individual Wires (in)	Complete Cable (in)	Complete Cable (cm)							DC at 20°C	AC at 75°C	DC at 20°C	AC at 75°C	
PEACHBELL	6/	7	0.0612	0.184	0.46736	0.0206	0.13290296	0.025	0.0372	563	2504.35	0.658	0.805	2.159	2.641	103
ROSE	4/	7	0.0772	0.232	0.58928	0.0328	0.21161248	0.039	0.0580	881	3918.88	0.414	0.506	1.358	1.660	138
IRIS	2/	7	0.0974	0.292	0.74168	0.0521	0.33612836	0.062	0.0923	1350	6005.10	0.26	0.318	0.853	1.043	185
PANSEY	1/	7	0.1093	0.328	0.83312	0.0657	0.42387012	0.079	0.1176	1640	7295.08	0.207	0.252	0.679	0.827	214
POPPY	1/0	7	0.1228	0.368	0.93472	0.0829	0.53483764	0.099	0.1473	1990	8851.96	0.164	0.202	0.538	0.663	247
ASTER	2/0	7	0.1379	0.414	1.05156	0.1045	0.6741922	0.125	0.1860	2510	11165.04	0.13	0.159	0.427	0.522	286
PHLOX	3/0	7	0.1548	0.464	1.17856	0.1318	0.85032088	0.158	0.2351	3040	13522.59	0.103	0.126	0.338	0.413	331
OXLIP	4/0	7	0.1739	0.522	1.32588	0.1662	1.07225592	0.199	0.2961	3830	17036.69	0.0817	0.0999	0.268	0.328	383
SNEEZEWART	/250	7	0.189	0.567	1.44018	0.1964	1.26709424	0.235	0.3497	4520	20105.96	0.0691	0.0846	0.227	0.278	425
VALERIAN	/250	19	0.1147	0.574	1.45796	0.1964	1.26709424	0.235	0.3497	4660	20728.71	0.0691	0.0846	0.227	0.278	426
DAISY	/266.8	7	0.1953	0.586	1.48844	0.2095	1.3516102	0.251	0.3735	4830	21484.91	0.0648	0.0793	0.213	0.260	443
LAUREL	/266.8	19	0.1185	0.593	1.50622	0.2095	1.3516102	0.251	0.3735	4970	22107.66	0.0648	0.0793	0.213	0.260	444
TULIP	/336.4	19	0.1331	0.666	1.69164	0.2642	1.70451272	0.316	0.4703	6150	27356.57	0.0514	0.063	0.169	0.207	513

DAFFODIL	/350	19	0.1357	0.679	1.72466	0.2749	1.77354484	0.329	0.4896	6390	28424.14	0.0494	0.0605	0.162	0.198	526
CANNA	/397.5	19	0.1447	0.724	1.83896	0.3122	2.01418952	0.373	0.5551	7110	31626.86	0.0435	0.0534	0.143	0.175	570
COSMOS	/477	19	0.1584	0.793	2.01422	0.3745	2.41676936	0.448	0.6667	8360	37187.14	0.0362	0.0472	0.119	0.155	639
ZINNIA	/500	19	0.1622	0.811	2.05994	0.3927	2.53354332	0.469	0.6979	8760	38966.42	0.0346	0.0425	0.114	0.139	658
HYACINTH	/500	37	0.1162	0.813	2.06502	0.329	2.1225764	0.469	0.6979	9110	40523.30	0.0346	0.0425	0.114	0.139	658
DAHLIA	/556.5	19	0.1711	0.856	2.17424	0.4371	2.81999436	0.522	0.7768	9750	43370.16	0.0311	0.0382	0.102	0.125	703
ORCHID	/636	37	0.1311	0.918	2.33172	0.4995	3.2225742	0.597	0.8884	11400	50709.73	0.0272	0.0335	0.089	0.110	765
PETUNIA	/750	37	0.1424	0.997	2.53238	0.5891	3.80063756	0.704	1.0477	13100	58271.71	0.023	0.0286	0.075	0.094	847
ARBUTUS	/795	37	0.1446	1.026	2.60604	0.6244	4.02837904	0.746	1.1102	13900	61830.29	0.0217	0.0271	0.071	0.089	878
MAGNOLIA	/954	37	0.1606	1.124	2.85496	0.7493	4.83418388	0.896	1.3334	16400	72950.84	0.0181	0.0226	0.059	0.074	982
HAWKWEED	/1,000	37	0.644	1.15	2.921	0.7854	5.06708664	0.939	1.3974	17200	76509.42	0.0173	0.0216	0.057	0.071	1010
BLUEBELL	/1,033.5	37	0.671	1.17	2.9718	0.8117	5.23676372	0.97	1.4435	17700	78733.53	0.0167	0.021	0.055	0.069	1031
LARKSPUR	/1,033.5	61	0.1302	1.172	2.97688	0.8117	5.23676372	0.97	1.4435	18300	81402.46	0.0167	0.021	0.055	0.069	1032
MARIGOLD	/1,113	61	0.1351	1.216	3.08864	0.8742	5.63998872	1.045	1.5551	19700	87629.97	0.0155	0.0195	0.051	0.064	1079

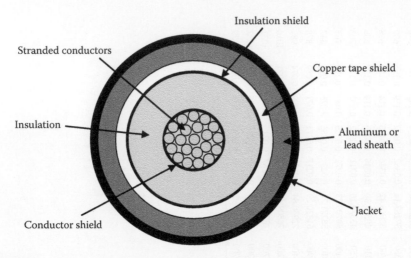

FIGURE 2.3
Single-phase cables.

Solid dielectric cables are used for applications up to 138 kV. A single-phase cable is shown in Figure 2.3. A typical cable has the following parts:

- Strands of aluminum or copper conductors to carry the current.
- Semiconducting conductor shield to protect the strands.
- Insulation surrounding the conductor to provide the dielectric strength needed. This is often made of cross-linked polyethylene or ethylene–propylene rubber.
- Semiconducting insulation shield to protect the insulation.
- Shield (screen) made up of copper tape or copper wire. This traps the electric field and reduces the magnetic field outside the cable. It can also be used for grounding purposes.
- Sheath made out of aluminum or lead alloy to insulate the cable from outside moisture.
- Jacket made out of polyolefine, polyvinyl chloride, or polyethylene to protect the cable from scratches.

The self-contained liquid-filled cables are basically the same as the solid dielectric cables with one exception—the cable has a hollow center filled with high dielectric strength liquid to remove all air pockets inside the sheath. This is because air can be ionized at high voltage, causing premature damage to the cable.

The high-pressure liquid-filled pipe cable is just three paper-insulated cables arranged in an equilateral configuration inside a buried steel pipe. The liquid is kept under pressure to as much 200 psi to ensure that all air

TABLE 2.2

Typical Data for Solid Dielectric Cable

Conductor Area (mm²)	Outer Diameter (mm)	Weight (kg/km)	Inductive Reactance (Ω/km at 60 Hz)	Capacitance (µF/km)
240	53.5	4070	0.128	0.180
300	55.6	4305	0.124	0.209
400	57.4	4635	0.118	0.238
500	60.5	5060	0.112	0.260
630	64.8	5665	0.108	0.291
800	69.2	6430	0.104	0.321
1000	73.8	7275	0.101	0.350

pockets are removed inside long cables. This cable is used for bulk power transmission at as much as 340 kV.

The submarine cables are made out of any of the aforementioned three types. However, because of the sensitive and harsh environment of the water, the lead alloy sheath is often used because of its resistance to corrosion. The sheath is usually covered by a number of outer layers made of polyethylene jackets in addition to metal wire armoring.

One of the key parameters of the cable is its capacitance, which is much higher than that for overhead transmission lines. This is because of the proximity of the cable conductor to the ground potential. Typical data for the solid dielectric cable are given in Table 2.2.

2.1.1.2 Bundled Conductors

The high voltage of the transmission line creates high electric field in its vicinity. This high electric field interacts with the electrons in the gas molecules of air (nitrogen, oxygen, carbon dioxide, etc.), creating what is known as the *corona region*. When the high voltage is positive, electrons are pulled out of the gas molecules and are attracted to the conductor. The gas molecules that have lost electrons now have a net positive charge, making them positive ions. The ionization is reversed for the high negative voltage. This exchange of electrons between the conductor and the air is known as *corona ionization*, which causes a number of problems, among which are

- *Energy loss*: The heat produced by the corona is a form of energy loss.
- *Conductor damage*: The current of the corona can create hot spots on the conductor, which can damage its surface.
- *Interference*: The corona produces a wide frequency spectrum that interferes with communication signals.
- *Noise*: Ionization at power frequency produces audible noise due to the vibration of air.

The corona is directly related to the electric field E at the surface of the conductor:

$$E = \frac{V}{d} \tag{2.1}$$

where
  V is the voltage of the conductor
  d is the diameter of the conductor
  E is the electric field at the surface of the conductor

To reduce the corona on power lines, we need to reduce the electric field. This can be done by reducing the voltage or increasing the diameter of the conductor. However, for bulk power transmission, we have to maximize the voltage to reduce the current carried by the conductor. In this way, the conductor size and overall weight is reduced, and the tower can be erected at longer spans. The second option is to increase the diameter, which seems to counter the advantage of using high voltage. To solve this problem, bundled conductors are used in high-voltage transmission lines.

Consider the three conductors in Figure 2.4. On the left side, we have a single solid conductor with a cross section determined by its current-carrying capability. Assume the voltage of this line is very high, causing excessive electric field that produces corona. As given in Equation 2.1, we should increase the diameter of the conductor to reduce the electric field at the surface of the conductor. To increase the diameter without increasing the cross-section area, we can build a conductor with a hollow core, as seen in Figure 2.4b where $d_2 > d_1$. But such a conductor would be very fragile and the core could collapse when stretched or when pressure is applied during installation. The third option is to use a few small subconductors with a total area equal to the area of the solid conductor. These subconductors are then separated by conductive spacers that increase the overall diameter of

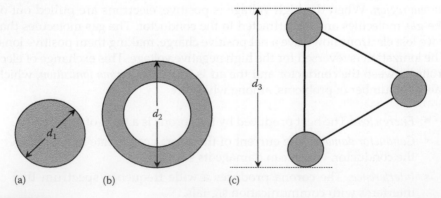

FIGURE 2.4
Concept of bundled conductors: (a) solid conductor, (b) hollow conductor, and (c) bundled conductor.

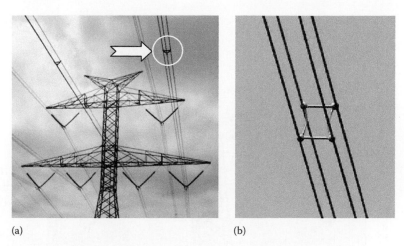

(a)   (b)

FIGURE 2.5
Bundled conductors: (a) three subconductors and (b) four subconductors.

the bundle to $d_3 > d_1$. These spacers are placed at short distances to prevent the conductors from getting closer to each other. Figure 2.5 shows two types of bundled conductors.

2.1.1.3 Static Wires

Static wires are mounted at the top of high-voltage towers in areas where lightning strikes are common. These wires are also called *shield wires* as they shield the power from lightning strikes. To be effective, they are located above the conductors and often grounded through the transmission line structure as well as at the substations. This is because lightning normally strikes the highest point with low potential. In wide towers, one static wire may not be effective in protecting all conductors; two static wires may be needed, as shown in Figure 2.6.

Without static wire, lightning that strikes a conductor can cause power line insulators to break down, creating permanent short inside them. With static wire, lightning strikes the static wire instead of the conductors. This way, the energy of the lightning bolt is dissipated along the transmission line at every tower and at the substations where the static wire is grounded.

In some designs, the static wire is isolated from the tower structure and is energized at low voltage to power communication equipment or obstruction lights near airport. The insulator withstands just a few hundreds of volts. In this small dielectric strength is too low to withstand lightning strikes. So when lightning hits the static wire, it will cause flashover across this low-voltage insulators, thus dissipating the lightning energy into the tower structures.

FIGURE 2.6
Static wires.

2.1.2 Insulators

Insulators mounted on towers and poles are designed to support the weight
of conductors in addition to all static and dynamic forces that could be exerted
on the conductor during windstorms, freezing rains, or earth movements. To
provide the needed dielectric strength, insulators are made of material such
as porcelain, glass, or glass-reinforced polymers. There are two categories
of insulators, pin or suspension, as shown in Figure 2.7. The pin-type insulators support the conductor above the insulator and the suspension-type

(a)   (b)

FIGURE 2.7
Insulators: (a) pin insulator and (b) suspension insulator.

FIGURE 2.8
Insulators with grading rings. (Courtesy of Adrian Pingstone through Wikipedia.)

insulators support the conductor below the insulator. For high-voltage transmission lines, suspension insulators are used. For distribution and subtransmission lines, both types are used. The pin insulator is manufactured as a single unit while the suspension insulator is made of multiple units that can be cascaded (chained). In this way, it can be used for a wide range of voltage as units can be added to provide the needed dielectric strength.

Some insulators are covered with a semiconductive glaze finish to allow for a few milliamps of current to flow on the surface of the insulator. The current warms the surface slightly, thus melting ice accumulation and evenly distributing voltage along the length of the chain.

Insulators used at more than 200 kV may have grading rings installed at their ends to reduce the electric field around the insulator, as shown in Figure 2.8. This reduces the corona discharge at the suspension ends.

The insulators in Figures 2.7 and 2.8 are made of disks to increase the flashover distance between the tower and the conductor. To protect against flashovers in high-voltage transmission lines, the insulators must be long enough to provide the needed dielectric strength as shown in Figure 2.9a, where *d* is the dielectric distance. Such a long insulator requires unrealistically long and wide towers to maintain the needed clearance between any two conductors, the clearance between any conductor and tower, and the clearance between any conductor and earth. However, if disk-shaped insulators are used, the distance between the tower and the conductor over the surface of the insulator can be made equal to *d* while the actual length of the insulator is reduced, as shown in Figure 2.9b. The jagged arrow shows the shortest distance along the surface of the insulation, which is longer than the actual height of the insulator. The jagged flashover distance is also known as the *creepage* distance.

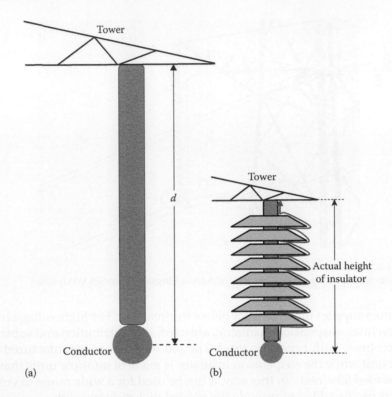

FIGURE 2.9
Power line insulator: (a) needed distance and (b) creepage distance.

2.2 Substations

Substations are the collection and distribution points of electrical energy. Generation substations receive energy from power plants and increase the voltage to the high values required by the transmission system. Subtransmission and distribution substations receive electrical energy from transmission lines and reduce the voltage to the distribution levels. The substation has a number of control, protection, and monitoring equipment as follows:

Transformers: to step up or step down the incoming and outgoing voltages

Switching equipment: automatic and manual disconnect switches, sectionalizers, isolators, bypass, load breakers, etc.

Protection equipment: circuit breakers, reclosers, fuses, surge protectors, current limiters, etc.

Measurement equipment: current transformers (CT), potential transformers (PT), etc.

Control equipment: power factor correction, voltage regulators, etc.

(a)   (b)

FIGURE 2.10
Transformers: (a) service transformer and (b) substation transformer.

2.2.1 Transformers

The main function of the transformer is to increase (step-up) or decrease (step-down) the voltage. The voltage of the transmission line must be high enough to reduce the current in its conductors. When the electric power is delivered to load centers, the voltage is stepped down to the distribution levels. When the power reaches customers' homes, the voltage is further stepped down to the household level of 100–240 V, depending on the various standards worldwide. Two main types of transformers are shown in Figure 2.10.

Large power transformers above 10 MVA often include tap changers to regulate the voltage on the load side. This is done by adjusting the turns ratio of a transformer. The tap changers can be on-load or off-load type. On-load tap changer has a diverter switch and a selector switch operating as a single unit to prevent arcing due to current interruptions.

2.2.2 Circuit Breakers

Circuit breakers (CBs) are among the most expensive equipment in substations. They are the front defense line for the power system against faults. The CB is an automatic interrupting device capable of opening (breaking) a circuit even under fault conditions. It is a highly accurate and very fast switch. For modern CBs, they can break a fault within a cycle. CB has two main settings: a tap setting and a time-dial setting. The tap setting is the level of current at which the breaker initiates the interruption. The time setting is the length of time before the breaker initiates the separation of its contactors. This is the length of time where the fault can be tolerated.

A typical application for the CB is given in Figure 2.11. The figure shows two substations connected by a transmission line. At substation A, the

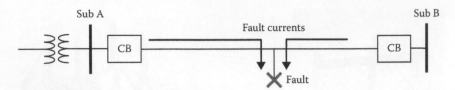

FIGURE 2.11
Faulted line with circuit breaker.

transformer feeds the transmission line through a CB. The CB is connected between the bus of substation A and the transmission line. At substation B, another CB is connected at the other end of the line. If a fault occurs somewhere in the transmission line, the CB on both ends of the line open to protect the line from excessive current that could damage its conductors and collapse the voltage of the power system.

To clear faults, the CB must extinguish the high-energy arc that is produced when the contacts of the breaker separate. The source of this arc energy is from the stored energy in all inductive elements in the power system. The presence of the arc could keep the current flowing in the circuit even after the contacts are fully separated. If not extinguished quickly, the tremendous amount of heat produced by the arc can destroy the contacts of the CB. To address this problem, we need to immerse the CB in a medium that has high dielectric strength and can provide cooling to the contacts. The most popular mediums are oil and SF_6 (gas). An SF_6 breaker is shown in Figure 2.12. At low rating, vacuum and air-blast CB can be used. In distribution systems, the breakers used are normally oil, vacuum, or air-magnetic.

CBs operate autonomously and are also remotely controlled. In the autonomous mode, the CB monitors the current passing through its contacts. When the current exceeds the tap setting and the time setting is elapsed, the CB

FIGURE 2.12
SF_6 circuit breaker.

opens its contacts. Sometimes, system operators operate CBs to reconfigure the power network or schedule maintenance job. This can be done from control centers or by substation operators.

2.2.3 Circuit Reclosers

Most distribution system faults are temporary due to events such as trees brushing power lines, bird and animal intrusion, or dust accumulation on insulators. When faults occur, utilities assume that the faults are temporary and can be cleared by themselves. Therefore, they reclose the faulted circuit automatically after a short time has lapsed, with the hope that the fault is cleared and customers' service is quickly restored. The device that automatically recloses the circuit is called the circuit recloser, which is a less accurate CB with lower interruption currents. They are used in medium- and low-voltage substations as well as distribution feeders. For single-phase distribution feeders, single-phase reclosers are used.

One of the key parameters of the recloser is its number of operations, also known as the number of shots. It is the number of times the recloser is allowed to automatically close after the fault is detected. Power system protection engineers set the number of shots based on the desired protection coordination of their system. However, most reclosers have two or three shots.

A typical application of the recloser is shown in Figure 2.13. The recloser (R) is connected to a feeder and protects the system downstream from the recloser. If a fault occurs, the recloser opens the circuit. After a set time interval, the recloser attempts to energize the system by reclosing. If the fault persists, the recloser breaks the circuit again. This process is repeated until the set time of shots elapses. In this case, the recloser assumes that the fault is permanent, and it breaks the circuit and locks itself in the open position until it is manually or remotely closed after the feeder is repaired.

2.2.4 Circuit Sectionalizers

Sectionalizers are programmable switches that can also be manually or remotely operated. They are not designed to break fault currents or even heavy normal currents. Instead, they are designed to open when there is no current in the circuit. This way, there is no need to have elaborate arc extinguish equipments. Typically, sectionalizers are used with reclosers in distribution

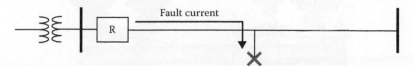

FIGURE 2.13
Feeder with recloser.

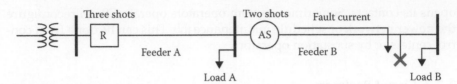

FIGURE 2.14
Feeder with recloser and sectionalizer.

systems, as shown in Figure 2.14. The figure shows a recloser (R) that is connected to feeder A and energizes loads A and B. The automatic sectionalizer (AS) energizes feeder B and load B. Assume that the recloser is set to have three shots and the sectionalizer is set to have two shots. If a permanent fault occurs in feeder B, the recloser breaks the circuit and both loads are left without power. The sectionalizer updates its counter from 0 to 1. The recloser then closes the circuit, but reopens again because the fault is still present. In this case, the sectionalizer updates its counter from 1 to 2. Since the sectionalizer has two shots, it will stay open and lock itself in the open position. The recloser still has one more shot, and it will close the circuit. But, because the sectionalizers are locked open, load A is now energized. This way, engineers limit the impact of the fault in distribution networks.

2.2.5 Isolators and Bypasses

Isolators are switches that are not designed to interrupt heavy currents, but are used mainly for maintenance purposes to isolate sections of a circuit or equipment. An isolator is shown in Figure 2.15. Isolators are remotely controlled for high-voltage applications. The isolator arms are visible to help workers verify their status from a distance.

FIGURE 2.15
Isolators in substation.

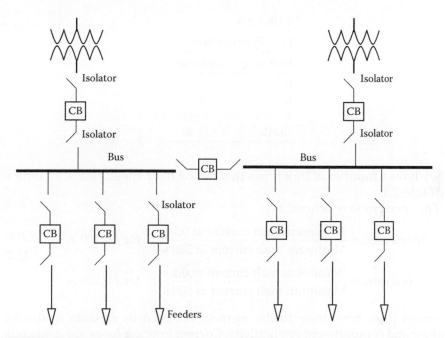

FIGURE 2.16
Circuit breakers with isolators.

Figure 2.16 shows a common use of the isolators in substations. The substation in the figure has two buses and nine CBs. Each CB is connected to two isolators, one on each side of the breaker. When a specific breaker is to be maintained, the breaker opens and the two isolators are opened next. The breaker is then isolated and safe to handle.

Bypasses are very similar to isolators, except that they are designed to short equipments. For example, when a CT is to be installed or goes through maintenance, the terminals of the CT are shorted before the work begins. This way, the transformer is not damaged when its secondary terminals are open.

2.2.6 Load Switches

Loads and capacitors in distribution networks are switched by air disconnect switches, oil switches, or vacuum switches. These switches can interrupt normal currents but cannot clear faults. They can also be used for synchronous closing applications and for isolating loads on distribution networks.

2.2.7 Fuses

Fuses are the most basic protective device in power networks. They are mainly used to protect equipment from being damaged due to overcurrent. The main types of fuses are expulsion, vacuum, and current-limiting.

TABLE 2.3

Link Type of Fuses

Link Type	Speed Ratio
H	4–6
K	6–8
T	10–13
DUAL	13–20

They have a number of link types that define their speed ratios, as shown in Table 2.3.

The speed ratio is defined as

$$\text{Speed ratio} = \frac{\text{Minimum melt current at 0.1 s}}{\text{Minimum melt current at 300 s}}; \quad \text{for } I < 100 \text{ A}$$

$$\text{Speed ratio} = \frac{\text{Minimum melt current at 0.1 s}}{\text{Minimum melt current at 600 s}}; \quad \text{for } I > 100 \text{ A} \tag{2.2}$$

Vacuum fuses have their fusible element enclosed in vacuum media for indoor and compartment applications. Current limiting fuses are nonexpulsion type and are used to limit the current into the equipment.

2.2.8 Surge Protectors

Transients in power systems are generated when the system experiences fast switching of highly inductive system or when system components are hit by a surge of energy from lighting. These transients can be very damaging for two reasons:

1. The magnitude of the voltage at several locations within the grid could reach excessive levels beyond the designed insulation levels. Thus, equipment could fail.
2. The frequency of the transients is often very high, causing the system capacitors to fail due to their low capacitive reactance at high frequency. Also, the internal parasitic capacitances of equipment such as transformers create internal short circuits.

To protect the equipment of power systems from these damaging transients, several devices that act as bypasses for the transient energy are employed. Among these devices are

- Surge protectors (surge arrester) such as the metal-oxide varistor (MOV)
- Spark gap discharge
- Overhead ground wire (OHGW), which is also known as shield wire
- Lightning poles

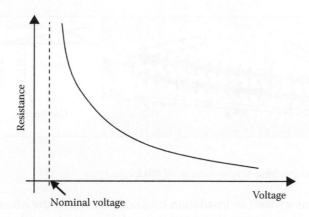

FIGURE 2.17
MOV characteristics.

The MOV is a variable resistance (varistor) that is inversely proportional to the applied voltage. The characteristic of the MOV is shown in Figure 2.17. When the system voltage is normal (nominal voltage), the MOV is essentially an open circuit because its internal resistance is very high. When the voltage is above the nominal value, the resistance of the MOV is reduced in inverse proportion to the voltage. For lightning bolts, when the voltage is excessive and could reach millions of volts, the MOV is almost a short circuit.

Figure 2.18 shows how the MOV is used to protect equipment such as transformers. The surge arrester is connected to the high-voltage side of a distribution transformer because almost all transients occur on high-voltage lines. Any large voltage spike reaching a transformer without a surge arrester will cause severe internal damage due to the parasitic capacitance

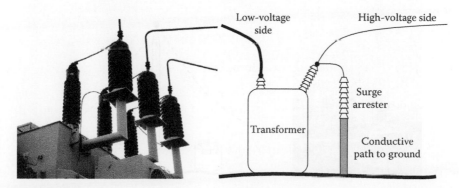

FIGURE 2.18
Transformer protection with MOV.

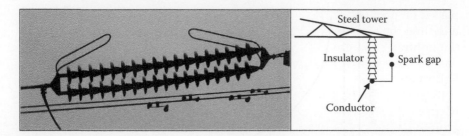

FIGURE 2.19
Spark gap across insulator. (Photo courtesy of Wikipedia.)

between turns as well as insulation breakdown. When the arrester is used, the energy in the surge is dissipated through the arrester to the ground, and the arrester prevents the voltage from reaching damaging levels.

Power line insulators are also vulnerable to large voltage surges. Most surges cause flashover on the surface of the insulators. This is a form of energy dissipation that would result in no damage to the insulators. However, if the voltage spike is very high and its frequency is very high as well, the discharge of energy could happen inside the insulator itself, causing permanent damage to the insulator as its dielectric strength is substantially reduced. The power line in this case is inoperable until the insulator is replaced. To prevent the failure of insulators, spark gap is used, as shown in Figure 2.19. The spark gap consists of two electrodes separated by a distance proportional to the needed breakdown voltage. When the surge voltage exceeds this value, an arc is created between the two electrodes, thus bypassing the insulator.

To protect equipment from being directly hit by lightning, grounded objects are placed higher than the equipment. The grounded object will attract the lightning because lightning tends to hit the highest grounded point in the area. Figure 2.20 shows a lightning rod that protects objects

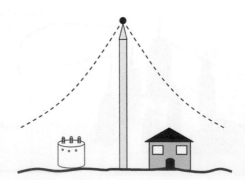

FIGURE 2.20
Lightning pole.

placed on ground. The lightning rod is solidly grounded and is capable of handling the high level of current in lightning bolts. As seen in Chapter 11, the lightning rod creates a tent-shaped protective zone represented by the dashed curves. Any object that is entirely inside the tent is protected from lightning.

To protect the transmission line from lightning strikes, static wires are used, as shown in Figure 2.6. In this case, instead of having a single point ground as with the lightning poles, we have a ground wire above the transmission line. The protective zone in this case is a tent that extends the length of the transmission line.

2.2.9 Measuring Equipment

Power system is monitored at various locations to assess the health of the grid and to identify customers' demands. Measuring devices come in a wide range of sophistication depending on the needed information and the speed by which the data are acquired. From electric safety viewpoints, the key devices are

- Voltage sensors
- Current sensors
- Frequency sensors

2.2.9.1 Voltage Sensors

Voltmeters are low dielectric strength devices. To measure the high voltage of the grid, the voltmeters are used in two main configurations: the voltage divider (VD) and the PT.

Two types of voltage dividers are shown in Figure 2.21. The one on the left is a capacitive divider and the other is a resistive divider. For the capacitive divider, two capacitors are connected in series, C_1 and C_2. The voltage of the

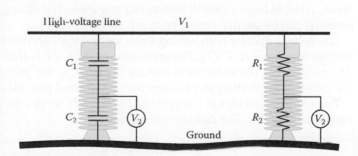

FIGURE 2.21
Voltage dividers.

high-voltage line is V_1 and the measured voltage is V_2. By placing the two capacitors in series, the total capacitance of the two is

$$C = \frac{C_1 C_2}{C_1 + C_2} \qquad (2.3)$$

Since the current in both capacitors is the same,

$$V_1 = IX_c = \frac{I}{\omega C}$$

$$\qquad (2.4)$$

$$V_2 = IX_{c2} = \frac{I}{\omega C_2}$$

where

X_c is the capacitive reactance of the two series capacitors
X_{c2} is the capacitive reactance of C_2 alone

The ratio of the line voltage V_1 to the measured voltage V_2 is

$$\frac{V_1}{V_2} = \frac{X_c}{X_{c2}} = \frac{C_2}{C} \qquad (2.5)$$

Hence, the line voltage is

$$V_1 = \frac{C_2}{C} V_2 = \frac{C_1 + C_2}{C_1} V_2 = \left(1 + \frac{C_2}{C_1}\right) V_2 \qquad (2.6)$$

To make this measurement method practical, the current passing through the capacitor must be very small so reasonable current rating capacitors can be used. This can be achieved by using capacitors with very small capacitance. Also, to have V_2 small enough so low dielectric strength voltmeters (or sensors) can be used, C_1 must be much smaller than C_2.

The voltage divider with tuning inductor and transformer is called capacitor voltage transformer (CVT). The schematic of the CVT is shown in Figure 2.22. The function of the inductor is to tune the CVT to the power frequency, and the transformer further steps down the voltage and provides isolation.

The resistive divider at the right of Figure 2.21 works the same way as the capacitive divider. The ratio of the voltages is

$$\frac{V_1}{V_2} = \frac{R_1 + R_2}{R_2} \qquad (2.7)$$

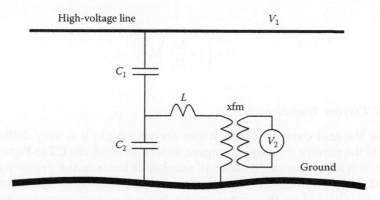

FIGURE 2.22
Capacitor voltage transformer.

Hence

$$V_1 = \left(1 + \frac{R_1}{R_2}\right)V_2 \qquad (2.8)$$

For this measurement method, the value of R_1 must be very large to limit the current, and $R_2 \ll R_1$ to limit the voltage across the voltmeter.

The PT method is shown in Figure 2.23. The device is just an auto-transformer of a total number of turns N, where

$$N = N_1 + N_2 \qquad (2.9)$$

The voltage ratio of the PT is

$$\frac{V_1}{V_2} = \frac{N_1 + N_2}{N_2} \qquad (2.10)$$

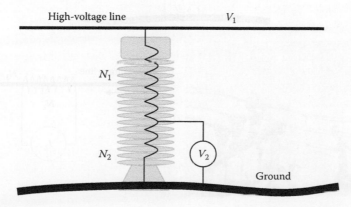

FIGURE 2.23
Potential transformer.

Hence

$$V_1 = \left(1 + \frac{N_1}{N_2}\right)V_2 \qquad (2.11)$$

2.2.9.2 Current Transformers

Because the grid currents and voltages are very high, it is very difficult to measure the current directly by ampere meters. Instead, the CT in Figure 2.24 is used. It is a transformer with small number of turns in the primary winding and a large number of turns in the secondary winding. The secondary winding is shorted by the ampere meter. Since the magnetomotive force of the transformer is constant,

$$I_1 N_1 = I_2 N_2 \qquad (2.12)$$

Hence

$$I_1 = I_2 \frac{N_2}{N_1} \qquad (2.13)$$

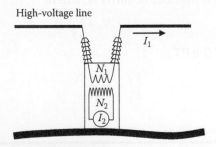

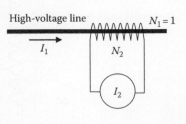

FIGURE 2.24
Current transformer.

In addition to measuring the line current, the CT provides electrical isolation between the high-voltage circuit of the grid and the measuring instrument. This is because the primary and secondary windings have their separate circuits that are only linked by the magnetic field.

The secondary winding of the CT must not open when the primary carries current; otherwise, excessive voltage across the open secondary will damage the CT. This is because the voltage/turn ratio of open secondary transformer is constant, that is,

$$\frac{V_1}{N_1} = \frac{V_2}{N_2}$$

$$V_2 = V_1 \frac{N_2}{N_1}$$

(2.14)

Since $N_2 \gg N_1$, V_2 is very high as compared to the already high voltage of the line being measured.

2.2.10 Reactive Power Control Equipment

Large- and medium-sized substations have reactive power compensators to stabilize and regulate the voltage. These devices can be manually or automatically operated. The main categories of these compensators are the fixed or switched capacitors, and the static reactive power compensators.

2.2.10.1 Fixed or Switched Capacitors

Capacitor racks similar to the one shown in Figure 2.25 are installed in substations. These capacitors are often switched mechanically only a few times a day to reduce their switching transients. These transients occur as the capacitor acts as a short circuit when it is closed and until the charge is built up. Moreover, the switching of capacitors produces voltage and current oscillations that can cause overvoltages in nearby areas. When the capacitor is switched off, the transients are minimal as the capacitor retains its charge and slowly dissipates its stored energy through its internal resistance.

2.2.10.2 Static Reactive Power Compensators

In major substations, electronic reactive power compensators may be installed to control the reactive power flow in transmission lines. The family of these devices is known as *static VAr compensator* (SVC). These devices do not have any moving parts, thus they are static. The applications of these devices include maintaining the voltage at the point of installation to constant levels, improving the power factor (pf), improving the system's stability, and correcting phase imbalance.

FIGURE 2.25
Substation capacitor racks.

The SVC family includes several designs, among which are

- Thyristor-controlled reactor (TCR): where air core reactors are switched electronically to adaptively consume excessive reactive powers.
- Thyristor-switched capacitor (TSC): where capacitors are switched electronically to adaptively deliver reactive power.
- Combinations of TCR and TSC: where reactive power can be either consumed or delivered.
- Static compensator (STATCOM): where a voltage source (battery or charged capacitor) is connected to the grid through dc/ac converter and transformer to control the reactive power. This is done by adjusting the voltage magnitude and its phase shift across the transformer winding on the converter side.

Exercises

2.1 Why are power line conductors made of strands?

2.2 Why are steel strands used in power line conductors?

2.3 What is the function of the sheath in cables?

2.4 What causes corona?

2.5 How can corona be reduced?

2.6 What is the main function of the static wire?

2.7 Why are tower insulators made as disk shapes?

2.8 What is the function of a circuit recloser?

2.9 What is the difference between isolator and bypass switches?

2.10 Compute the capacitance that limits the current in a capacitor divider to 100 mA when it is connected to a 500 kV conductor.

2.11 A 500 kV capacitor divider has 5 nf total capacitance. If the measuring instrument is rated at 200 V, compute C_2.

2.12 A CT has $N_1 = 1$ and $N_2 = 10,000$. If the measured current is 10 A, compute the current in the transmission line.

2.13 If the voltage across the primary winding of the CT in the previous example is 1 V, compute the voltage across the secondary winding if the measurement circuit is opened.

2.14 What would happen to the transformer in Exercise 2.14?

2.15 What are the expected transients when a capacitor is switched on?

2.16 What are the main advantages of static reactive power compensators?

3

Physiological Effects of Electricity

An electric shock occurs when a person becomes part of an electrical circuit causing electrical current to pass through his or her body. This current can cause a wide range of harmful effects from minor skin sensations to death.

Generally, electric shocks are divided into two categories: *secondary shocks* and *primary shocks*. A secondary shock is due to low currents; this could be painful but without major physical harm. A primary shock, however, is due to higher currents that can cause severe physical harm or even death.

Current passing through a body can cause two main problems:

- Overheating of cells leading to internal and external burns
- Interfering with body rhythmic signals that control the heart, lungs, muscles, etc.

The magnitude of the current passing through the body, the duration of the shock, and the path of the current are the three main factor that determine the severity of the electric shock. Human tolerance to electric current is not consistent. For example, people with large body mass or more fat cells tend to tolerate more current than those who are lean.

One of the unique characters of electric shocks is what is known as the *let-go current*. Small amounts of current can cause muscle spasms, causing the person to grip the energized object and not be able to release his or her grip. This prolongs the period of the electric shock and can lead to more damage to the person.

3.1 Classifications of Electric Shocks

Electric shocks are often classified based on their impacts. Secondary or primary shocks can generally be grouped into the following categories:

- *Slight sensation on skin*: This is the level at which a person may become aware of the existence of an electric current. This occurs at a very low level of currents, less than 0.5 mA in most cases.

- *Perception threshold*: It is the level of current that makes a person aware of the existence of current in his body. At this level, the current does not cause any discomfort to the person. For most people, a current of 1 mA would induce this effect.

- *Mild shock*: Current between 1 and 2 mA, a person experiences shock, but the shock is not painful and muscle control is not lost. This is considered secondary shock.

- *First-stage shock*: Current less than 10 mA can cause painful shock, which has been described to feel like a person's muscle is hit by a sledgehammer. At this level of current, the person is still able to release his or her grip. This stage is considered a secondary shock.

- *Second-stage shock*: Current between 15 and 23 mA, most people will not be able to release their grip of the energized object. This is the level of current above the let-go threshold for most people. The current also causes painful and severe shock and may cause breathing difficulty. This shock is considered secondary unless the breathing difficulty causes physiological damage.

- *Third-stage shock*: This is a primary shock. Current around 75 mA, a person may experience *ventricular fibrillation* (VF), which is the uncoordinated twitching or quivering of the heart. Contractions of the heart become chaotic and the heart no longer pumps blood effectively, and blood could even become stagnant. A few seconds of VF can lead to fainting because the brain does not receive adequate amount of blood. Eventually, VF can lead to cardiac arrest. The initial symptoms of VF are chest pain, dizziness, nausea, rapid heartbeat, and shortness of breath. Because VF continues even after the electric shock has ceased, it is a medical emergency and must be treated immediately to save the life of the person. The normal synchronous operation of the heart can be restored by a defibrillator, which produces a pulse that is strong enough to depolarize all of the heart nerve and muscle tissues at once. This is similar to computer reboot.

- *Fourth-stage shock*: This is a primary shock. At about 200 mA, for as little as 5 s, the shock is primary and VF is fatal.

- *Fifth-stage shock*: This is a primary shock. At about 4 A, the shock is primary and the heart could be paralyzed.

The Institute of Electrical and Electronics Engineers (IEEE) compiled statistics on electric shocks, which are given in Table 3.1 for secondary shocks and Table 3.2 for primary shocks. The US Department of Defense published more detailed data for secondary and primary shocks. These are shown in Table 3.3. Table 3.1 shows the levels of dc and ac that produce various types

TABLE 3.1

Effects of AC and DC Secondary Shock Current

	Current (mA)			
	DC		AC	
Reaction	**Men**	**Women**	**Men**	**Women**
No sensation on hand	1.0	0.6	0.4	0.3
Tingling (threshold of perception)	5.2	3.5	1.1	0.7
Shock: uncomfortable, muscular control not lost	9.0	6.0	1.8	1.2
Painful shock, muscular control not lost	62.0	41.0	9.0	6.0

Source: IEEE Standard 524a, 1993.

TABLE 3.2

Threshold Limit of AC Primary Shock Current

	Current (mA)			
	0.5% of Population		50% of Population	
Threshold	**Men**	**Women**	**Men**	**Women**
Let-go level: worker cannot release gripped energized object	9	6	16	10.5
Respiratory tetanus: breathing is arrested			23	15
Ventricular fibrillation	100	67		

Source: IEEE Standard 1048, 1990.

TABLE 3.3

Primary Shock Currents from the US Department of Defense (DoD)

	Current (mA)	
Effect	**Men**	**Women**
Painful and severe shock, muscular contractions, breathing difficulties	23	15
Ventricular fibrillation, threshold	75	75
Ventricular fibrillation, fatal (usually fatal for shock duration of 5 s or longer)	235	235
Heart paralysis (no ventricular fibrillation), threshold (usually not fatal; heart often restarts after short shocks)	4000	4000
Tissue burning (usually not fatal unless vital organs damaged)	5000	5000

Source: UFC 3-560-01, 2006.

of secondary shocks on men and women. Although secondary shocks can be painful, they are not life threatening. As seen in the table, women are more sensitive to electric shocks than men. This is because the body mass of women is generally less than that for men. The table also shows that the ac is more dangerous than the dc.

Table 3.2 shows the primary shocks and their corresponding ac limits. The population in the table is divided into sensitive and average sets. The sensitive population, about 0.5% of the total population, is harmed at lower currents than the rest of the population.

People have wide variation in their physical conditions. The threshold and limits given in the tables are average numbers; thus, a deviation of up to ±50% should be expected. When using the numbers in these tables, keep in mind the following two facts:

- Most of the primary shock data were collected based on animal tests and some bizarre experiments done on condemned inmates during the early days of the twentieth century. Therefore, the data should be considered as a general guideline and should not be used with scientific certainties.
- The shock classifications are associated with the total current entering the human body. This current is likely to travel through multiple paths and the amount of current reaching a specific organ such as the heart or lungs is almost impossible to predict. In some lab tests, VF was induced on animal hearts when just a few microamperes passed directly into the heart chambers.

3.2 Factors Determining the Severity of Electric Shocks

The severity of electric shocks is determined by mainly six factors:

1. Voltage level of the touched equipment
2. Amount of current passing through the person
3. Resistance of the person's body
4. Pathway of the current inside the body
5. Duration of the shock
6. Frequency of the source

3.2.1 Effect of Voltage

Electric shocks are due to currents passing through tissues. However, the magnitude of the voltage difference between any two points determines the amount of current passing through these points. The higher the voltage, the higher is the current.

Most electric shocks occur at household voltage levels between 100 and 240 V, because it is accessible by everyone, and the voltage is high enough to cause significant current to flow inside the body. At this voltage level, muscles may contract tightly to the energized object and the person may not be able to let go of the energized object.

At the transmission and distribution voltage levels, electric shocks are even more lethal because they produce high currents inside bodies. In addition to severe electrical shocks, the current may cause fierce involuntary muscle contractions that could throw the person off balance. Fortunately, the general public does not have access to high-voltage conductors, unless a person at elevated platform touches the conductor or a conductor is downed. Therefore, statistics show that deaths from high-voltage shocks are much less than that from the household level.

3.2.2 Effect of Current

Electric currents can interfere with the normal operation of the heart and lungs, causing the heart to beat out of step and the lungs to function irregularly. Moreover, electric currents passing through tissues can produce heat inside the body causing severe burns. This heat is a form of energy that can be represented by

$$E = Pt = I^2 Rt \tag{3.1}$$

where
P is the electric power caused by the current inside the body
E is the heat energy inside the body
I is the current inside the body
R is the body resistance
t is the duration of the shock

Example 3.1

Compute the electric shock energy received by a person with a body resistance of 1 kΩ. Assume the shock lasts 1, 2, 5, or 10 s.

Solution:

Use Equation 3.1:

$$E = I^2 Rt = 1000 \times I^2 t$$

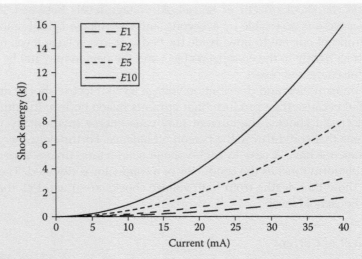

FIGURE 3.1
Energy of electric shocks for various exposure times.

We can plot the energy versus shock current, as given in Figure 3.1. *E*1, *E*2, *E*5, and *E*10 are the energy for 1, 2, 5, and 10 s shocks, respectively.

Note that the energy of the electric shock is proportional to the square of the current. A 40 mA for 10 s will produce 16 kJ of energy. This level of energy is extremely hazardous.

3.2.3 Effect of Body Resistance

Body resistance determines the amount of shock current based simply on Ohm's law. The human body is a complex system with a large number of paths of nonlinear and variable resistances. The resistance of any single path is not the same for any two people. Tissue and organ resistances depend on several conditions such as the weight of the person, the amount of body fat, the hydration and the amount of body minerals, skin condition, time of the day, etc. Moreover, the resistance of any given person fluctuates depending on the level of voltage, the frequency of the current, and the time of the day. Therefore, it is impossible to know precisely the body resistance under all possible conditions. Alternatively, a range of resistances can be obtained for a given population under given conditions. This range is used as a guideline for electrical safety analysis. However, for the design of safety equipment and work conditions, the minimum resistance in the range should be used.

A simple way to determine body resistance is to aggregate the resistances of the current paths. For example, the resistance of the torso is the aggregated resistance of organs such as the liver, the gallbladder, and the intestines in addition to skin. In general, there are six major groups of resistances:

- Skin resistance
- Internal resistance of chest
- Internal resistance of torso
- Hand resistance
- Leg resistance
- Footwear resistance

For each of these groups, the resistance could vary widely. For example, the skin may have a resistance ranging from a few ohms to 100 kΩ. Dry skin has high resistance, and broken or wet skin has low resistance. Nerves, arteries, and muscles have low resistance, while bone, fat, and tendons have high resistance.

The resistance for a given shock scenario depends entirely on the entry and exit points of the electric current. Hand-to-hand shock will involve the resistance of the hands, skin, and chest. A hand-to-foot shock will involve the resistances of the hands, skin, chest, torso, and feet.

The IEEE established the ranges for body resistances that are used as guidelines. The data are shown in Table 3.4. These numbers are used to roughly estimate the current passing through the human body. However, the variability in human body resistance could make the results inaccurate for people with body resistances outside the specified range. Therefore, when selecting electric safety equipment, the lower values of body resistance should always be used.

Because of the variability in body resistance, the values in Figure 3.2 are suggested for noninductive resistances of the human body. These values are used in most electric safety studies. Each arm is about 500 Ω, each leg is about 500 Ω, and the torso is about 100 Ω. When using these values, keep in mind the following key points:

- During the shock, damages to internal tissues and skin will lower the total body resistance, which causes the current to be even higher.
- Shoe resistance should not be considered because it is uncertain and is very low for damped leather or steel plates.

TABLE 3.4

Body Resistance in Ohms

	Hand-to-Hand		Hand-to-Feet
Resistance	Dry Condition	Wet Condition	Wet Condition
Maximum	13,500	1,260	1,950
Minimum	1,500	610	820
Average	4,838	865	1,221

Source: IEEE Standard 1048, 1990.

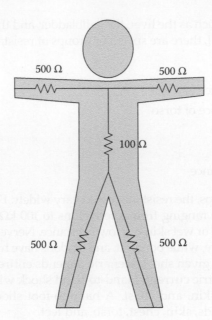

FIGURE 3.2
Suggested values for body resistances.

- Hand resistance is very low when wet, so it should be assumed to be zero.
- The resistance of the vital organs is often very low. The resistance of the human heart can be as low as 25 Ω.

Other approximate resistances that are widely used assume

- 1 kΩ between the two hand
- 1 kΩ between a hand and a foot
- 1 kΩ between the two feet

3.2.4 Effect of Current Pathway

A current passing through the skin is not as harmful as a current passing through the vital organs. The hazardous currents are the ones that pass through the heart, spinal cord, lungs, and brain and the ones that cause severe burns to the internal and external tissues.

A mere 10 μA applied directly to the hearts of animals via pacemaker catheter induced VF in lab tests. A small current in the spinal cord may also alter the respiratory control mechanism.

Another shock path is when a person walks on a surface that carries electric current, such as the case when a fault current is injected into the ground or during lightning strikes. In this case, the current passes between the two legs of the person. The path is often less hazardous as most of the current does not reach the heart, lungs, or brain. However, the current may cause the person to lose balance and fall on the surface. In this case, a more hazardous current could pass through his or her vital organs.

3.2.5 Effect of Shock Duration

The longer the duration of the shock, the higher the likelihood of death. This is because the thermal heat inside the tissues is a form of energy, and therefore proportional with time, as shown in Equation 3.1. Also, if the current interferes with the operation of the heart or lungs, the longer the duration, the higher the chance of death from respiratory failure or cardiac arrest.

Charles Dalziel carried out research on the time–current relationship for VF at 60 Hz. He, and other researchers, obtained their results by experimenting on animals of weights and organ sizes similar to humans. Although inconclusive, the data of these experiments are the best available information to date. Based on these studies, Dalziel developed the empirical formula for 99.5% of the test subjects:

$$I_{vf} = \frac{K}{\sqrt{t_{shock}}} \tag{3.2}$$

where
I_{vf} is the VF threshold current in mA
t_{shock} is the duration of exposure in seconds
K is a constant that depends on the weight of the test subject; K is also known as the energy constant (the unit of K^2 is $(mA)^2$ s)

The value of K is dependent on the weight of the person. Table 3.5 gives some average values based on tests performed on animals. The sensitive population (0.5%) has a much lower K than the average population (50%). The last column in the table is for the maximum nonfibrillation current (MNC). This is the maximum current that did not induce VF in all subjects. Because of the

TABLE 3.5

Values of K for Various Weights

	K ([mA]² s)		
Body Weight (kg)	Sensitive Population 0.5%	Average Population 50%	Maximum Nonfibrillation Current
20	78	177	61
50	185	368	116
70	260	496	157

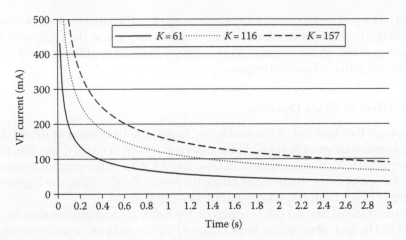

FIGURE 3.3
Ventricular fibrillation current as a function of shock duration.

variability of data even within the same weight population, the value of K to be used in electric safety studies must be the ones given by the MNC.

The Dalziel formula was based on tests limited to a time range of 0.03–3.0 s. Outside this range, the formula is not valid. Figure 3.3 depicts the Dalziel formula for different body weights. As seen in the figure, it takes a very short time to induce VF; for 100 mA, it takes about 4 ms to induce VF in a child, 1.4 s for a person weighing up to 70 kg, and 2.4 s for a person weighing more than 70 kg.

Some studies have suggested that for a 3 s shock, the VF threshold current is 100 mA. Other studies suggest the existence of two distinct thresholds based on the shock duration relative to the time of a single heartbeat. In this study, the threshold was 500 mA for a duration less than one heartbeat, and 50 mA for longer shocks. In all these studies, the survival duration is very short and cannot be utilized as a design parameter for safety equipment; rather, it should be used for academic curiosity.

The energy acquired by the body during an electric shock is a key factor that determines the severity of the shock:

$$E = \int_0^{t_{shock}} P\, dt = \int_0^{t_{shock}} I^2 R_b\, dt = I^2 R_b\, t_{shock} \tag{3.3}$$

where
P is the electric power during the shock
t_{shock} is the duration of the shock
I is the shock current
R_b is the body resistance

TABLE 3.6

Shock Energy and Its Impact

Impact	Threshold of Shock Energy
Perceptible	4 µJ
Painful	50 mJ
Harmful	408 mJ
VF	13 J

The shock energy is another indicator determining the severity of an electric shock. Table 3.6 shows general threshold values for several impacts. To put the data into perspective, the static energy from carpet shock is normally less than 10 mJ. The Taser can produce between 360 mJ and 1.6 J. The defibrillator injects 150–360 J.

For VF, the energy threshold E_{fib} can be calculated as

$$E_{fib} = \int_{0}^{t_{shock}} P\, dt = \int_{0}^{t_{shock}} I_{vf}^2 R_b\, dt = I_{vf}^2 R_b\, t_{shock} \qquad (3.4)$$

where I_{vf} is the VF current threshold. Since the Dalziel formula is

$$I_{vf}^2 t_{shock} = K^2 \qquad (3.5)$$

the VF energy is

$$E_{fib} = K^2 R_b \qquad (3.6)$$

Example 3.2

Compute the threshold of the VF energy for a lineman. Assume his body resistance is 1 kΩ.

Solution:

For linemen, $K = 157$.

$$E_{fib} = K^2 R_b = 157^2 \times 10^{-6} \times 1000 = 24.65 \text{ J}$$

Keep in mind that this is the threshold for 99.5% of the population. This means that 99.5% can withstand this level of energy. There are more sensitive subjects that can experience VF at lower energy levels. This is why IEEE suggests lowering the threshold to almost half (13 J).

3.2.6 Effect of Frequency

The frequency that impacts humans can be divided into two categories:

- Nonionizing frequency (0–100 PHz), which includes electric power, radio, microwave, and infrared frequencies
- Ionizing frequency (>100 PHz), which includes x-rays and gamma rays

From the electric safety point of view, we focus on the low end of the nonionizing frequencies (up to 10 kHz). In this range, the research indicates that human tolerance to electric shocks is a parabolic-shaped function with respect to the frequency, as shown in Figure 3.4. Unfortunately, at 50–60 Hz, humans are very vulnerable to electric current as the let-go level is the lowest. At the dc level or the high-frequency range (3–10 kHz), the let-go level is relatively high.

Table 3.7 shows additional information for 0, 60 Hz, and 10 kHz shocks. The table includes the dc and ac data given in Tables 3.1 and 3.2. Notice that while the perception threshold for men is 1.1 mA at 60 Hz, it is 12 mA at 10 kHz (more than 10 times). The let-go threshold for men at 60 Hz is 16 mA, while it is 75 mA at 10 kHz (almost 6 times).

From the electric safety perspective, humans are not really exposed to the ionizing frequency. These frequencies are from radiations such as x-rays and gamma rays. They are extremely dangerous because they can break molecules apart and can damage the DNA of the cells. Fortunately, these types of radiations are not a concern for work related to power systems.

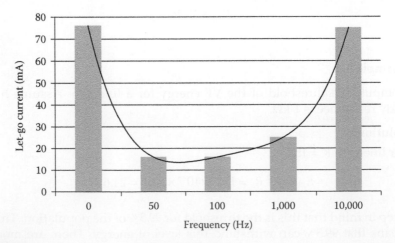

FIGURE 3.4
Let-go current as a function of frequency for men.

TABLE 3.7

Effect of Frequency on Electric Shocks

	Current (mA)					
	Direct Current		Alternating Current			
	0 Hz		60 Hz		10 kHz	
Effect	Men	Women	Men	Women	Men	Women
Slight sensation on hand	1.0	0.6	0.4	0.3	7	5
Perception threshold, median	5.2	3.5	1.1	0.7	12	8
Shock: not painful and no loss of muscular control	9	6	1.8	1.2	17	11
Painful shock: muscular control is not lost	62	41	9	6	55	37
Painful shock: let-go threshold, median	76	51	16	10.5	75	50
Painful and severe shock: breathing difficulty, muscular control lost by 99.5%	90	60	23	15	94	63

3.2.7 Effect of Impulse versus Continuous Current

Hazards from equipment capable of producing impulse discharge have been a concern for a long time. One source of these types of shocks is from the power system capacitors. The capacitors can retain energy even after the source is disconnected. Anyone who comes in contact with the terminals of the charged capacitor will receive an electric shock that is short in duration but with a high peak. Switching transients is another example. The transient occurs when the current in inductive element is interrupted, causing the voltage to rapidly surge.

George Lichtenberg discovered that when an impulse charge of enough strength is applied on dielectric objects, interesting patterns appear. These patterns are called Lichtenberg figures; one of them is shown in Figure 3.5. The pattern looks like an inverted tree with several branches. The base of the tree is the entry point of the impulse current.

Since the human skin is a sort of dielectric material, an impulse discharge through the skin exhibits the same Lichtenberg figure, as shown on the subject in Figure 3.6. These patterns are a clear indication that the injury is due to impulse discharge. This is in contrast with the excessive destruction of tissues due to shocks that are high voltage and low frequency.

In a detailed study made Dalziel in the middle of the twentieth century, several impulse charge injuries in the United States, Europe, and Japan were surveyed. Dalziel related impulse discharge energy to the injuries of the subjects. Table 3.8 shows a summary of his results.

FIGURE 3.5
Lichtenberg figure in dielectric material. (Courtesy of Bert Hickman through Wikipedia.)

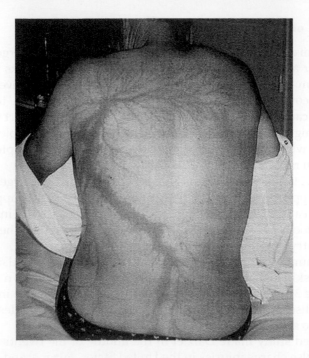

FIGURE 3.6
Lichtenberg figure on human skin. (Courtesy of the United States Department of Health and Human Services.)

TABLE 3.8

Impact of Impulse Energy Discharge

Impulse Energy (J)	Injury
25,000	Lichtenburg figures, intense muscular reaction and pain, deep burns, current pathway from abdomen to feet, temporary paralysis
5,000	Loss of sight in one eye, pain, no trace of discharge on body
720	Unconsciousness, wounds on arms and hands
520	Unconsciousness and temporary paralysis
429	Lichtenburg figures, intense muscular reaction, temporary paralysis

3.3 Symptoms and Treatments of Electric Shock

Skin burns and Lichtenberg figures are easy indicators of electric shocks. However, not all shocks produce these indicators. In such cases, people rely on other indicators, such as the following:

- Victim is unconscious but could still be breathing.
- Victim may stop breathing though the heart may continue to beat for a short time and the blood may still be circulated to the body cells after breathing has stopped.
- Victim may have cardiac arrest as well as loss of breathing.
- Victim's skin may be pale white or blue due to lack of oxygen. For people with dark skin, the color inside the mouth, under the eyelid, or under the nail can be checked.
- Victim's body may become stiff in a few minutes due to muscles reacting to the shock.

A person who has stopped breathing is not necessarily dead but is in immediate danger. This is because of the lack of oxygen feeding the cells through the blood. Since the blood contains small amount of oxygen after breathing is arrested, the body cells do not die immediately, and the victim's life may be saved by artificial ventilation (AV). AV is also known as respiration. With VP, air is forced rhythmically in and out of the lungs until natural breathing happens. If AV starts within 3 min of the shock, 70% of the victims are revived.

If the victim suffers cardiac arrest, AV alone is not enough to save the person. In this case, a cardiopulmonary resuscitation (CPR) has to be administered very quickly. If the victim's heart is in VF state, a defibrillator must be used.

3.4 Microshocks

Microshock is defined as currents between 1 and 100 μA. This level of current is not perceptible. However, it could be harmful to a person attached to medical equipment such as needles, catheters, and ECG and EEG electrodes. The effect of a microshock depends greatly on its point of entry. If the current passes through an arm or a leg, little harm, if any, would occur. However, some medical procedures require the intrusion of medical instruments, such as intravenous catheters or biopotential electrode, into the skin of the patient. Since the outer layer of the skin is the first line of defense against electric shock as it provides 10 kΩ–1 MΩ resistance, these procedures reduce or even eliminate skin resistance. This increases the risk of electrical shock. If catheters are used, fibrillation can occur for as little as 50 μA. Thus, the safety limit in hospitals is set at 10 μA.

Exercises

3.1 State seven factors that determine the severity of the electric shock when a person comes in contact with an energized conductor.

3.2 Compute the survival time of a heavy man receiving an electric shock of 100 mA.

3.3 Compute the energy received by the victim in the previous example until VF is induced.

3.4 What is the value commonly used for human resistance in electric safety studies?

3.5 At what value can current injected directly into the heart cause VF?

3.6 What is the Lichtenberg figure?

3.7 State some of the symptoms of electric shocks.

3.8 If a person stops breathing after an electric shock, can you consider him dead?

3.9 How to treat a person who stops breathing?

3.10 Can respiration revive a person experiencing cardiac arrest?

3.11 How will you treat a person having cardiac arrest?

3.12 How will you treat a person experiencing VF?

3.13 What is the current range for microshocks?

3.14 Where are microshocks likely to occur?

4

Ground Resistance

The center of the earth is the only absolute zero potential spot; any other location inside or on the earth has a nonzero potential. Hence, objects on the surface of the earth such as water pipes, building foundations, and steel structures have nonzero potential. Figure 4.1 shows a simple schematic of the earth with an object at the earth's surface. The object, which is in contact with the soil, has a potential higher than zero. Hence, the potential difference V between the object and the center of the earth can be represented by Ohm's law

$$V = IR \tag{4.1}$$

where
  I is the current flowing from the object to the center of the earth
  R is the resistance between the object and the center of the earth
  V is the potential of the object with respect to the center of the earth

The resistance R is called the *ground resistance* of the object. It plays a major role in the ability of the object to attain charges on its surface. For metallic equipment near or on the ground surface, the equipment must be grounded through an object with low ground resistance to ensure that anyone who comes in contact with the surface of the equipment is not exposed to excessively charged object with high potential.

From the safety point of view, the design of any grounding system has the following three objectives:

1. To ensure that a person in the vicinity of grounded facilities is not exposed to excessive charges, and thus is not exposed to the hazards of electric shocks.

2. To provide means to carry electric currents into earth under normal and fault conditions.

3. Under fault conditions, the grounding system must withstand the fault current without being damaged until the overcurrent protection devices isolate the fault.

There are two terminologies often used in utilities to describe two different grounding systems:

- *Intentional grounding,* which consists of ground electrodes, wires, ground grids, or any combination buried at some depth below the earth's surface. They could be connected to the exteriors of equipment

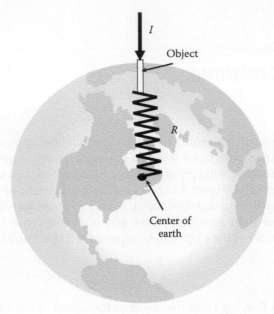

FIGURE 4.1
Ground resistance.

near ground levels. The intentional grounding system is designed to protect people and equipment by providing low resistance paths for the charges to go to the center of the earth (remote earth). This way, the charges (current) go through the system instead of the people who are touching the equipment.

- *Accidental grounding,* which is caused by unintentional contact between a charged object and the ground. This is an abnormal condition that could be hazardous. Examples of accidental grounding include downed lines and a person touching an energized object while standing on the ground.

Unfortunately, people often assume that any grounded object can be safe to touch. This assumption can be hazardous. The potential of a grounded objet is a function of the ground resistance of the object and the current passing through it to the ground. So it is possible that a high ground resistance with little current can have less voltage than a low ground resistance with high current.

4.1 Ground Resistance of Objects

The ground resistance of an object is a function of several factors such as the shape of the object, the type of ground soil, and the dampness of the soil. An exact computation of the ground resistance of an object is a very tedious task that requires the knowledge of highly varying parameters that are hard to measure under all possible conditions. Nevertheless, an approximate computation of the ground resistance of an object can be made with a reasonable degree of accuracy as discussed in the following subsections.

4.1.1 Ground Resistance of Hemisphere

Assume the simple case in Figure 4.2 of a hemisphere buried in soil. Assume that the soil is homogeneous with a resistivity ρ. If the hemisphere is connected to a conductor that carries a current I, the current enters the hemisphere and is then dispersed uniformly to earth. The current density at the surface of the hemisphere J is the current leaving the hemisphere divided by the surface area of the hemisphere:

$$J = \frac{I}{2\pi r^2} \tag{4.2}$$

Assuming the soil to be homogeneous, the current density at any point x from the center of the hemisphere and outside the hemisphere can be computed as

$$J(x) = \frac{I}{2\pi x^2}; \quad x \geq r \tag{4.3}$$

Ohm's law states that the current in a medium creates an electric field intensity E that is equal to the current density multiplied by the resistivity

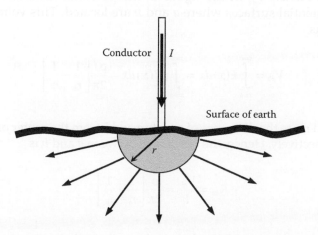

FIGURE 4.2
Current dispersion of a hemisphere.

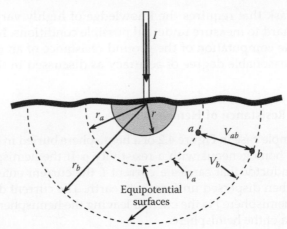

FIGURE 4.3
Equipotential surface.

of the medium ρ. Hence, the electric field intensity $E(x)$ at any distance x outside the hemisphere is

$$E(x) = \rho J(x); \quad x \geq r \tag{4.4}$$

For homogeneous soil, the potentials of all points located at the same distance from the center of the hemisphere are equal. The surface of these points is known as the *equipotential surface*, as shown in Figure 4.3. The potential difference between any two points (such as a and b) inside earth can be computed by integrating the electric field intensity between these two points. Hence, V_{ab} in the figure is the potential difference between two equipotential surfaces where a and b are located. This voltage can be computed as

$$V_{ab} = \int_{x=a}^{x=b} E(x)\,dx = \int_{x=a}^{x=b} \rho J(x)\,dx = \frac{\rho I}{2\pi}\left[\frac{1}{r_a} - \frac{1}{r_b}\right] \tag{4.5}$$

where r_a and r_b are the distances of points a and b from the center of the hemisphere, respectively. Hence, the resistance between a and b is

$$R_{ab} = \frac{V_{ab}}{I} = \frac{\rho}{2\pi}\left[\frac{1}{r_a} - \frac{1}{r_b}\right] \tag{4.6}$$

The ground resistance of the hemisphere R_g is the resistance between the surface of the hemisphere and the center of the earth. This can be computed using Equation 4.6 by setting r_a equal to the radius of the hemisphere, and r_b to ∞:

$$R_g = \frac{\rho}{2\pi r} \tag{4.7}$$

The soil resistivity is sometimes called *specific resistance*. In generic terms, the resistivity of a given material is defined as the resistance measured between the opposite sides of a 1 m long cube of the material. For soil, the resistivity is defined as the resistance measured between opposite sides of a cube of soil with a side length of 1 m.

The soil resistivity varies widely with the type of soil, the soil condition (temperature, moisture, salt content, and compactness), and the homogeneity of the soil material. Table 4.1 shows the approximate resistivity of various soil compositions. The table divides the soil into four general types: wet organic, moist, dry, and bedrock. As seen in the table, the value of the resistivity is substantially reduced if the soil is wet or organic.

Soil resistivity can vary widely within any one of the types in Table 4.1. For a more accurate value of soil resistivity, one could measure the resistivity or use more detailed data based on industrial practice such as the one in Table 4.2.

TABLE 4.1

Resistivity of Four Categories of Soil

	Soil Composition			
	Wet Organic	**Moist**	**Dry**	**Bedrock**
Resistivity ρ (Ω-m)	10	100	1000	10,000

Source: IEEE Standard 1048-1990, *Guide for Protective Grounding of Power Lines.*

TABLE 4.2

Resistivity of Soil with Specific Characteristics

Soil Type	**Soil Resistivity (Ω-m)**
Very moist, swamp like	30
Farming and clay	100
Sandy clay	150
Moist sandy	300
Concrete 1:5	400
Moist gravel	500
Dry sandy	1,000
Dry gravel	1,000
Stony soil	30,000
Rock	10^7

Example 4.1

A hemisphere 2 m in diameter is buried in wet organic soil. Compute the ground resistance of the hemisphere. Also compute the ground resistance 2, 10, and 100 m away from the hemisphere.

Solution:

$$R_g = \frac{\rho}{2\pi r} = \frac{10}{2\pi 1} = 1.6\,\Omega$$

The resistance between two points can be computed by using Equation 4.6:

$$R_{ab} = \frac{\rho}{2\pi}\left[\frac{1}{r_a} - \frac{1}{r_b}\right]$$

At 2 m

$$R_{ab2} = \frac{10}{2\pi}\left[\frac{1}{1} - \frac{1}{2}\right] = 0.8\,\Omega$$

At 10 m

$$R_{ab10} = \frac{10}{2\pi}\left[\frac{1}{1} - \frac{1}{10}\right] = 1.43\,\Omega$$

At 100 m

$$R_{ab100} = \frac{10}{2\pi}\left[\frac{1}{1} - \frac{1}{100}\right] = 1.57\,\Omega$$

Figure 4.4 shows the ground resistance between the hemisphere and points at various distances. Notice that the change in ground resistance is insignificant when the distance from the center of the hemisphere increases beyond 10 m. So practically, the ground resistance of an object is a function of the immediate distance rather than the distance to the center of the earth.

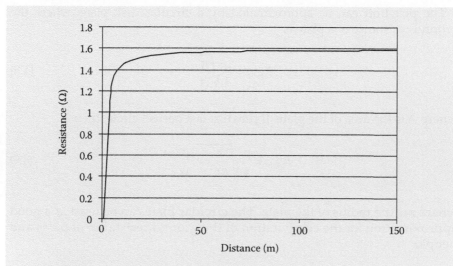

FIGURE 4.4
Ground resistance between the center of the hemisphere and points at various distances.

4.1.2 Ground Resistance of Circular Plate

Wooden poles often have copper plates at the bottom used as grounding electrode, as shown in Figure 4.5. The ground wire runs along the pole and is often covered by insulating tube. The ground wire is bonded to the plate. The plate may have cups to retain moisture to reduce the ground resistance.

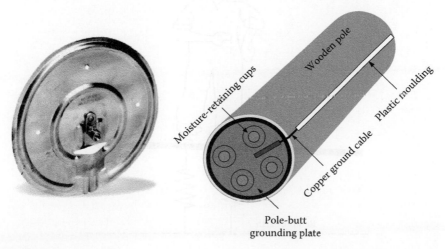

FIGURE 4.5
Pole-butt grounding plate.

The pole-butt can be approximated by a circular disk plate, where the ground resistance of a plate is

$$R_{plate} = \frac{\rho}{4}\sqrt{\frac{\pi}{A}} \tag{4.8}$$

where A is the area of the plate. If the disk is a perfect circle, then

$$R_{plate} = \frac{\rho}{4}\sqrt{\frac{\pi}{\pi r^2}} = \frac{\rho}{4r} \tag{4.9}$$

where r is the radius of the plate. The circular plate can be used as a good approximation for the computation of the ground resistance of poles and people.

4.1.3 Ground Resistance of People

A person standing on the ground has a ground resistance under his or her feet, as shown in Figure 4.6. Each foot has a ground resistance R_f between the bottom of the foot and the center of the earth. The trick is to know how to model the bottom of a shoe which has irregular shape. An approximation is often made by using a circular plate that has the same area as the footprint of an average person. Using the ground resistance

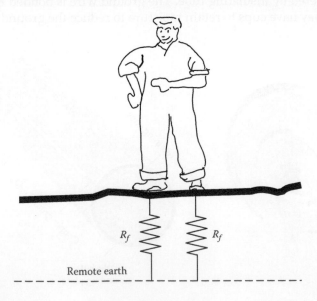

FIGURE 4.6
Feet resistance.

equation of the circular plate given in Equation 4.9, the ground resistance of a single foot R_f is

$$R_f = \frac{\rho}{4r} \qquad (4.10)$$

The area of the plate is

$$A = \pi r^2 \qquad (4.11)$$

Assuming an approximate area for a footprint of about 0.02 m², we can approximately compute the ground resistance of a foot as

$$R_f = \frac{\rho}{4\sqrt{\dfrac{A}{\pi}}} = \frac{\rho}{4\sqrt{\dfrac{0.02}{\pi}}} \approx 3\rho \qquad (4.12)$$

The total ground resistance of a walking person and a standing person are different. If you assume that the standing person has his or her feet close enough to each other, then the total ground resistance R_g is the parallel combination of two R_f:

$$R_g = \frac{R_f \times R_f}{R_f + R_f} = 0.5R_f \approx 1.5\rho \qquad (4.13)$$

Equation 4.10 assumes a uniform soil resistivity below the feet. However, it is a common practice in substations and industrial installations to spread a shallow depth (5–50 cm) of gravel on the surface to increase the contact resistance between the feet and the ground soil. Gravel is mostly a nonconductive natural stone. The foot resistance in this case depends on the resistivity of the soil, the resistivity of the gravel, as well as the thickness of the gravel.

The resistivity of the surface material must be higher than that for the ground soil. In this case, most of the ground currents flow inside the soil and very little go through the surface material. Thus, the step potential is reduced. In addition, the increase in feet resistance reduces the touch potential.

IEEE standards have recommended modifications to Equation 4.10 to include the effect of the surface material:

$$R_f = 3\rho_s C_s \qquad (4.14)$$

where
  ρ_s is the surface material resistivity
  C_s is a derating factor for the surface layer, which is a function of the thickness of the surface material in meters

The computation of C_s is tedious, but can be obtained from the curves in Figure 4.7 or by the approximate formula in Equation 4.16. h_s refers to the

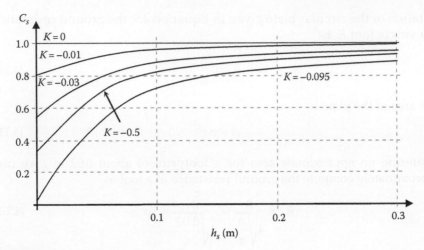

FIGURE 4.7
Derated factor for surface material.

depth of the surface material. K is the reflection factor between different materials, which is given by

$$K = \frac{\rho - \rho_s}{\rho + \rho_s} \qquad (4.15)$$

where ρ is the soil resistivity under the surface material.

The curves in Figure 4.7 can be approximated by the empirical formula

$$C_s = 1 - \frac{0.09\left(1 - \dfrac{\rho}{\rho_s}\right)}{2h_s + 0.09} \qquad (4.16)$$

Table 4.3 shows the range of resistivity of commonly used surface material. Note that the resistivity of the surface material is high even when wet with the exception of concrete; wet concrete can have extremely low resistivity.

TABLE 4.3

Resistivity of Surface Material

Surface Material	Resistivity (Ω-m)	
	Dry	**Wet**
Crusher run granite	4×10^3–1.6×10^8	$(1.2$–$1.3) \times 10^3$
Washed granite	$(2$–$190) \times 10^6$	$(5$–$10) \times 10^3$
Asphalt	$(2$–$30) \times 10^6$	$(0.01$–$6) \times 10^6$
Concrete	10^6–10^9	20–100
Limestone	7×10^6	$(2$–$3) \times 10^3$

Source: IEEE Standard 80.

Example 4.2

Compute the foot resistance for a homogeneous soil of 100 Ω-m. Repeat the calculations assuming that 0.2 m of surface material with a resistivity of 1000 Ω-m is used.

Solution:

Without surface material, the foot resistance is given by Equation 4.12:

$$R_f = 3\rho = 300 \ \Omega$$

With the surface material, the foot resistance is given by Equation 4.14:

$$R_f = 3\rho_s C_s$$

C_s can be obtained using Equation 4.16:

$$C_s = 1 - \frac{0.09\left(1 - \dfrac{\rho}{\rho_s}\right)}{2h_s + 0.09} = 1 - \frac{0.09\left(1 - \dfrac{100}{1000}\right)}{2 \times 0.2 + 0.09} = 0.834$$

The foot resistance is then

$$R_f = 3\rho_s C_s = 3 \times 1000 \times 0.834 = 2.5 \ \text{k}\Omega$$

This value is 12 times the foot resistance without surface material.

Example 4.3

Compute the foot resistance when the soil is 100 Ω-m covered with 30 cm of concrete. When dry, the concrete resistivity is 1 MΩ-m. When wet, it is 50 Ω-m.

Solution:

C_s for dry concrete using Equation 4.16 is

$$C_s = 1 - \frac{0.09\left(1 - \dfrac{\rho}{\rho_s}\right)}{2h_s + 0.09} = 1 - \frac{0.09\left(1 - \dfrac{100}{1,000,000}\right)}{2 \times 0.3 + 0.09} = 0.88$$

The foot resistance is then

$$R_f = 3\rho_s C_s = 3 \times 10^6 \times 0.88 = 2.64 \ \text{M}\Omega$$

C_s for the wet concrete using Equation 4.16 is

$$C_s = 1 - \frac{0.09\left(1 - \dfrac{\rho}{\rho_s}\right)}{2h_s + 0.09} = 1 - \frac{0.09\left(1 - \dfrac{100}{50}\right)}{2 \times 0.3 + 0.09} = 1.13$$

The foot resistance is then

$$R_f = 3\rho_s C_s = 3 \times 50 \times 1.13 = 169.5 \ \Omega$$

Note the substantial drop in foot resistance when the concrete is wet.

4.1.4 Ground Resistance of Rod

Ground rod is a simple device that provides quick low ground resistance. The rod is made out of a solid tube made of copper-clad rod, zinc-clad steel, solid copper rod, galvanized iron rod, or galvanized iron pipe. The diameter of the ground rod is at least 1.6 cm (5/8 in.). Figure 4.8 shows a schematic of a ground rod. Besides the rod itself, it has a clamp to bond the ground rod with a ground wire. The ground wire is connected to the object that needs to be grounded.

The computation of the ground resistance of a solid cylindrical rod is very involved and requires complex analysis. If we assume a uniform

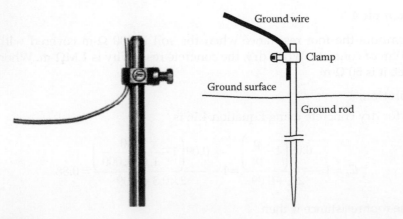

FIGURE 4.8
Ground rod.

current density outside the rod, the formula for a homogeneous soil is given by

$$R_{rod} = \frac{\rho}{2\pi L_{rod}} \left[\ln\left(\frac{4L_{rod}}{r} \right) - 1 \right] \tag{4.17}$$

where
R_{rod} is the ground resistance of the rod
L_{rod} is the depth of the buried section of the rod
r is the radius of the rod

As seen in Equation 4.17, the ground resistance of the rod is dependent on three parameters:

1. Soil type and moisture content
2. Radius of the rod
3. Depth of the rod section inserted into the ground

Since the soil resistivity at a given site is fixed and the radii of the rods are made out of a few cross sections, the main factor in reducing the ground-rod resistance is the depth of the ground rod below the surface of the earth.

Example 4.4

Compute the ground resistance of a 4 cm diameter rod. The rod is driven into a soil of 100 Ω-m resistivity at various depths ranging from 0.1 to 3 m.

Solution:

By directly substituting in Equation 4.17,

$$R_{rod} = \frac{\rho}{2\pi L_{rod}} \left[\ln\left(\frac{4L_{rod}}{r} \right) - 1 \right] = \frac{100}{2\pi L_{rod}} \left[\ln\left(\frac{4L_{rod}}{0.02} \right) - 1 \right]$$

By changing L_{rod} from 0.1 to 3 m, we can obtain the curve in Figure 4.9. Note that the ground resistance of the rod is high if the rod is not inserted the full 3 m into the ground.

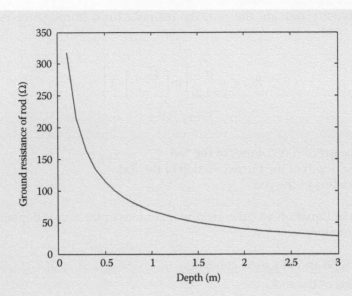

FIGURE 4.9
Ground resistance of rod as a function of depth.

As seen in Example 4.4, the ground resistance of the rod is highly dependent on the depth of the rod into the ground. Ground rods are commonly used to ground objects such as equipment and trucks. If the ground rod is not driven deep enough into the soil, its ground resistance could be high enough to create hazards to personnel touching the object.

4.1.4.1 Ground Resistance of Rods Inserted in Concrete

In permanent installations for buildings and concrete structures, the ground rods could be inserted in concrete shells. In some cases, the reinforcement steel bars of the structure are used as ground electrodes.

If a single ground rod is in concrete shell inserted into the ground, the resistance of the ground rod given in Equation 4.17 is modified to take into account the concrete resistivity:

$$R_{rod-c} = \frac{1}{2\pi \, L_{rod}} \left[\rho \left(\ln \left(\frac{4 L_{rod}}{r_c} \right) - 1 \right) + \rho_c \ln \left(\frac{r_c}{r} \right) \right] \qquad (4.18)$$

where
ρ is the resistivity of the soil
ρ_c is the resistivity of the concrete
L_{rod} is the length of the rod below ground surface
r is the radius of the ground rod
r_c is the radius of the concrete shell

4.1.4.2 Ground Resistance of a Cluster of Rods

When a single ground rod is not enough to provide low ground resistance, multiple ground electrodes can be used, as shown in Figure 4.10. The electrodes form a cluster of parallel connected rods, and the cluster must satisfy the following conditions:

1. All electrodes must equally share the fault currents.
2. No two electrodes share the spear of influence.

To achieve the first condition, the following conditions must be satisfied:

- All electrodes must be of the same type.
- All electrodes must be inserted in the same soil.
- All electrodes must be driven to the same depth.

To achieve the second condition, the distance between electrodes must be long enough so that the current density at the midpoint is negligibly low. As a rule of thumb, the distance must be at least equal to the depth of the driven rod.

If the ground rods are clustered linearly, the total resistance ($R_{cluster}$) is just the parallel resistance of all rods. In this case, Equation 4.17 can be modified as

$$R_{cluster} = \frac{\rho}{2\pi \, n \, L_{rod}} \left[\ln\left(\frac{4L_{rod}}{r} \right) - 1 \right] \tag{4.19}$$

where n is the number of rods in the linear cluster.

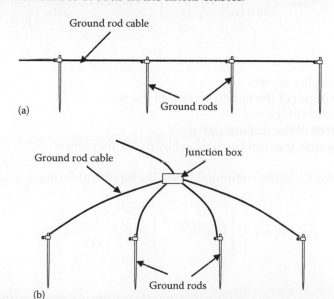

FIGURE 4.10
Ground rods in linear cluster: (a) cascade structure and (b) parallel structure.

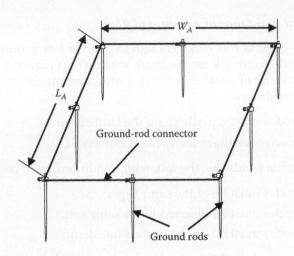

FIGURE 4.11
Ground rods in rectangular cluster.

If the ground rods are clustered in a rectangular form as shown in Figure 4.11, Equation 4.17 is modified as

$$R_{cluster} = \frac{\rho}{2\pi n L_{rod}} \left[\ln\left(\frac{4L_{rod}}{r}\right) - 1 + 2L_{rod} K_1 \frac{\left(\sqrt{n}-1\right)^2}{\sqrt{A}} \right] \qquad (4.20)$$

where
n is the number of rods
L_{rod} is the depth of the buried section of the rod
r is the radius of the rod
A is the area of the rectangular array
K_1 is a variable that can be obtained by experimentation

Alternatively, K_1 can be computed using the empirical formula

$$K_1 = 1.37 \left[1 - e^{-\left(\sqrt{A}/h\right)^{0.25}} \right] - 0.045 \left(\frac{L_A}{W_A} - 1 \right) \qquad (4.21)$$

where
h is the depth of the top of the rod if the rod is fully inserted into the ground
L_A is the length of the array area
W_A is the width of the array area

If the rods are not fully inserted into the soil, the depth of the top of the rods is zero. Hence

$$K_1 = 1.37 - 0.045\left(\frac{L_A}{W_A} - 1\right) \tag{4.22}$$

and if the arrangement is a perfect square,

$$K_1 = 1.37 \tag{4.23}$$

Example 4.5

A cluster of rods is arranged in a rectangular shape of 10×15 m. The separation between any two adjacent rods is 5 m. Each rod is 4 cm in diameter and is 3 m long. The top of each rod is 1 m below the ground surface. Assume a soil resistivity is 100 Ω-m, compute the ground resistance of the cluster.

Solution:

The area of the cluster is

$$A = L_A \times W_A = 15 \times 10 = 150 \text{ m}^2$$

If the separation between any two adjacent rods is 5 m, then the number of rods along the peripheral of the cluster is 10.

Use Equation 4.21 to compute K_1:

$$K_1 = 1.37\left[1 - e^{-\left(\sqrt{A}/h\right)^{0.25}}\right] - 0.045\left(\frac{L_A}{W_A} - 1\right)$$

$$= 1.37\left[1 - e^{-\left(\sqrt{150}/1\right)^{0.25}}\right] - 0.045\left(\frac{15}{10} - 1\right) = 1.137$$

Use Equation 4.20 to compute the resistance of the cluster:

$$R_{cluster} = \frac{\rho}{2\pi n L_{rod}}\left[\ln\left(\frac{4L_{rod}}{r}\right) - 1 + 2L_{rod}K_1\frac{\left(\sqrt{n} - 1\right)^2}{\sqrt{A}}\right]$$

$$= \frac{100}{2\pi 10 \times 3}\left[\ln\left(\frac{4 \times 3}{0.02}\right) - 1 + 2 \times 3 \times 1.137\frac{\left(\sqrt{10} - 1\right)^2}{\sqrt{150}}\right] = 4.245\ \Omega$$

Notice that the ground resistance of such effective grounding system is very low.

If the cluster is arranged in circular shape, the area in Equations 4.20 and 4.21 is the area of the circle, and the ratio $L_A/W_A = 1$ in Equation 4.21. In this case, K_1 is

$$K_1 = 1.37\left[1 - e^{-(\sqrt{A}/h)^{0.25}}\right] \tag{4.24}$$

4.1.5 Ground Resistance of Buried Wires

Buried wires are often used to form a ground grid in substations and permanent installations. The equation of the ground resistance of a buried wire is

$$R_{wire} = \frac{\rho}{2\pi L_c}\left(\ln\left(\frac{L_c}{r}\right) + \ln\left(\frac{L_c}{2d}\right)\right) \tag{4.25}$$

where
L_c is the length of the conductor (wire)
r is the radius of the wire
d is the depth at which the wire is buried

In areas where ground is hard and digging is difficult, ground rods can be installed horizontally at relatively shallow depth. In this case, we can treat the ground rod as a buried wire and the ground resistance of the horizontal rod can be computed by Equation 4.25.

Example 4.6

A 2.54 cm (1 in.) diameter ground rod is inserted 2.44 m (8 ft) into ground soil of 100 Ω-m resistivity. Compute the ground resistance of the rod.

If the ground rod is installed horizontally, compute the depth of the rod's trench that produces the same ground resistance as the vertically installed ground rod.

Solution:

For the first part, let us compute the ground resistance of the rod when it is fully inserted vertically. To do this, we need to use Equation 4.17.

$$R_{rod} = \frac{\rho}{2\pi L_{rod}}\left[\ln\left(\frac{4L_{rod}}{r}\right) - 1\right] = \frac{100}{2\pi \times 2.44}\left[\ln\left(\frac{4 \times 2.44}{0.0127}\right) - 1\right] = 36.82\ \Omega$$

To achieve the same ground resistance by installing the ground rod horizontally, we need to use Equation 4.25 for buried wire:

$$R_{wire} = \frac{\rho}{2\pi\,L_c}\left(\ln\left(\frac{L_c}{r}\right) + \ln\left(\frac{L_c}{2d}\right)\right)$$

$$36.82 = \frac{100}{2\pi\,2.44}\left(\ln\left(\frac{2.44}{0.0127}\right) + \ln\left(\frac{2.44}{2d}\right)\right)$$

$$36.82 = \frac{100}{2\pi\,2.44}\left(\ln\left(\frac{2.44}{0.0127}\right) + \ln(2.44) - \ln(2d)\right)$$

$$\ln(2d) = 0.505$$

Hence

$$d = \frac{1}{2}e^{0.505} = 0.83 \text{ m}$$

As seen in this example, two different orientations of the ground rod can produce the same ground resistance if the correct depths are maintained. The real issue is whether digging a 2.44 m trench at 0.83 m depth is a simpler job than driving the rod 2.44 m vertically into the ground.

4.1.5.1 Ground Resistance of Buried Wires in Grid Arrangement

The ground grid is a common system for substations and permanent installations. It consists of a group of ground wires connected in mesh structure, as shown in Figure 4.12. The wires are bonded together at every intersection

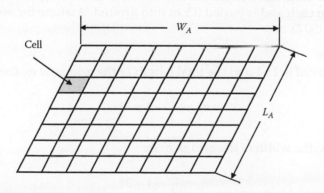

FIGURE 4.12
Ground grid.

point to form a number of cells. The entire grid is buried underground at a given depth that determines the ground resistance of the grid.

The equation that determines the ground resistance of the grid is given by

$$R_{grid} = \rho \left[\frac{1}{L_{grid}} + \frac{1}{\sqrt{20A}} \left(1 + \frac{1}{1 + d\sqrt{20/A}} \right) \right] \qquad (4.26)$$

where

L_{grid} is the combined length of all conductors forming the grid
A is the grid area
d is the depth at which the grid is buried

A and L_{grid} can be computed by

$$A = L_A \times W_A \qquad (4.27)$$

$$L_{grid} = n_L \times L_A + n_W \times W_A \qquad (4.28)$$

where

L_A is the length of the grid
W_A is the width of the grid
n_L is the number of conductors with length L_A
n_W is the number of conductors with length W_A

Example 4.7

Compute the ground resistance of a 30×20 m grid. The grid forms cells of 1×1 m each and is buried 0.5 m into ground. Assume the soil resistivity is 150 Ω-m.

Solution:

Since the cell is 1×1 and the total length of the grid is 30 m, then

$$n_W = \frac{30}{1} + 1 = 31$$

Similarly, the width of the grid is 20 m:

$$n_L = \frac{20}{1} + 1 = 21$$

Using Equations 4.26 through 4.28, we can compute the ground resistance of the grid

$$A = L_A \times W_A = 30 \times 20 = 600 \text{ m}^2$$

$$L_{grid} = n_L \times L_A + n_W \times W_A = 21 \times 30 + 31 \times 20 = 1250 \text{ m}$$

$$R_{grid} = \rho \left[\frac{1}{L_{grid}} + \frac{1}{\sqrt{20A}} \left(1 + \frac{1}{1 + d\sqrt{20/A}} \right) \right]$$

$$R_{grid} = 150 \left[\frac{1}{1250} + \frac{1}{\sqrt{20 \times 600}} \left(1 + \frac{1}{1 + 0.5\sqrt{20/600}} \right) \right] = 2.74 \ \Omega$$

The number of conductors determines to a large extent the resistance of the grid. As seen in Equation 4.28, the higher the number, the larger is the combined length of the grid conductors. When L_{grid} increases, the ground resistance of the grid decreases. Figure 4.13 shows a case of a square grid with different number of conductors along the length and width of the grid.

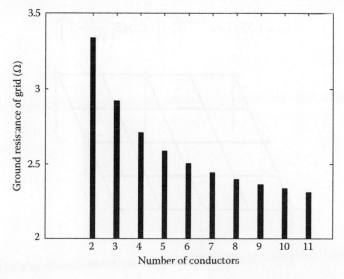

FIGURE 4.13
Grid resistance as a function of the number of conductors.

4.1.5.2 Ground Resistance of Combined Grids and Rods

As seen in Example 4.7, the grid structure has very low ground resistance as compared to the ground rods and hemispheres. Therefore, it is one of the best methods for grounding and is used in substations and permanent installations. To lower the ground resistance even further, the grid is combined with ground rods, as shown in Figure 4.14.

The ground resistance of the grid–rod combination is given by

$$R_{grid+rod} = \frac{R_1 R_2 - R_m^2}{R_1 + R_2 - 2R_m} \tag{4.29}$$

where

$$R_1 = \frac{\rho}{\pi L_{grid}}\left[\ln\left(\frac{2L_{grid}}{\alpha}\right) + \frac{L_{grid}}{\sqrt{A}}K_1 - K_2\right] \tag{4.30}$$

$$R_2 = \frac{\rho}{2\pi n L_{rod}}\left[\ln\left(\frac{4L_{rod}}{r_{rod}}\right) - 1 + \frac{2L_{rod}}{\sqrt{A}}K_1\left(\sqrt{n}-1\right)^2\right] \tag{4.31}$$

$$R_m = \frac{\rho}{\pi L_{grid}}\left[\ln\left(\frac{2L_{grid}}{L_{rod}}\right) + \frac{L_{grid}}{\sqrt{A}}K_1 - K_2 + 1\right] \tag{4.32}$$

K_1 and K_2 are coefficients that can be approximated by the empirical formulas

$$K_1 = 1.37\left[1 - e^{-\left(\sqrt{A}/h\right)^{0.25}}\right] - 0.045\left(\frac{L_A}{W_A} - 1\right) \tag{4.33}$$

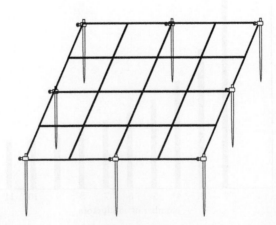

FIGURE 4.14
Ground grid with rods.

$$K_2 = 5.65 + 0.14\left(\frac{L_A}{W_A} - 1\right) \qquad\qquad \text{when } h = 0$$

(4.34)

$$K_2 = 5.65\left[1 - e^{-(\sqrt{A}/h)^{0.25}}\right] + 0.055\left(\frac{L_A}{W_A} - 1\right) \quad \text{when } h > 0$$

where

ρ is the soil resistivity
L_{grid} is the total length of all grid conductors
A is the area covered by the grid
r_c is the radius of the conductor
d is the depth at which the conductor is buried

$$\alpha = \begin{cases} r_c & \text{for conductors on earth's surface} \\ \\ \sqrt{2r_c d} & \text{for conductors buried at depth } d \end{cases}$$

h is the depth of the top of the rod
L_A is the length of the grid area
W_A is the width of the grid area
n is the number of rods used in the grid
L_{rod} is the depth of the buried section of the rod
r_{rod} is the radius of the ground rod

Example 4.8

For the grid in Example 4.7, 10 ground rods are added to the grid. Assume that each rod is 2.54 cm (1 in.) in diameter and 2.44 m (8 ft) in length. Each rod is inserted 3 m into the ground. Compute the ground resistance of the rod.

Solution:

From Example 4.7, we have

$$A = 600 \text{ m}^2 \quad \text{and} \quad L_{grid} = 1250 \text{ m}$$

Since the length of the rod is 2.44 m and the rod is inserted 3 m into the ground,

$$h = 3.0 - 2.44 = 0.56 \text{ m}$$

$$K_1 = 1.37\left[1 - e^{-(\sqrt{A}/h)^{0.25}}\right] - 0.045\left(\frac{L_A}{W_A} - 1\right)$$

$$K_1 = 1.37\left[1 - e^{-(\sqrt{600}/0.56)^{0.25}}\right] - 0.045\left(\frac{30}{20} - 1\right) = 1.24$$

$$K_2 = 5.65\left[1 - e^{-\left(\sqrt{A}/h\right)^{0.25}}\right] + 0.055\left(\frac{L_A}{W_A} - 1\right)$$

$$K_2 = 5.65\left[1 - e^{-\left(\sqrt{600}/0.56\right)^{0.25}}\right] + 0.055\left(\frac{30}{20} - 1\right) = 5.25$$

$$R_1 = \frac{\rho}{\pi L_{grid}}\left[\ln\left(\frac{2L_{grid}}{\sqrt{2r_c\,d}}\right) + \frac{L_{grid}}{\sqrt{A}}K_1 - K_2\right]$$

$$R_1 = \frac{150}{\pi 1250}\left[\ln\left(\frac{2\times1250}{\sqrt{2\times0.00635\times0.5}}\right) + \frac{1250}{\sqrt{600}}1.24 - 5.25\right] = 2.61\ \Omega$$

$$R_2 = \frac{\rho}{2\pi n L_{rod}}\left[\ln\left(\frac{4L_{rod}}{r_{rod}}\right) - 1 + \frac{2L_{rod}}{\sqrt{A}}K_1\left(\sqrt{n} - 1\right)^2\right]$$

$$R_2 = \frac{150}{2\pi\times10\times2.44}\left[\ln\left(\frac{4\times2.44}{0.0127}\right) - 1 + \frac{2\times2.44}{\sqrt{600}}1.24\left(\sqrt{10} - 1\right)^2\right] = 6.65\ \Omega$$

$$R_m = \frac{\rho}{\pi L_{grid}}\left[\ln\left(\frac{2L_{grid}}{L_{rod}}\right) + \frac{L_{grid}}{\sqrt{A}}K_1 - K_2 + 1\right]$$

$$R_m = \frac{150}{\pi 1250}\left[\ln\left(\frac{2\times1250}{2.44}\right) + \frac{1250}{\sqrt{600}}1.24 - 5.25 + 1\right] = 2.52\ \Omega$$

$$R_{grid+rod} = \frac{R_1 R_2 - R_m^2}{R_1 + R_2 - 2R_m}$$

$$R_{grid+rod} = \frac{2.61\times6.65 - 2.52^2}{2.61 + 6.65 - 2\times2.52} = 2.6\ \Omega$$

Note that even when the grid is as extensive as the one in Example 4.7, adding ground rods reduces the ground resistance even further.

4.2 Soil Resistivity

The resistance of a material is the opposition of the material to the flow of electric current when a voltage is applied. The resistance is the ratio of the applied voltage V to the resulting current flow I as defined by Ohm's law:

$$V = I R \tag{4.35}$$

The resistance of a conductor depends on the atomic structure of the material, or the resistivity ρ, and can be computed by

$$R = \frac{\rho l}{A} \tag{4.36}$$

where
  l is the length of the conductor
  A is the cross-sectional area of the conductor

4.2.1 Measuring Ground Resistance

The ground resistance can be measured by using a technique known as the fall-of-potential method. This method, which is also known as the three-point test, is explained in Figure 4.15. The setup consists of the object, whose ground resistance is to be determined, a current electrode, and a potential probe. The electrode is a small copper rod that is driven into the soil. The distance between the current electrode and the object is normally set at or greater than 40 m. A voltage source is connected between the object and the current electrode. A voltmeter is connected between the object and the potential probe. The potential probe moves along the line between the object and the current electrode. At each position, the current (of the current electrode) and the voltage (of the potential probe) are recorded and plotted as shown at the bottom of Figure 4.15. When the potential probe touches the object under test, the measured voltage is zero; when it touches the current electrode, the voltage is equal to the source voltage. The magnitude of the measured voltage is a nonlinear function with respect to the distance x. However, the voltage is often unchanged for a wide range of x, as shown in the flat region in the figure. At the middle of this region, V_f is recorded at the distance x_f. The ratio V_f/I is the ground resistance of the object. Note that the curve in Figure 4.15 is similar to that obtained in Figure 4.4.

The fall-of-potential method can be useful in calculating the ground resistance of transmission line towers. In this case, the tower footage is the object under test, as shown in Figure 4.16.

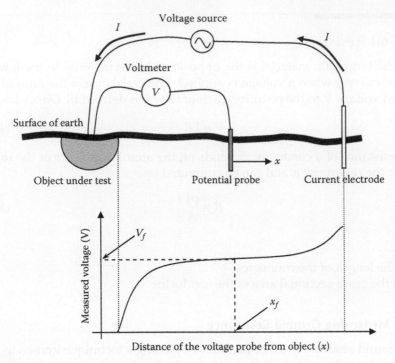

FIGURE 4.15
Fall-of-potential method to measure ground resistance.

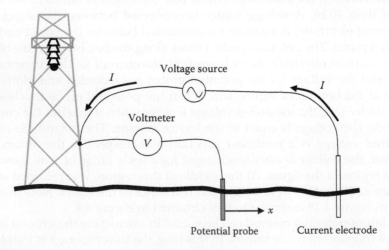

FIGURE 4.16
Fall-of-potential method to measure footage resistance of towers.

4.2.2 Measuring Soil Resistivity

If the soil is uniform and the ground resistance of the buried object is known, the soil resistivity can be calculated. For example, to calculate the soil resistivity surrounding a hemisphere, we first perform the fall-of-potential measurement described in the previous section to find the ground resistance of the hemisphere. Then, use Equation 4.7 to compute the soil resistivity:

$$\rho = 2\pi x_f R_g \qquad (4.37)$$

Besides electric safety studies, measuring soil resistivity is a very important task for subsurface geophysical surveyors. It is used to locate ores, identify the depth of bedrocks, and estimate the degree of corrosion for underground pipelines. Geophysical scientists have developed several methods to measure soil resistivities. Three of them are commonly used in electric safety studies: the Wenner four-pin method, the Schlumberger four-pin method, and the driven-rod method (also known as three-pin method).

4.2.2.1 Wenner Four-Pin Method

The Wenner method, which is shown in Figure 4.17, requires the use of four identical pins (electrodes or probes) placed at equal distances a along a straight line and inserted into earth at depth d. The voltage between the two inner potential electrodes and the current of the outer electrodes is measured.

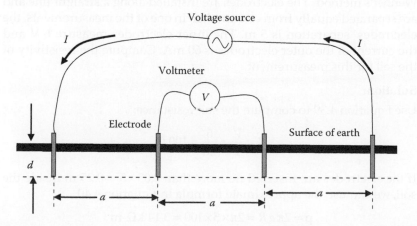

FIGURE 4.17
Wenner four-pin method.

The soil resistivity can then be computed using Equation 4.38:

$$\rho = \frac{4\pi a R}{1 + \dfrac{2a}{\sqrt{a^2 + 4d^2}} - \dfrac{a}{\sqrt{a^2 + d^2}}} \tag{4.38}$$

The resistance R is

$$R = \frac{V}{I} \tag{4.39}$$

where
  V is the measured voltage of the inner electrodes
  I is the measured current of the outer electrodes

If the electrodes are not inserted deep into the soil (i.e., $d < 0.05a$), we can approximate Equation 4.38 to

$$\rho = 2\pi a R \tag{4.40}$$

Example 4.9

Four electrodes are used to measure the resistivity of a soil using Wenner's method. The electrodes are installed along a straight line and are separated equally from one another. In one of the measurements, the electrodes' separation is 5 m. The inner electrodes measure 6 V and the current of the outer electrodes is 60 mA. Compute the resistivity of the soil for this measurement.

Solution:

Use Equation 4.39 to compute the soil resistance:

$$R = \frac{V}{I} = \frac{6}{60} = 100\ \Omega$$

If the electrodes are inserted no more than $0.05 \times 5 = 0.25$ m inside the soil, we can use the approximate formula in Equation 4.40:

$$\rho = 2\pi a R = 2\pi \times 5 \times 100 = 3.14\ \text{k}\Omega\text{-m}$$

Keep in mind that one measurement is not enough to accurately identify the soil resistivity. This point is explained later in this chapter.

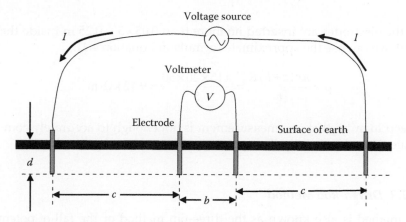

FIGURE 4.18
Schlumberger four-pin method.

4.2.2.2 Schlumberger Four-Pin Method

The Schlumberger method, which is shown in Figure 4.18, is similar to the Wenner's method, but the inner electrodes can be set at a fixed separation b and the outer electrodes move along a straight line at variable separation c.

If the electrodes are not inserted deep into the soil ($d < 0.05b$), the soil resistivity is

$$\rho = \frac{\pi c(c+b)R}{b} \tag{4.41}$$

Example 4.10

Four electrodes are used to measure the resistivity of a given soil using Schlumberger's method. The inner electrodes are separated by 5 m. In one of the measurements, the outer electrodes are separated from the inner electrodes by 15 m. The inner electrodes measure 5 V and the current of the outer electrodes is 100 mA. Compute the resistivity of the soil for this measurement.

Solution:

Use Equation 4.39 to compute the soil resistance:

$$R = \frac{V}{I} = \frac{5}{0.1} = 50 \ \Omega$$

If the electrodes are inserted no more than $0.05 \times 5 = 0.25$ m inside the soil, we can use the approximate formula in Equation 4.41:

$$\rho = \frac{\pi c (c+b) R}{b} = \frac{\pi 15 \times 20 \times 50}{5} = 9.42 \text{ k}\Omega\text{-m}$$

Keep in mind that one measurement is not enough to accurately compute the soil resistivity. This point is explained later in this chapter.

4.2.2.3 Driven-Rod Method

This method is also known as the three-pin method or the fall-of-potential method. The setup of this method is shown in Figure 4.19. The method uses a long rod that can be placed at various depths. The ground resistance of the rod is computed by the fall-of-potential method

$$R_{rod} = \frac{V(x)}{I(x)} \tag{4.42}$$

where
$V(x)$ is the voltage measured between the potential probe and the rod
$I(x)$ is the current in the driven rod
x is the distance between the rod and the potential probe

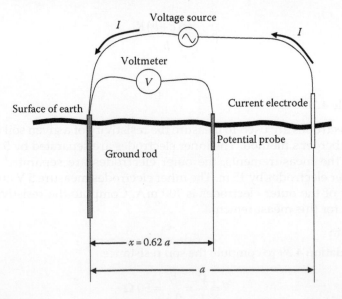

FIGURE 4.19
Three-pin method.

It is recommended that $x \approx 0.62a$, where a is the distance between the current electrode and the rod.

The resistivity of the soil can be computed using Equation 4.17 for the ground rod:

$$\rho = R_{rod} \frac{2\pi L_{rod}}{\left[\ln\left(\dfrac{4L_{rod}}{r}\right)-1\right]} \qquad (4.43)$$

Example 4.11

The three-pin method is used to measure the resistivity of soil. In one of the measurements, the current electrode is 10 m away from the rod, and the potential probe is 6.2 m away from the rod. The rod is driven 3 m into the ground and is 4 cm in diameter. The potential probe measure is 2 V and the current of the current electrode is 50 mA. Compute the resistivity of the soil for this measurement.

Solution:

Use Equation 4.42 to compute soil resistance:

$$R_{rod} = \frac{V(x)}{I(x)} = \frac{2}{0.05} = 40 \ \Omega$$

Use the approximate formula in Equation 4.43 to compute soil resistivity:

$$\rho = R_{rod} \frac{2\pi L_{rod}}{\left[\ln\left(\dfrac{4L_{rod}}{r}\right)-1\right]} = 40 \frac{2\pi 3}{\left[\ln\left(\dfrac{12}{2}\right)-1\right]} = 950 \ \Omega\text{-m}$$

Keep in mind that one measurement is not enough to accurately compute soil resistivity. This point is explained later in this chapter.

4.2.2.4 Nonuniform Soils

For temporary grounding, it is reasonable to assume that the soil is homogeneous and uniform. However, for permanent grounding systems, more elaborate analysis of the soil is needed as most soil is made of materials with widely different resistivities.

The change in resistivity occurs in the horizontal and vertical directions. The change of resistivity in the horizontal direction is often

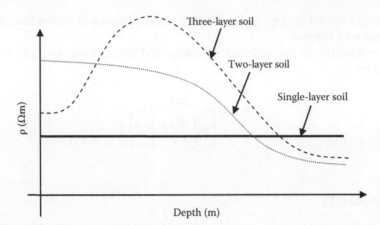

FIGURE 4.20
Soil resistivity measurement.

gradual and small. But, the resistivity in the vertical direction could change widely and abruptly due to the layering of the ground material. In most studies, however, only the vertical change in soil resistivity is considered.

Figure 4.20 shows three resistivity curves of hypothetical soils composed of single, double, and triple layers. As seen in the figure, the resistivity of the soil changes with depth. The important question is how to account for the variation in resistivity? Can we find an "equivalent" resistivity for the layered soil? IEEE in its Standard 81-1983 (*Guide for Earth Resistivity, Ground Impedance, and Earth Surface Potentials of a Ground System*) suggests the use of the *Apparent Resistivity* method assuming two layers of soil. This is a good approximation that would work for most permanent grounding systems. The configuration of the soil is given in Figure 4.21. The top layer occupies an earth crust of depth h, and the bottom layer is rest of the ground material.

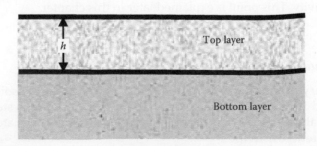

FIGURE 4.21
Two-layer soil.

For the soil in Figure 4.21, the *reflection coefficient K* of the resistivities is defined as

$$K = \frac{\rho_2 - \rho_1}{\rho_2 + \rho_1} \tag{4.44}$$

where
ρ_1 is the soil resistivity of the top layer
ρ_2 is the soil resistivity of the bottom layer

The reflection coefficient gives information on how wide is the variation of soil resistivities between layers. The value of K can be positive or negative: when the top soil resistivity is higher than that for the bottom layer, K is negative.

The *apparent* (equivalent) resistivity ρ_a of the two-layer soil can be computed by using the modified Wenner equation:

$$\rho_a = \rho_1 \left(1 + 4 \sum_{n=1}^{n_{max}} \left(\frac{K^n}{\sqrt{1 + \left(\frac{2nh}{a} \right)^2}} - \frac{K^n}{\sqrt{4 + \left(\frac{2nh}{a} \right)^2}} \right) \right) \tag{4.45}$$

where
h is the depth of top layer
a is the separation between adjacent electrodes
n_{max} is the maximum number of summation terms

In theory, $n_{max} = \infty$, but after the first few terms, the change in the apparent resistivity is insignificant as seen in the following example.

Example 4.12

A soil is composed of two layers. The top layer has a resistivity of 100 Ω-m and the lower layer's resistivity is 1000 Ω-m. The depth of the top soil is 5 m. The four-pin method is used where the separation between adjacent electrodes is 20 m. Compute the apparent resistivity of the soil.

Solution:

Compute the reflection coefficient using Equation 4.44:

$$K = \frac{\rho_2 - \rho_1}{\rho_2 + \rho_1} = \frac{1000 - 100}{1000 + 100} = 0.82$$

Use Equation 4.45 to compute the apparent resistivity of the two-layer soil:

$$\rho_a = \rho_1 \left(1 + 4 \sum_{n=1}^{n_{max}} \left(\frac{K^n}{\sqrt{1 + \left(\frac{2nh}{a} \right)^2}} - \frac{K^n}{\sqrt{4 + \left(\frac{2nh}{a} \right)^2}} \right) \right)$$

$$= 100 \left(1 + 4 \sum_{n=1}^{n_{max}} \left(\frac{0.82^n}{\sqrt{1 + (0.5n)^2}} - \frac{0.82^n}{\sqrt{4 + (0.5n)^2}} \right) \right)$$

But what is the value of n_{max}? If the above equation is plotted against n, we obtain Figure 4.22.

Note that after six terms, the change in the apparent resistivity is insignificant. In this case, $n_{max} = 10$ is adequate enough.

The apparent resistivity of the two-layer soil is 373.18 Ω-m.

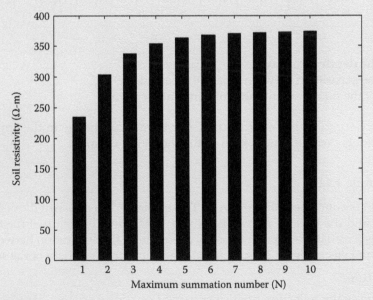

FIGURE 4.22
Soil resistivity versus n_{max}.

The accuracy of the measured resistivity depends largely on the separation between the electrodes. To explain this point, let us modify Equation 4.45:

$$\Gamma = \frac{\rho_a}{\rho_1} = 1 + 4 \sum_{n=1}^{n_{max}} \left(\frac{K^n}{\sqrt{1 + \left(\frac{2n}{r}\right)^2}} - \frac{K^n}{\sqrt{4 + \left(\frac{2n}{r}\right)^2}} \right) \qquad (4.46)$$

where $r = a/h$, the ratio of the separation to the depth of the top layer.

If you plot Γ with respect to r for various values of K, you obtain a family of curves known as the *standard curves* or *resistivity sounding curves*. An example of these curves is shown in Figure 4.23. The sounding curves are plotted in log scale to show the symmetry in the characteristics. For small electrode separation, the apparent resistivity is very close to the resistivity of the surface layer ($\Gamma = 1$). Hence, the effect of the bottom layer is not measured. If the separation between electrodes increases, the apparent resistivity changes as the lower layer affects the measurement. Every curve in the figure has two asymptotes: the first is when $\rho_a \approx \rho_1$ and the second when ρ_a approaches ρ_2. But which value of apparent resistivity to use? The obvious

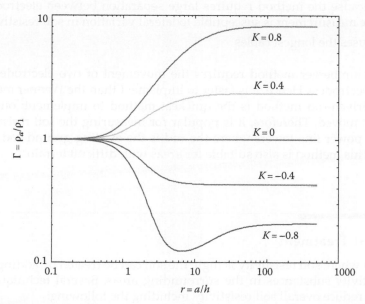

FIGURE 4.23
Sounding curves of two-layer soil.

answer is the value that includes the effect of the lower layer at the second asymptote. To achieve this asymptote, the following need to be considered:

- In most sites, the depth of the top layer is unknown. Hence, the separation of the electrode that would lead to measurements in the second asymptote is also unknown.
- Because of the previous point, several measurements at various separations are needed to reach the second asymptote. In some cases, you may want to start with wide separation to minimize the number of measurements. However, very wide separation may lead to inaccuracies if the soil resistivity changes significantly in the lateral direction.
- In all measurements, the depth of the electrode should be much less than the electrode spacing.

4.2.2.5 Comments on the Three Methods

Since we have to repeat the measurement at various spacing, the Wenner method requires more time and manpower than the other two methods. This is because

- For every additional measurement, all four electrodes must be moved
- Because the method requires large separation between electrodes, the measurement is susceptible to lateral variation in soil resistivity
- It uses the longest cables

The Schlumberger method requires the movement of two electrodes only (outer electrodes). Hence, it is faster to implement than the Wenner method.

The driven-rod method is the quickest method to implement; only one probe is moved. Therefore, it is popular for measuring the soil resistivities around power line towers where the rod is the existing ground rod of the tower. This method is also suitable for areas with difficult terrain.

4.3 Soil Treatment

In areas where soil resistivity is high, the soil can be treated by adding high-conductivity substances in the surrounding areas. Several techniques are used to reduce overall soil resistivity including the following:

- *Soil Infusion*: The soil surrounding the electrode is infused with conductive mineral such as sodium chloride, magnesium, copper sulfates, and calcium chloride. This infusion increases the conductivity of the soil. Although effective, the treatment must be renewed periodically as water dissolves and washes away minerals. This method is not suitable in wet environments.

- *Bentonite*: Some volcanic material such as bentonite is known for its low resistivity and its resistance to corrosion. The minerals in bentonite are microscopic crystals mixed with clay. There are several types of bentonites such as potassium (K), sodium (Na), calcium (Ca), and aluminum (Al). When the bentonite is mixed with water, an electrolytic process occurs that reduces the resistivity of the bentonites to less than 3 Ω-m. Furthermore, when water is added, the bentonite increases in volume by more than 10 times, and thus adheres to the entire surface of the electrode. This method is suitable for wet environments.

- *Hollow electrode*: The grounding electrodes are made of copper tubes filled with conductive mineral such as salt. The tube has tiny holes to allow water to enter and slowly dissolve the salts. The salt solution seeps out of the tube and into the surrounding ground. To keep the electrodes and surrounding areas at low resistivity, the tubes are refilled periodically.

- *Concrete*: Concrete is used as a grounding medium for soils with high resistivities. Because concrete is a hydroscopic material that absorbs moisture, its resistivity is 30–100 Ω-m. If the soil resistivity is high, a grounding electrode encased in concrete will have lower resistance than the identical electrode buried directly in the soil. The only drawback of this method is the corrosion of the steel reinforcement bars inside the concrete due to the presence of ground currents, especially the dc component.

4.4 Factors Affecting Ground Resistance

The ground resistance of any object can be affected by a number of factors including voltage gradient, current, moisture, temperature, and chemical contents in soil and surface material.

4.4.1 Effect of Voltage Gradient

When the voltage gradient (voltage/distance) is high (several kV/cm), an arc in the soil could occur. The arc, which is a low-resistance path for current, starts at the surface of the ground electrode and travels through the ground. The arc continues until the voltage gradient is reduced below the arcing level. This is similar to the flashover that could happen on an insulator surface.

Several factors can contribute to the increase in the voltage gradient of the ground soil. These include fault currents and lightning strikes.

4.4.2 Effect of Current

When currents pass through ground electrodes, they tend to dry the surrounding soil. This could substantially increase the soil resistivity, thus increasing the ground resistance of the electrodes. High ground resistance could pose safety hazards. IEEE Standard 80 suggests a conservative value of current density that does not exceed 200 A/m² for more than 1 s.

4.4.3 Effect of Moisture, Temperature, and Chemical Content

The ground resistance is highly dependent on the moisture content in soil, as shown in Figure 4.24a. The moisture content in soil is dependent on two factors: (1) the availability of rain or ground water and (2) the capability of the soil to retain water. The resistivity of most soils rises abruptly whenever the moisture content accounts for less than 15% of the soil weight. The capability of the soil to retain water is a function of soil composition and soil temperature. Soil composition is the grain size and compactness; large grains and loose soil tend to retain less water than small grains and compact soil. High temperatures evaporate water in soils. To lessen the effect of high temperatures, it is common to use gravel or surface material coverings, 0.08–0.15 m (3–6 in.) in depth, to prevent the evaporation of moisture in the soil.

Pure water (H_2O alone) is a nonconductive fluid, so its resistance is very high. However, when soluble salts, acids, or alkali are added to the pure water, the mixture is conductive. The resistivity of the soil is substantially reduced when salt is present, as shown in of Figure 4.24b.

The effect of temperature on soil resistivity is negligible for temperatures above freezing point. If the temperature of the soil is below the freezing point, the water in the soil freezes and the soil resistivity increases substantially, as shown in Figure 4.24c.

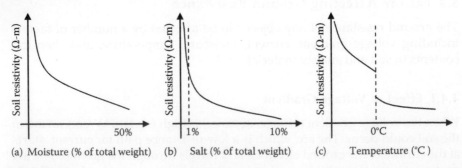

(a) Moisture (% of total weight)   (b)   Salt (% of total weight)   (c)   Temperature (°C)

FIGURE 4.24
Soil resistivity as a function of (a) moisture, (b) salt, and (c) temperature.

4.4.4 Surface Material

Surface material such as gravel, sod, sand, or stone can be useful in two ways: (1) it reduces the evaporation of moisture especially during prolonged dry weather periods and (2) it reduces shock currents if the material covering the surface is of high resistivity. Normally, about 8–15 cm (3–6 in.) in depth of surface material is needed. IEEE Standard 80 published a table of resistivity for some common surface material (see Table 4.3 for a summary).

Exercises

4.1 What are the objectives of grounding systems?

4.2 What is intentional grounding?

4.3 What is accidental grounding?

4.4 A grounding hemisphere with 1 m diameter is inserted into a soil of 200 Ω-m resistivity. The hemisphere receives 100 A. Compute the current density at its surface.

4.5 For the hemisphere in Exercise 4.4, compute the potential difference between two points on the surface of the soil that are separated by 1 m. Assume the nearest point to the center of the hemisphere is 5 m away.

4.6 What is the ground resistance of the hemisphere in Exercise 4.5?

4.7 A pole-butt is 0.5 m in diameter. Compute its ground resistance assuming the soil resistivity is 120 Ω-m.

4.8 A person is standing on a wet organic soil. Compute his or her ground resistance.

4.9 A ground rod is inserted 2 m into a soil of 100 Ω-m resistivity. Compute its ground resistance.

4.10 What conditions must be satisfied when ground rods are connected in a cluster?

4.11 Under what conditions do two electrodes share fault currents?

4.12 How to prevent parallel electrodes from sharing their sphere of influence?

4.13 Four pins were separated by 5 m while using the Wenner method. The pins were inserted 2 m into the ground. The current probe measures 2 A and the voltage probe measures 50 V. Compute the soil resistivity.

4.14 State four common methods for soil treatment.

4.15 State several factors that impact solid resistance.

5

General Hazards of Electricity

Electricity cannot be detected by human senses before the shock occurs. This makes electric hazards less evident than others such as fires and crashes. Electric shocks are unique in two aspects:

1. Electric hazards often produce no warning signs. The calm and serene environment that often surrounds energized stationary objects may give a false sense of safety.
2. Electric shocks are always quick leaving no time for corrective actions. Injury or death often occurs in a fraction of a second.

Because of these aspects, electric equipment must always be considered hazardous and exceptional precautions must be employed when working on or near them. The United States through the Occupational Safety and Health Administration (OSHA) and the Institute of Electrical and Electronics Engineers (IEEE) has established strict safety rules for electrical works. These rules are designed to protect workers and the public from electric shocks. Almost all electric shock fatalities and injuries occur when workers do not follow these established rules due to time pressure, lack of knowledge, or poor supervision.

Electric shocks occur due to high touch and step potentials. The *touch potential* occurs when a person touches a charged metallic object, causing a current to flow through his or her body. The *step potential* is when a person is stands or walks on ground or a surface that carries electric current.

5.1 Touch Potential

Touch potential occurs when a person touches a conductive object carrying electric charges while part of his or her body is touching another conductive object with different amounts of charges. The charges on a conductive object determine the potential of the object. In this case, the person's hands and the other parts of his or her body are at two different potentials. Since the human body is not a perfect insulator, the potential difference causes the charges to move from one object to the other. The rate of flow of these charges through the person's body is the electric current that causes harm.

IEEE Standard 1048 defines touch potential (voltage) as "The voltage difference between a grounded metallic structure or equipment and a point on the earth's surface separated by a distance equal to one normal maximum horizontal reach, approximately 1 m."

More general examples of touch potential and contact voltage hazards are as follows:

- A person touches an energized object while standing on a lower potential conductive object (such as ground). The severity of this touch potential depends on several factors including the potential difference between the objects and the resistances of the various components of the circuit, particularly the ground resistance.

- A person touches two different de-energized conductive objects with separate grounds in areas with ground currents. The severity of this touch potential depends on the distance between the two objects, the magnitude of the ground current, and the difference in the ground resistances.

In the first example, the object being touched is designed to carry electric current. In the second example, the object is not designed to carry electric current, but the surrounding environment causes the difference in potentials.

5.1.1 Touch Potential of Energized Objects

Assume a person touches an energized object while standing on conductive surface such as ground, as shown in Figure 5.1. The left part of the figure shows the man touching the energized conductor while standing on the ground. On the right side of Figure 5.1, the man is replaced by his body resistance and the earth by his feet ground resistance. Because neither the

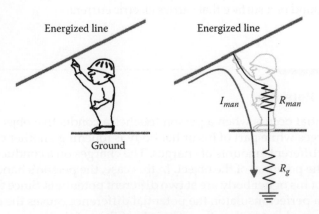

FIGURE 5.1
Man touching an energized line while standing on the ground.

resistance of the human body nor the feet ground resistance is infinite, a current will flow through the body based on Ohm's law:

$$I_{man} = \frac{V}{R_{man} + R_g} \qquad (5.1)$$

where
R_{man} is the body resistance of the person
R_g is the feet ground resistance of the person
V is the voltage of the energized line

Example 5.1

A man standing on wet soil touches a 120 V wire. Assume the body resistance (hand to feet) of the man is 1000 Ω. Also assume the man is standing with his feet close to each other. Compute the current passing through the man. Repeat the solution assuming the man is standing on bedrock soil.

Solution:

The case in this example is depicted in Figure 5.1. As shown in Chapter 4, the wet soil has a resistivity of 100 Ω-m and the ground resistance of the standing man is

$$R_g \approx 1.5\rho = 1.5 \times 100 = 150 \ \Omega$$

The current passing through the man is

$$I_{man} = \frac{V}{R_g + R_{man}} = \frac{120}{150 + 1000} = 104.3 \ \text{mA}$$

This is a lethal level of current.

For the bedrock case, the soil resistivity is 10,000 Ω-m. If the man stands on bedrock soil, his ground resistance is

$$R_g \approx 1.5\rho = 1.5 \times 10,000 = 15 \ \text{k}\Omega$$

The current passing through the man is

$$I = \frac{V}{R_g + R_{man}} = \frac{120}{15,000 + 1,000} = 7.5 \ \text{mA}$$

Note that the current passing through the man is much lower when he is standing on bedrock as compared to wet soil. The man may not survive the shock for the wet soil case, but he may survive the shock for the case of bedrock soil.

5.1.2 Touch Potential of Unintentionally Energized Objects

Touch potential hazards can also exist for de-energized objects due to unintentional energization. Consider, for example, the high-voltage towers shown in Figure 5.2. The tower structure is often made of steel alloy to provide the needed strength. The tower is not always at zero potential as it could carry currents due to a number of reasons such as the following:

- *Electromagnetic coupling*: Because the energized lines are in the vicinity of the tower's metallic objects (structure, OHGW, etc.), a voltage is induced on these objects. The tower dissipates its discharges through all adjacent grounds including the tower ground. Since the tower ground resistance is not zero, the tower will always be at some potential.

- *Insulator leakage*: In humid environments, especially close to seashores, the surface of the insulator could become contaminated by salty moisture and dust. This provides a path for current from the energized line to the tower structure, as shown in Figure 5.3. This is known as insulator leakage or flashover. Also, insulators can develop cracks that can hold water. When filled with water and exposed to freezing temperatures, the cracks expand. When water flows through the cracks, it increases the flashover current.

- *Accident*: Accidents could occur in various forms, causing the energized conductor to come in contact with the tower.

Under any of the conditions mentioned earlier, electric shock hazards occur if a person touches the steel tower while standing on the ground. Figure 5.4 depicts this scenario. If the structure carries a current I, most of this current passes through the structure and goes to the ground of the tower I_{gt}; the rest

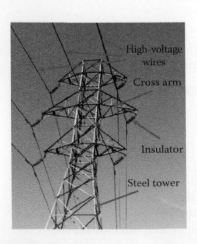
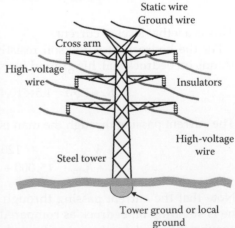

FIGURE 5.2
Main components of a steel power line.

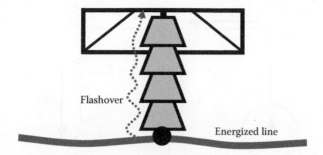

FIGURE 5.3
Insulation leakage or flashover due to conductive depositions on the insulator.

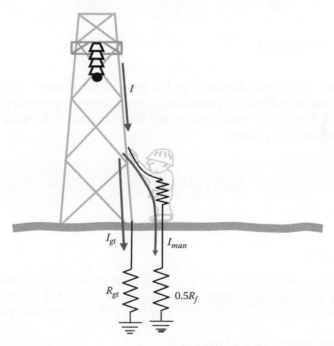

FIGURE 5.4
Touch potential.

of the current goes through the man touching the tower I_{man}. The potential of the structure without the person touching the tower is known as the *ground potential rise* (GPR). It is the voltage that the grounded object attains with respect to the remote earth:

$$GPR = I\,R_{gt} \tag{5.2}$$

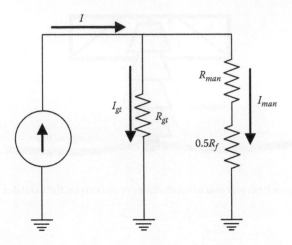

FIGURE 5.5
Circuit diagram of Figure 5.4.

Assume that the man is standing, so his ground resistance is $0.5R_f$, as discussed in Chapter 4. Figure 5.5 shows a circuit diagram representing the case in Figure 5.4. The current passing through the man can be obtained by using the current divider equation:

$$I_{man} = I \frac{R_{gt}}{R_{gt} + R_{man} + 0.5R_f}$$

(5.3)

Example 5.2

A power line insulator partially fails and 1.0 A passes through the tower structure to the ground. Assume that the tower ground is a hemisphere with a radius of 0.5 m, and the soil surrounding the hemisphere is moist.

  a. Compute the GPR of the tower.
  b. Assume that a man with a body resistance of 1.0 kΩ touches the tower while standing on the ground. Compute the current passing through the man.

Solution:

  a. Compute the ground resistance of the hemisphere:

$$R_{gt} = \frac{\rho}{2\pi r} = \frac{100}{2\pi \times 0.5} = 32 \ \Omega$$

The GPR of the tower is

$$\text{GPR} = IR_{gt} = 1.0 \times 32 = 32 \text{ V}$$

b. To compute the current passing through the man, first compute R_f:

$$R_f = 3\rho = 300 \ \Omega$$

The current passing through the man is given in Equation 5.3:

$$I_{man} = I \frac{R_{gt}}{R_{gt} + R_{man} + 0.5R_f} = 1.0 \frac{32}{32 + 1000 + 150} = 27 \text{ mA}$$

This is a hazardous level of current.

Example 5.3

Repeat Example 5.2 assuming that a ground rod is used instead of the hemisphere to ground the tower. Assume that the rod is 4 cm in diameter and is driven 1 m into the ground.

Solution:

a. Compute the ground resistance of the rod:

$$R_{gt} = \frac{\rho}{2\pi l} \ln\left(\frac{2l+r}{r}\right) = \frac{100}{2\pi \times 1} \ln\left(\frac{2+0.02}{0.02}\right) = 73.45 \ \Omega$$

The GPR of the tower is

$$\text{GPR} = IR_{gt} = 1.0 \times 73.45 = 73.45 \text{ V}$$

Notice that the GPR of a ground rod is much higher than the GPR of a hemisphere.

b. $I_{man} = I \dfrac{R_{gt}}{R_{gt} + R_{man} + 0.5R_f} = 1.0 \dfrac{73.45}{73.45 + 1000 + 150} = 60 \text{ mA}$

If the ground rod is driven 1 m into the soil, it provides less protection than the hemisphere. Repeat the problem by assuming that the ground rod is driven 2 m into the soil. Can you draw a conclusion?

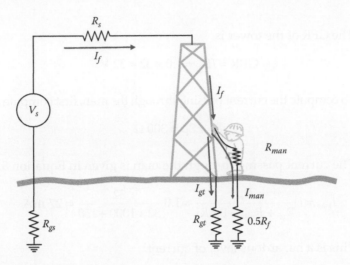

FIGURE 5.6
Structure carrying fault current.

To compute the touch potential for more complex circuits, it is convenient to use Thevenin's theorem. Consider the case in Figure 5.6 where a line is faulted by touching the tower structure while a man on the ground is in contact with the structure. Assume that the source voltage is V_s and the resistance between the source and the structure is R_s. Also, assume that the ground resistance of the source is R_{gs}. To compute the current passing through the man, we remove the man from the circuit, as shown in Figure 5.7,

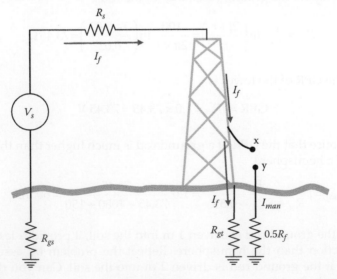

FIGURE 5.7
Circuit without man touching the structure.

and compute Thevenin's resistance, R_{th}, and Thevenin's voltage, V_{th}, between point x and y (the contact points of the man). To compute the Thevenin's resistance, we short all voltage sources and open all current sources. Then, we compute the equivalent resistance between x and y. In the case in Figure 5.6, R_s and R_{gs} are in series; hence

$$R_1 = R_{gs} + R_s \qquad (5.4)$$

Then, R_1 is in parallel with R_{gt}:

$$R_2 = \frac{R_1 R_{gt}}{R_1 + R_{gt}} \qquad (5.5)$$

Finally, R_2 is in series with $0.5R_f$. This is Thevenin's resistance between points x and y:

$$R_{th} = R_2 + 0.5R_f \qquad (5.6)$$

Thevenin's voltage is the voltage between x and y without the man. This is also the GPR of the structure:

$$V_{th} = \text{GPR} = I_f R_{gt} = \frac{V_s}{R_1 + R_{gt}} R_{gt} \qquad (5.7)$$

The touch potential V_{touch} is defined as the potential between points x and y before the person is in contact with these two points. This is the same as Thevenin's voltage:

$$V_{touch} = V_{th} \qquad (5.8)$$

The final step is to compute the current passing through the man using Thevenin's equivalent circuit in Figure 5.8:

$$I_{man} = \frac{V_{th}}{R_{th} + R_{man}} \qquad (5.9)$$

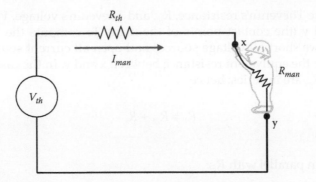

FIGURE 5.8
Current through man due to touch potential using Thevenin's theorem.

Example 5.4

An energized line is faulted into a metallic structure. The circuit consists of an equivalent voltage of 120 V and a source resistance of 1 Ω. The ground resistance of the source is 20 Ω. Assume the ground resistance of the structure is 30 Ω, the foot resistance of the man is 50 Ω, and his body resistance is 1000 Ω. Compute the touch potential and the current passing through the man.

Solution:

The first step is to compute Thevenin's resistance using Equations 5.4 through 5.6:

$$R_1 = R_{gs} + R_s = 20 + 1 = 21\ \Omega$$

$$R_2 = \frac{R_1\,R_{gt}}{R_1 + R_{gt}} = \frac{21 \times 30}{21 + 30} = 12.4\ \Omega$$

$$R_{th} = R_2 + 0.5R_f = 12.4 + 0.5 \times 50 = 37.4\ \Omega$$

The second step is to compute Thevenin's voltage using Equation 5.7:

$$V_{th} = \frac{V_s}{R_1 + R_{gt}}\,R_{gt} = \frac{120}{21 + 30}\,30 = 70.6\ \text{V}$$

The touch potential is the same as V_{th}.
   To compute the current passing through the man, use Equation 5.9:

$$I_{man} = \frac{V_{th}}{R_{th} + R_{man}} = \frac{70.6}{37.4 + 1000} = 68\ \text{mA}$$

5.1.3 Touch Potential of De-Energized Objects

Touch potentials can exist between two separated de-energized metallic objects touching ground. This is known as the metal-to-metal touch potential. It is a common problem for metallic pipes, rails, and fences. Substantial metal-to-metal touch voltages could be present when a person touches one grounded object and comes into contact with another metallic object grounded at a different location. The severity of the touch potential depends on several factors such as the following:

- *Magnitude of ground current*: The ground current creates a different GPR for each object.
- *Separation between the two objects*: The wider the separation, the higher the difference between the GPRs of the two objects.
- *Proximity to high-voltage lines*: Electromagnetic coupling between the power line and a metallic object elevates the voltage of the object.
- *Poorly grounded object*: Insulated or poorly grounded objects retain their charges, and thus maintain their elevated voltage.
- *Bonding of the objects*: Bonding the two objects equalized their potential. However, the touch potential between one object and the ground could still be hazardous.

Consider the case in Figure 5.9. A person is standing on insulated material (e.g., gravel) and touches two grounded objects located at *a* and *b*. Each of these objects has its own ground resistance (R_{ga} and R_{gb}). Assume that a current *I* is injected into the ground by a nearby structure. To compute the metal-to-metal touch potential and the current passing through the person, we need to compute Thevenin's voltage and impedance.

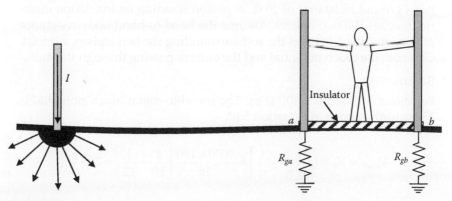

FIGURE 5.9
Metal-to-metal touch potential.

Thevenin's voltage is the touch voltage between the two structures. It is computed by using the voltage difference between points *a* and *b* given by Equation 4.5:

$$V_{th} = V_{touch} = \frac{I\rho}{2\pi}\left[\frac{1}{r_a} - \frac{1}{r_b}\right] \tag{5.10}$$

Thevenin's impedance is the sum of the two ground impedances of the object:

$$R_{th} = R_{ga} + R_{gb} \tag{5.11}$$

The current passing through the person is computed using Equation 5.9:

$$I_{man} = \frac{V_{th}}{R_{th} + R_{man}} \tag{5.12}$$

The resistance of the man is his hand-to-hand resistance.

Example 5.5

A short-circuit current of 1000 A passes through a hemisphere ground object. Two conductive objects are separated from the hemisphere by 10 and 12 m. The closer object has a ground resistance of 30 Ω and the other has a ground resistance of 20 Ω. A person standing on insulation material touches the two objects. Assume the hand-to-hand body resistance of the person is 1 kΩ and the soil surrounding the hemisphere is moist. Compute the touch potential and the current passing through the man.

Solution:

For moist soil, use ρ = 100 Ω-m. The metal-to-metal touch potential is Thevenin's voltage in Equation 5.10:

$$V_{touch} = V_{th} = \frac{I\rho}{2\pi}\left[\frac{1}{r_a} - \frac{1}{r_b}\right] = \frac{1000 \times 100}{2\pi}\left[\frac{1}{10} - \frac{1}{12}\right] = 265 \text{ V}$$

Thevenin's resistance is

$$R_{th} = R_{ga} + R_{gb} = 30 + 20 = 50 \ \Omega$$

The current passing through the man can be computed by Equation 5.12:

$$I_{man} = \frac{V_{th}}{R_{th} + R_{man}} = \frac{265}{50 + 1000} = 252 \text{ mA}$$

The voltage between the person's hands

$$V_{man} = I_{man} \times R_{man} = 252 \times 1000 = 252 \text{ V}$$

Can you tell why the voltage across the man is slightly lower than the metal-to-metal touch potential?

Another common touch potential problem is when a person standing on the ground is touching a grounded object at a distance. A common scenario is depicted in Figure 5.10. The person is touching a gate mounted on a metallic post. The potential of the gate is the GPR of the post. If the gate is wide, the separation between points *a* and *b* is wide as well. In this case, the touch potential could be substantial.

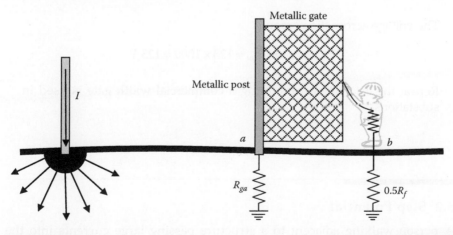

FIGURE 5.10
Touch potential during fault.

Example 5.6

A short-circuit current of 1000 A passes through a hemisphere ground object. A conductive post is separated from the hemisphere by 10 m. The post has a ground resistance of 30 Ω. A person standing on the ground touches a 1 m wide gate mounted on the post. Assume the hand-to-foot body resistance of the person is 1 kΩ and the soil surrounding the hemisphere is moist. Compute the touch potential and the current passing through the man.

Solution:

For moist soil, $\rho = 100$ Ω-m. The touch potential is Thevenin's voltage in Equation 5.10:

$$V_{touch} = V_{th} = \frac{I\rho}{2\pi}\left[\frac{1}{r_a} - \frac{1}{r_b}\right] = \frac{1000 \times 100}{2\pi}\left[\frac{1}{10} - \frac{1}{11}\right] = 145 \text{ V}$$

Since the man is standing, the ground resistance of the man is $0.5R_f$ or 1.5ρ. The Thevenin's resistance between points a and b is

$$R_{th} = R_{ga} + 1.5\rho = 30 + 1.5 \times 100 = 180 \text{ }\Omega$$

The current passing through the man is

$$I_{man} = \frac{V_{th}}{R_{th} + R_{man}} = \frac{145}{180 + 1000} = 123 \text{ mA}$$

The voltage across the person is

$$V_{man} = I_{man} \times R_{man} = 123 \times 1000 = 123 \text{ V}$$

Repeat the calculations assuming a commercial width gate is used in substations.

5.2 Step Potential

A person walking adjacent to a structure passing large currents into the ground can be vulnerable to electric shocks. This is particularly hazardous during faults, especially near substations, near downed lines, near faulted power line towers, or even during lightning storms.

FIGURE 5.11
Current passing through a person due to step potential.

IEEE Standard 1048 defines step potential (voltage) as "The potential difference between two points on the earth's surface separated by a distance of one pace (assumed to be 1 m in the direction of maximum potential gradient). This potential difference could be dangerous when current flows through the earth or material upon which a worker is standing, particularly under fault conditions."

Figure 5.11 shows a person walking near a structure that discharges current into the ground. The ground current may cause a high enough potential difference between the person's feet, resulting in current passing through his legs and abdomen. This current may not be fatal since it is not likely to pass through the heart, lungs, or brain. However, it could cause the person to lose his balance and fall on the ground, where a lethal current could flow through his vital organs.

The step potential V_{step} can be computed using Thevenin's theorem, where Thevenin's voltage V_{th} is the open circuit voltage between point a and b in Figure 5.11. This is the same as V_{ab} without the person. As given in Equation 5.10, the voltage for a hemisphere grounding object is

$$V_{th} = V_{ab} = V_{step} = \frac{I\rho}{2\pi}\left[\frac{1}{r_a} - \frac{1}{r_b}\right] \tag{5.13}$$

where
r_a is the distance from the center of the hemisphere to the person's rear foot, assuming the grounding hemisphere is behind the walking person
r_b is the distance from the center of the hemisphere to his front foot
I is the total current into the ground by the hemisphere

FIGURE 5.12
Thevenins's impedance.

Thevenin's resistance R_{th} is the open circuit resistance between point a and b (without the person), as shown in Figure 5.12.

$$R_{th} = 2R_f \qquad (5.14)$$

Figure 5.13 shows the equivalent circuit for step potential calculations. The body resistance of the person R_{man} is the leg-to-leg resistance. The current passing through the man is

$$I_{man} = \frac{V_{th}}{R_{th} + R_{man}} \qquad (5.15)$$

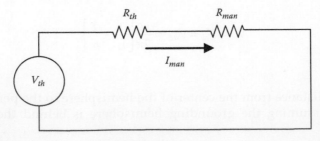

FIGURE 5.13
Equivalent circuit for step potential.

Example 5.7

A short-circuit current of 1000 A passes through a hemisphere ground. A person is walking 5 m away from the center of the hemisphere. Assume that the leg-to-leg resistance of the person is 1.0 kΩ and the soil surrounding the hemisphere is moist. Compute the step potential, the current passing through the person, and the voltage across the man.

Solution:

For moist soil, $\rho = 100$ Ω-m. Assume that the step of the person is about 0.6 m. Thevenin's voltage can be computed using Equation 5.13:

$$V_{step} = V_{th} = \frac{I\rho}{2\pi}\left[\frac{1}{r_a} - \frac{1}{r_b}\right] = \frac{1000 \times 100}{2\pi}\left[\frac{1}{5} - \frac{1}{5.6}\right] = 341 \text{ V}$$

$$R_f = 3\rho = 300 \text{ }\Omega$$

The current passing through the man can be computed by Equation 5.15:

$$I_{man} = \frac{V_{th}}{2R_f + R_{man}} = \frac{341}{600 + 1000} = 213.13 \text{ mA}$$

The voltage between the person's feet is

$$V_{man} = I_{man} \times R_{man} = 213.13 \times 1000 = 213.13 \text{ V}$$

Example 5.8

Repeat Example 5.7 assuming that a ground rod is used instead of the hemisphere. Assume the rod is inserted 1 m into the ground.

Solution:

Using the rod equation in Chapter 5, we can obtain Thevenin's voltage as

$$V_{step} = V_{th} = \frac{I\rho}{2\pi l}\left[\ln\left(\frac{2l + r_a}{r_a}\right) - \ln\left(\frac{2l + r_b}{r_b}\right)\right]$$

$$V_{th} = \frac{1000 * 100}{2\pi}\left[\ln\left(\frac{2+5}{5}\right) - \ln\left(\frac{2+5.6}{5.6}\right)\right] = 495 \text{ V}$$

Notice that the voltage on the ground is higher for the ground rod as compared to the hemisphere.

The current passing through the man is given in Equation 5.15:

$$I_{man} = \frac{V_{th}}{2R_f + R_{man}} = \frac{495}{600+1000} = 246.88 \text{ mA}$$

Notice that the ground rod provides less protection than the ground hemisphere.

The voltage across the man is

$$V_{man} = I_{man} \times R_{man} = 246.88 \times 12,000 = 246.88 \text{ V}$$

The severity of the step potential depends on the distance between the person and the location where current is injected into ground. Consider the case in Figure 5.14 where the GPR is plotted as a function of distance from the point of injection. The GPR is reduced rapidly near the injection point, which is known as the maximum gradient region. Hence, the closer the person is to the injection point, the higher is his step potential. IEEE Standard 1048 requires the computation of the step potential to be in the *maximum potential gradient* region.

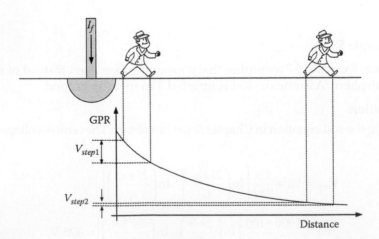

FIGURE 5.14
Step potential in the direction of maximum gradient.

Exercises

5.1 A person is working on a steel structure while standing on the ground. An accident occurs, in which 1 A pass through the structure to the ground. The structure is grounded by a metal rod 6 cm in diameter. The rod is dug 2 m into the ground. The surrounding soil is dry. Assume that the resistance of the man's body is 1 kΩ. Compute the current passing through the man.

5.2 Repeat Exercise 5.1 for wet organic soil.

5.3 During a weather storm, an atmospheric discharge hits a lightning pole. The pole is grounded through a hemisphere. The maximum value of the lightning current through the pole is 10 kA. The soil of the area is moist. A man who is walking 20 m away from the center of the hemisphere experiences an excessive step potential. The man's body resistance is 1 kΩ. Compute the current passing through his legs and the potential between his legs.

5.4 During a weather storm, an atmospheric discharge hits a lightning pole that is grounded through a hemisphere. The maximum value of the lightning current passing through the rod is 20 kA. The soil of the area is moist. A man is playing golf 50 m away from the center of the hemisphere. At the moment of the lightning strike, the distance between his two feet is 0.4 m. Compute the current passing through the person assuming that the resistance between his legs is 1 kΩ.

5.5 Repeat Exercise 5.4 but assume the person is 5 m away from the center of the hemisphere. What is the effect of the proximity of the man to the grounding hemisphere?

5.6 A power line insulator partially fails and 10 A passes through the structure to the tower's ground. The tower's ground is a hemisphere with a radius of 0.5 m. The soil resistivity is 100 Ω-m. Assume that a man touches the tower while standing on the ground. Compute the current passing through the man assuming his body resistance is 1000 Ω.

6

Induced Voltage due to Electric Field

The National Electrical Safety Code (NESC) defines *de-energized* line as a conductor that is "disconnected from all sources of electrical supply by open switches, disconnectors, jumpers, taps, or other means."

A de-energized circuit is *not* a dead circuit. It is a fatal mistake for anyone to assume that de-energized lines are safe to touch; they are not. De-energized metallic objects acquire charges from adjacent high-voltage lines due to the electromagnetic field produced by the energized lines. These charges elevate the potentials of the de-energized objects. This problem is particularly severe when the de-energized conductors share the right-of-way (ROW) with energized circuits. Figure 6.1 shows the electromagnetic field (electric field and magnetic field) produced by the energized line. If these fields reach the de-energized line, a voltage is induced on the de-energized conductor. The magnitude of this induced voltage depends on several factors:

- The voltage level of the energized conductor
- The current of the energized conductor
- The distance between the two lines
- The length for which the two lines are run adjacent to each other

The NESC further states that

> De-energized conductors or equipment could be electrically charged or energized through various means, such as induction from energized circuits, portable generators, lightning, etc.

The IEEE Standard 524a-1993 defined the induced voltage due to electrical and magnetic fields as follows:

> A de-energized line in close proximity to an energized line will have a voltage induced on it through electric field (capacitive coupling) induction. This voltage will have a magnitude somewhere between zero voltage (ground) and the voltage of the energized line. In practical circumstances, induced voltage can be as high as 30% of the energized line voltage.
>
> Magnetic fields can induce hazardous open circuit voltages in partially grounded loops of de-energized lines adjacent to energized lines. This voltage can be as high as 300 V/mile under normal load conditions or as high as 5000 V/mile under short circuit conditions. In addition, hazardous current levels can be induced in grounded loops to levels that create hazardous step or touch voltages at or near ground electrodes.

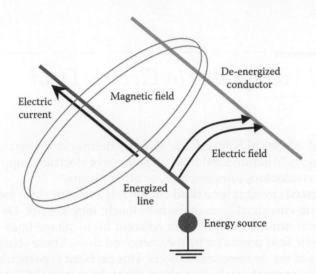

FIGURE 6.1
Electromagnetic coupling between energized and de-energized lines.

In this chapter, we shall study induced voltage due to the electric field alone. Induced voltage due to magnetic field is covered in the next Chapter 7.

6.1 Equipotential Surface

To understand the effect of electric field, we start with the simple case shown in Figure 6.2. This is for a charged sphere located high enough above ground. This way, the electric field is uniformly distributed and is free from ground distortions. The charge Q produces an electric field in the radial direction from the surface of the sphere. Any two points outside the sphere that are located at the same distance from the center of the sphere will have identical potential. Thus, any imaginary sphere in the air that is concentric with the charged sphere will have no voltage difference between any two points on its surface. This is known as the *equipotential surface*.

The electric field is dense at the surface of the sphere and becomes less dense when we move away. The electric field density D_x at an equipotential surface whose radius x can be expressed by

$$D_x = \frac{Q}{A_x} = \frac{Q}{4\pi x^2} \tag{6.1}$$

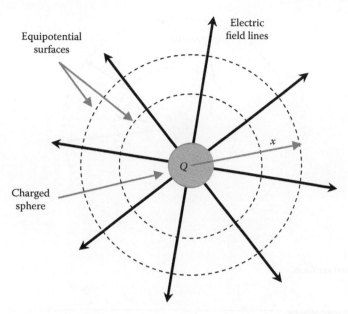

FIGURE 6.2
Electric field and equipotential surface.

where
    Q is the total charge of the sphere in coulomb (C or A s)
    A_x is the area of the equipotential surface at distance x from the center of
        the sphere

The electric field strength E_x at a distance x from the center of the sphere is proportional to the electric field density

$$E_x = \frac{D_x}{\varepsilon_0} = \frac{Q}{4\pi\varepsilon_0 \, x^2} \tag{6.2}$$

where ε_0 is a proportionality constant known as *air permittivity or absolute permittivity* (8.85 × 10⁻¹² F/m). The unit of E_x is V/m.

The voltage between any two points outside the sphere, see Figure 6.3, is the integration of the electric field strength between these two points:

$$V_{12} = \int_{x_1}^{x_2} E_x \, dx = \int_{x_1}^{x_2} \left(\frac{Q}{4\pi\varepsilon_0 \, x^2} \right) dx = \frac{Q}{4\pi\varepsilon_0} \left(\frac{1}{x_1} - \frac{1}{x_2} \right) \tag{6.3}$$

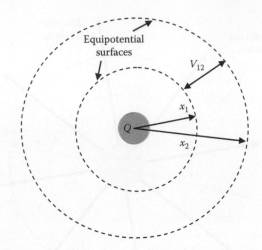

FIGURE 6.3
Equipotential surfaces.

6.2 Induced Voltage on a Conductor without Field Distortion

In the analysis of this section, it is assumed that the energized transmission line is leveled, cylindrical, and surrounded by air (see Figure 6.4). To ignore the electric field distortion, the line is assumed to be high above ground. The total charge of the energized line Q produces an electric field in the radial direction from the surface of the conductor. Any two points outside the conductor that are located at the same distance from the center of the conductor will have identical voltage. Thus, the imaginary cylinders in the air that are concentric with the conductor are the *equipotential surfaces*.

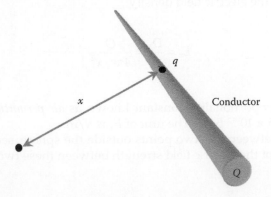

FIGURE 6.4
Charged conductor.

The electric field density at a normal distance x from the center of the conductor is

$$D_x = \frac{Q}{A_x} = \frac{Q}{2\pi x L} \tag{6.4}$$

where
Q is the total charge on the conductor in coulomb
A_x is the area of the cylindrical equipotential surface at distance x from the center of the conductor
L is the length of the conductor

Define the charge per unit length q as

$$q = \frac{Q}{L} \tag{6.5}$$

Hence,

$$D_x = \frac{q}{2\pi x} \tag{6.6}$$

The electric field strength E_x at a distance x from the center of the conductor is

$$E_x = \frac{D_x}{\varepsilon_0} = \frac{q}{2\pi\varepsilon_0 x} \tag{6.7}$$

The voltage difference between any two points in space is the integration of the electric field strength between these two points:

$$V_{12} = \int_{x_1}^{x_2} E_x \, dx = \int_{x_1}^{x_2} \left(\frac{q}{2\pi\varepsilon_0 x} \right) dx = \frac{q}{2\pi\varepsilon_0} \ln\left(\frac{x_2}{x_1} \right) \tag{6.8}$$

Equation 6.8 can also be used to compute the potential of any de-energized object near energized conductors. Consider the case in Figure 6.5. The energized conductor is placed above a plane at a height h. Point p, which represents a de-energized object, is placed at a distance d from the center of the energized conductor. Assume that the plane is not distorting the electric field. This is why the equipotential surfaces are perfectly cylindrical.

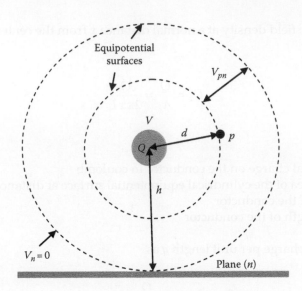

FIGURE 6.5
Induced voltage at point *p* with no ground distortion.

The potential at point *p* with respect to the plane V_{pn} can be computed by setting $x_1 = d$, and $x_2 = h$ in Equation 6.8.

$$V_{pn} = \int_d^h E_x \, dx = \frac{q}{2\pi\varepsilon_0} \ln\left(\frac{h}{d}\right) \qquad (6.9)$$

where
  d is the distance between point *p* and the center of the conductor
  h is the height of the center of the conductor above the plane
  n is the subscript indicating a point on the plane

Equation 6.2 can similarly be used to compute the potential of the energized conductor *V* with respect to the plane. In this case, we integrate the electric field from the surface of the conductor to the plane:

$$V = \int_r^h E_x \, dx = \frac{q}{2\pi\varepsilon_0} \ln\left(\frac{h}{r}\right) \qquad (6.10)$$

where
  r is the radius of the conductor
  h is the height of the conductor above the plane

The relationship between the voltage and the charge of the conductor is known as the line *capacitance C*:

$$C = \frac{q}{V} = \frac{2\pi\varepsilon_0}{\ln\left(\dfrac{h}{r}\right)} \tag{6.11}$$

Example 6.1

A 1 km, 140 kV conductor is placed 50 m above a plane. Assume the conductor diameter is 1 cm. Compute the charge of the conductor and its capacitance.

Solution:

By direct substitution in Equation 6.10, we can obtain the charge of the conductor per unit length:

$$V = \frac{q}{2\pi\varepsilon_0} \ln\left(\frac{h}{r}\right)$$

$$140 \times 10^3 = \frac{q}{2\pi \times 8.85 \times 10^{-12}} \ln\left(\frac{50}{0.005}\right)$$

$$q = 0.845\,\mu C/m$$

The total charge of the conductor is q multiplied by the length of the conductor:

$$Q = ql = 0.845 \times 1000 = 0.845\,mC$$

The capacitance of the conductor is

$$C = \frac{q}{V} = \frac{0.845 \times 10^{-6}}{140 \times 10^3} = 6.0357\,pF/m$$

Example 6.2

The radius of an energized conductor is 1 cm and its height above a plane is 25 m. The voltage of the energized line is 160 kV. Compute the induced voltage at point p that is 20 m in height and at a horizontal distance of 1 m from the energized conductor.

Solution:

Use Equation 6.9 to compute the induced voltage:

$$V_{pn} = \frac{q}{2\pi\varepsilon_0} \ln\left(\frac{h}{d}\right)$$

The distance d can be computed by the Pythagorean theorem:

$$d = \sqrt{(25-20)^2 + 1} = 5.1\,\text{m}$$

Now, we need to calculate the charge of the energized conductor using Equation 6.10:

$$V = \frac{q}{2\pi\varepsilon_0} \ln\left(\frac{h}{r}\right)$$

$$160 \times 10^3 = \frac{q}{2\pi \times 8.85 \times 10^{-12}} \ln\left(\frac{25}{0.01}\right)$$

$$q = 1.137\,\mu\text{C/m}$$

Hence, the induced voltage at point p is

$$V_{pn} = \frac{q}{2\pi\varepsilon_0} \ln\left(\frac{h}{d}\right)$$

$$V_{pn} = \frac{1.137 \times 10^{-6}}{2\pi \times 8.85 \times 10^{-12}} \ln\left(\frac{25}{5.1}\right) = 32.5\,\text{kV}$$

6.2.1 Effect of Line Length on Induced Voltage

The length of the conductor can impact the calculations of the induced voltage, especially for short conductors. Consider the arrangement in Figure 6.6, which shows two parallel conductors: conductor 1 carries a positive charge and conductor 2 carries the corresponding negative charge. Keep in mind that charges are formed in pairs; any positive charge will have a negative charge associated with it. An arbitrary point is placed at point p. d_1 and d_2 are the minimum distances (normal distances) between point p and the center of conductors 1 and 2, respectively. Assume z_1 and z_2 are arbitrary points on conductors 1 and 2, respectively. For the same

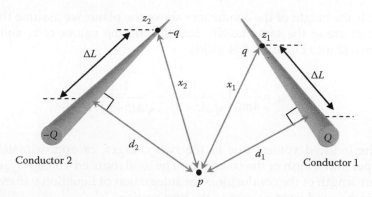

FIGURE 6.6
Parallel conductors.

segment of the line ΔL, the distances between point p and z_1 and p and z_2 using the Pythagorean theorem are

$$x_1 = \sqrt{\Delta L^2 + d_1^2}$$
$$x_2 = \sqrt{\Delta L^2 + d_2^2}$$
(6.12)

Assume the charges at z_1 and z_2 are $+q$ and $-q$, respectively. These charges can be viewed as the charges per unit length of the conductors. We can also assume that they are clusters of charges forming spheres. The electric fields at point p due to either charge is

$$E^+ = \frac{q}{4\pi\varepsilon_0 \, x_1^2}$$

$$E^- = \frac{-q}{4\pi\varepsilon_0 \, x_2^2}$$
(6.13)

where
E^+ is the electric field due to the positive charge at z_1
E^- is the electric field due to the negative charge at z_2

The induced voltage at point p with respect to a plane due to both charges is

$$v_{pn} = \int_{x_1}^{h} E^+ \, dx + \int_{x_2}^{h} E^- \, dx = \frac{q}{4\pi\varepsilon_0 \, x_1} - \frac{q}{4\pi\varepsilon_0 \, x_2}$$
(6.14)

where h is the height of the conductors above the plane; we assume the two conductors are at the same height. Substituting the values of x_1 and x_2 in Equation 6.12 into Equation 6.14 yields

$$v_{pn} = \frac{q}{4\pi\varepsilon_0}\left(\frac{1}{\sqrt{\Delta L^2 + d_1^2}} - \frac{1}{\sqrt{\Delta L^2 + d_2^2}} \right) \qquad (6.15)$$

v_{pn} is the induced voltage due to the two charges, or approximately the charge per unit length of the conductor. The total induced voltage V_{pn} due to the entire length of the conductor is the integration of Equation 6.15 over the length of the conductor, as seen in the next section.

6.2.2 Generalized Method for Short Lines

Assume that point z_1 or z_2 is located at a distance $-L_1$ from one end of the line and L_2 from the other end, as shown in Figure 6.7.

Integrating Equation 6.15 for the entire length of the conductor yields the total induced voltage V_{pn} at point p with respect to a plane:

$$V_{pn} = \frac{q}{4\pi\varepsilon_0} \int_{\Delta L=-L_1}^{L_2} \left(\frac{1}{\sqrt{\Delta L^2 + d_1^2}} - \frac{1}{\sqrt{\Delta L^2 + d_2^2}} \right) d\Delta L \qquad (6.16)$$

The integral of Equation 6.16 is

$$V_{pn} = \frac{q}{4\pi\varepsilon_0}\left(\ln\left(\frac{L + \sqrt{L^2 + d_1^2}}{L + \sqrt{L^2 + d_2^2}} \right) \right)\Bigg|_{-L_1}^{L_2} \qquad (6.17)$$

$$V_{pn} = \frac{q}{4\pi\varepsilon_0}\left(\ln\left(\frac{L_2 + \sqrt{L_2^2 + d_1^2}}{L_2 + \sqrt{L_2^2 + d_2^2}} \right) - \ln\left(\frac{-L_1 + \sqrt{L_1^2 + d_1^2}}{-L_1 + \sqrt{L_1^2 + d_2^2}} \right) \right) \qquad (6.18)$$

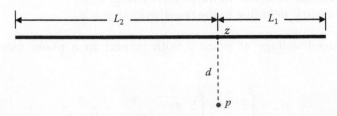

FIGURE 6.7
Induced voltage at point z near the long conductor.

$$V_{pn} = \frac{q}{4\pi\varepsilon_0} \ln\left(\left(\frac{L_2 + \sqrt{L_2^2 + d_1^2}}{L_2 + \sqrt{L_2^2 + d_2^2}}\right)\left(\frac{-L_1 + \sqrt{L_1^2 + d_2^2}}{-L_1 + \sqrt{L_1^2 + d_1^2}}\right)\right) \tag{6.19}$$

Equation 6.19 is general and can be used for any line length. Note that when point p is placed on the middle plane between the two conductors, where $d_1 = d_2$, the voltage at point p is zero.

Equation 6.19 can also be used to compute the potential of any of the two conductors with respect to a plane. This is done by placing point p at the surface of that conductor. Hence, the potential of conductor 1 with respect to a plane is

$$V_{1n} = \frac{q}{4\pi\varepsilon_0} \ln\left(\left(\frac{L_2 + \sqrt{L_2^2 + r_1^2}}{L_2 + \sqrt{L_2^2 + (D - r_2)^2}}\right)\left(\frac{-L_1 + \sqrt{L_1^2 + (D - r_2)^2}}{-L_1 + \sqrt{L_1^2 + r_1^2}}\right)\right) \tag{6.20}$$

where

r_1 is the radius of conductor 1
r_2 is the radius of conductor 2
D is the separation between the center points of the two conductors

For transmission lines, the separation between the conductors is often much larger than the radii of the conductors. Hence, we can simplify Equation 6.20 as

$$V_{1n} = \frac{q}{4\pi\varepsilon_0} \ln\left(\left(\frac{L_2 + \sqrt{L_2^2 + r_1^2}}{L_2 + \sqrt{L_2^2 + D^2}}\right)\left(\frac{-L_1 + \sqrt{L_1^2 + D^2}}{-L_1 + \sqrt{L_1^2 + r_1^2}}\right)\right) \tag{6.21}$$

Similarly, we can compute the potential of conductor 2 as

$$V_{2n} = \frac{q}{4\pi\varepsilon_0} \ln\left(\left(\frac{L_2 + \sqrt{L_2^2 + D^2}}{L_2 + \sqrt{L_2^2 + r_2^2}}\right)\left(\frac{-L_1 + \sqrt{L_1^2 + r_2^2}}{-L_1 + \sqrt{L_1^2 + D^2}}\right)\right) \tag{6.22}$$

Then, the potential difference between the two conductors is

$$V_{12} = V_{1n} - V_{2n} \tag{6.23}$$

6.2.3 Approximate Method for Long Lines

For long lines, we can approximate Equation 6.19 by assuming the lengths L_1 and L_2 are very long compared to the separation between the conductors. In this case, the first ratio inside the **ln** function is approximately equal to 1:

$$\left(\frac{L_2 + \sqrt{L_2^2 + d_1^2}}{L_2 + \sqrt{L_2^2 + d_2^2}} \right) \approx \left(\frac{2L_2}{2L_2} \right) = 1 \tag{6.24}$$

Hence, the potential at point p with respect to a plane is

$$V_{pn} = \frac{q}{4\pi\varepsilon_0} \ln\left(\frac{-L_1 + \sqrt{L_1^2 + d_2^2}}{-L_1 + \sqrt{L_1^2 + d_1^2}} \right) \tag{6.25}$$

The ratio inside the **ln** function in Equation 6.25 is undefined when L_1 is much larger than the distances d_1 and d_2. So we need to analyze it algebraically:

$$\lim_{L_1 \to \infty} \left(\frac{-L_1 + \sqrt{L_1^2 + d_2^2}}{-L_1 + \sqrt{L_1^2 + d_1^2}} \right) = \lim_{L_1 \to \infty} \left(\frac{-1 + \sqrt{1 + \left(\dfrac{d_2}{L_1}\right)^2}}{-1 + \sqrt{1 + \left(\dfrac{d_1}{L_1}\right)^2}} \right) \tag{6.26}$$

The term under the square root can be expanded into a polynomial where

$$\sqrt{1 + \left(\frac{d_1}{L_1}\right)^2} = 1 + 0.5\left(\frac{d_1}{L_1}\right)^2 + \text{HOT}$$

$$\sqrt{1 + \left(\frac{d_2}{L_1}\right)^2} = 1 + 0.5\left(\frac{d_2}{L_1}\right)^2 + \text{HOT} \tag{6.27}$$

HOT are the higher-order terms in the polynomial. They are much smaller than the first two terms for the long line because

$$\left(\frac{d_2}{L_1}\right)^2 \gg \left(\frac{d_2}{L_1}\right)^3 \gg \left(\frac{d_2}{L_1}\right)^4 \gg \cdots \tag{6.28}$$

Ignoring the HOT, Equation 6.26 can be written as

$$\lim_{L_1 \to \infty} \left(\frac{-1+\sqrt{1+\left(\dfrac{d_2}{L_1}\right)^2}}{-1+\sqrt{1+\left(\dfrac{d_1}{L_1}\right)^2}} \right) \approx \left(\frac{-1+1+0.5\left(\dfrac{d_2}{L_1}\right)^2}{-1+1+0.5\left(\dfrac{d_1}{L_1}\right)^2} \right) = \left(\frac{d_2}{d_1}\right)^2 \tag{6.29}$$

Substituting Equation 6.29 into Equation 6.25 yields

$$V_{pn} = \frac{q}{2\pi\varepsilon_0} \ln\left(\frac{d_2}{d_1}\right) \tag{6.30}$$

Equation 6.30 can be used to construct the equipotential surfaces. If we move point p in the space between the two conductors while maintaining the ratio d_2/d_1 constant, the potential of point p is unchanged. The contour that forms the points of unchanged voltage is the *equipotential surface*. Figure 6.8 shows some of the equipotential surfaces around energized conductors. Keep in mind that the equipotential surfaces are orthogonal to the electric field lines.

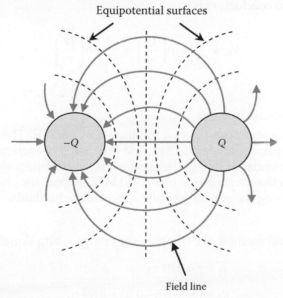

FIGURE 6.8
Equipotential surfaces.

Equation 6.30 can also be used to compute the voltage of an energized conductor. The potential of conductor 1 can be computed by placing point p at the surface of the conductor. In this case, d_1 is the radius of the conductor r_1, and d_2 is the distance between the centers of the two conductors D:

$$V_{1n} = \frac{q}{2\pi\varepsilon_0} \ln\left(\frac{D}{r_1}\right) \tag{6.31}$$

Similarly, we can compute the potential of conductor 2 by setting $d_1 = D$ and $d_2 = r_2$:

$$V_{2n} = \frac{q}{2\pi\varepsilon_0} \ln\left(\frac{r_2}{D}\right) \tag{6.32}$$

The potential difference between the two conductors is then

$$V_{12} = V_{1n} - V_{2n} = \frac{q}{2\pi\varepsilon_0}\left(\ln\left(\frac{D}{r_1}\right) - \ln\left(\frac{r_2}{D}\right)\right) \tag{6.33}$$

$$V_{12} = \frac{q}{2\pi\varepsilon_0} \ln\left(\frac{D^2}{r_1 r_2}\right)$$

If we assume that the radii of the two conductors are equal, then the voltage between the two conductors is

$$V_{12} = \frac{q}{2\pi\varepsilon_0} \ln\left(\frac{D}{r}\right)^2 = \frac{q}{\pi\varepsilon_0} \ln\left(\frac{D}{r}\right) \tag{6.34}$$

Example 6.3

A 200 m line is composed of two conductors in parallel. The separation between the two conductors is 5 m, and the radius of each conductor is 1 cm. The potential of one of the conductors with respect to a plane at the middle of the transmission line is 140 kV. Compute the charge on the conductors using the general and the approximate methods.

Solution:

For the general method, we compute the charge using Equation 6.19:

$$d_1 = r = 0.01\,\mathrm{m}$$

$$d_2 = D - r = 5 - 0.05 = 4.99\,\mathrm{m}$$

$$V = \frac{q}{4\pi\varepsilon_0}\left(\ln\left(\frac{L_2 + \sqrt{L_2^2 + d_1^2} \; -L_1 + \sqrt{L_1^2 + d_2^2}}{L_2 + \sqrt{L_2^2 + d_2^2} \; -L_1 + \sqrt{L_1^2 + d_1^2}} \right) \right)$$

$$140,000 = \frac{q}{4\pi\times 8.85\times 10^{-12}}\left(\ln\left(\frac{100 + \sqrt{100^2 + 0.01^2} \; -100 + \sqrt{100^2 + 4.99^2}}{100 + \sqrt{100^2 + 4.99^2} \; -100 + \sqrt{100^2 + 0.01^2}} \right) \right)$$

$$q = 1.253201\times 10^{-6} \; \text{C}$$

Using the long line approximation in Equation 6.30,

$$V = \frac{q}{2\pi\varepsilon_0}\ln\left(\frac{d_2}{d_1} \right)$$

$$140,000 = \frac{q}{2\pi\times 8.85\times 10^{-12}}\ln\left(\frac{4.99}{0.01} \right)$$

$$q = 1.253075\times 10^{-6} \; \text{C}$$

Note that both methods produce very comparable results even for such a short line.

Example 6.4

A transmission line is composed of two conductors in parallel forming a single circuit. The radius of each conductor is 1 cm. The charge of the conductor is 1 μC/m. Compute the induced voltage at a point located at $d_1 = 20$ m and $d_2 = 40$ m using the general method. Assume various line lengths and $L_1 = L_2$. Compare the results with the approximate method.

Solution:

For the general method, we use Equation 6.19:

$$V_{pn} = \frac{q}{4\pi\varepsilon_0}\left(\ln\left(\frac{L_2 + \sqrt{L_2^2 + d_1^2} \; -L_1 + \sqrt{L_1^2 + d_2^2}}{L_2 + \sqrt{L_2^2 + d_2^2} \; -L_1 + \sqrt{L_1^2 + d_1^2}} \right) \right)$$

$$V_{pn} = \frac{10^{-6}}{4\pi 8.85\times 10^{-12}}\left(\ln\left(\frac{L + \sqrt{L^2 + 20^2} \; -L + \sqrt{L^2 + 40^2}}{L + \sqrt{L^2 + 40^2} \; -L + \sqrt{L^2 + 20^2}} \right) \right)$$

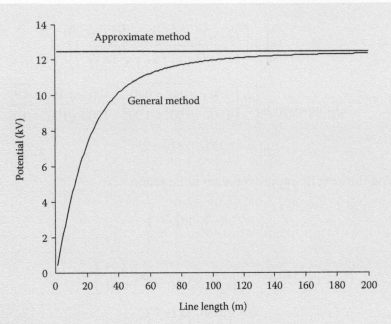

FIGURE 6.9
Comparison between the general and the approximate methods.

For the long line approximation, we use Equation 6.30:

$$V_{pn} = \frac{q}{2\pi\varepsilon_0} \ln\left(\frac{d_2}{d_1}\right)$$

$$V_{pn} = \frac{10^{-6}}{2\pi 8.85\times10^{-12}} \ln\left(\frac{40}{20}\right) = 12.47 \text{ kV}$$

A plot for the general method with respect to the line length is given in Figure 6.9. The figure also shows the result of the approximate method. Note that when the line is 200 m in length, the two methods provide very similar results.

6.3 Electric Field Distortion due to Earth

Electric field lines are perpendicular to the surface of the charged object. In space, they form straight lines leaving the positive charge, as shown in Figure 6.2. When a ground is near a charged conductor, the electrical fields leave the conductor in the radial direction, but eventually curve to reach the

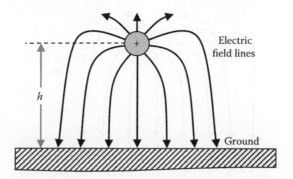

FIGURE 6.10
Electric field of charged conductor above earth.

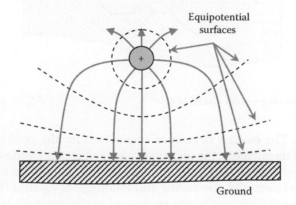

FIGURE 6.11
Equipotential surfaces with ground effect.

low voltage earth's surface, as shown in Figure 6.10. Because the equipotential surfaces are perpendicular to the field lines, the curved field lines distort the equipotential surfaces, making them imperfect cylinders around the charged conductor, as shown in Figure 6.11.

To include the earth effect in the induced voltage calculations, elaborate and complex analyses are needed. However, we can greatly simplify the calculations if we use the image charge theory and assume that the earth is an equipotential horizontal plane with permittivity equal to that for air.

6.3.1 Image Charge

The image charge method assumes that for any charge above ground, there is a fictitious opposite charge image inside the earth. If the earth's relative permittivity is close to that for air, the opposite charge is at a distance under the surface of the earth equal to the height of the conductor above earth, as shown

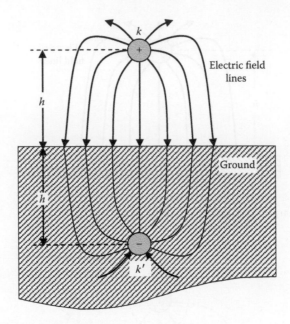

FIGURE 6.12
Charged conductor above earth and its image.

in Figure 6.12. The electric field lines originate from the positively charged conductor and terminate at the negatively charged image inside earth.

Figure 6.13 shows conductor k at height h and its image k' at depth h. The voltage between the conductor and its image is $V_{kk'}$ and the voltage of the conductor with respect to the surface of earth (neutral) is V_{kn}.

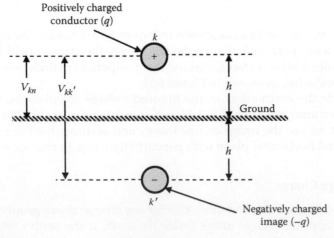

FIGURE 6.13
Potentials of conductor and its image.

The voltage $V_{kk'}$ is the integration of the electric field strength between the conductor and its image. This process is the same as that used for Equation 6.34. The separation between the two conductors $D = 2h$:

$$V_{kk'} = \frac{q}{\pi\varepsilon_0}\ln\left(\frac{2h}{r}\right)$$ (6.35)

where
  r is the radius of the conductor
  h is the height of the center of the conductor above the ground level

Because the earth's relative permittivity is assumed equal to that of air, the voltage between the conductor and the surface of the earth V_{kn} is

$$V_{kn} = \frac{V_{kk'}}{2} = \frac{q}{2\pi\varepsilon_0}\ln\left(\frac{2h}{r}\right)$$ (6.36)

Figure 6.14 shows the voltage of the energized conductor with respect to its charge, height above ground, and its radius. Let us place a de-energized conductor at point p that is located at distance d from the center of an energized conductor and d' from the center of the image of the energized conductor, as shown in Figure 6.15. Using Equation 6.9, we can develop the general equation to compute the potential of point p with respect to the ground surface. This can be done by the superposition theorem. The first step is to compute the contribution of the positive charge on the de-energized object at point p

$$V_{p+} = \frac{q}{2\pi\varepsilon_0}\ln\left(\frac{h}{d}\right)$$ (6.37)

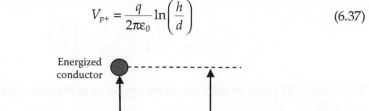

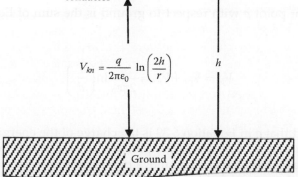

FIGURE 6.14
Voltage of energized conductor.

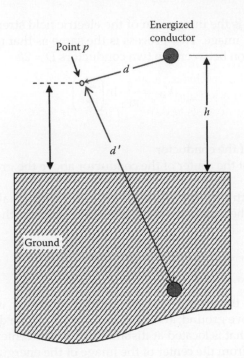

FIGURE 6.15
Voltage at a point above ground near energized conductor.

Then, compute the voltage at point p due to the image charge:

$$V_{p-} = \frac{-q}{2\pi\varepsilon_0}\ln\left(\frac{h}{d'}\right) = \frac{q}{2\pi\varepsilon_0}\ln\left(\frac{d'}{h}\right) \tag{6.38}$$

The voltage at point p with respect to ground is the sum of Equations 6.37 and 6.38:

$$V_{pn} = V_{p+} + V_{p-} = \frac{q}{2\pi\varepsilon_0}\ln\left(\frac{d'}{d}\right) \tag{6.39}$$

Keep in mind that q in Equation 6.39 is the charge of the energized conductor. The equation can be used to compute the voltage at any point in space near energized conductors. If you move point p to the surface of conductor k, you obtain the potential of the conductor with respect to the surface of the earth. In this case, $d' = 2h$, and d is the radius of the conductor r. This would lead to the same result as in Equation 6.36.

Example 6.5

Repeat Example 6.2, but include the earth effect.

Solution:

The first step is to use the voltage of the energized conductor to compute its charge. This is accomplished by using Equation 6.36:

$$V_{kn} = \frac{q}{2\pi\varepsilon_0} \ln\left(\frac{2h}{r}\right)$$

$$160 \times 10^3 = \frac{q}{2\pi \times 8.85 \times 10^{-12}} \ln\left(\frac{2 \times 25}{0.01}\right)$$

$$q = 1.0446\ \mu C/m$$

To compute the voltage at point p, we need to compute the distances d and d' using the Pythagorean theorem:

$$d = \sqrt{1^2 + (25 - 20)^2} = 5.1\ m$$

$$d' = \sqrt{1^2 + (25 + 20)^2} \approx 45\ m$$

Use Equation 6.39 to compute the potential at point p with respect to ground:

$$V_{pn} = \frac{q}{2\pi\varepsilon_0} \ln\left(\frac{d'}{d}\right)$$

$$V_{pn} = \frac{1.0446 \times 10^{-6}}{2\pi \times 8.85 \times 10^{-12}} \ln\left(\frac{45}{5.1}\right) = 40.9\ kV$$

Example 6.6

Repeat Example 6.5 but assume point p is placed at a horizontal distance of 10 m from the energized conductor and 20 m above the ground,

$$d = \sqrt{10^2 + (25 - 20)^2} = 11.2\ m$$

$$d' = \sqrt{10^2 + (25 + 20)^2} = 46.1\ m$$

Use Equation 6.39 to compute the potential at point p with respect to the ground:

$$V_{pn} = \frac{1.0446 \times 10^{-6}}{2\pi \times 8.85 \times 10^{-12}} \ln\left(\frac{46.1}{11.2}\right) = 26.58 \text{ kV}$$

Note that the induced voltage is reduced when we move away from the energized conductor.

6.3.2 Error When Earth Effect Is Ignored

Comparing Equations 6.9 through 6.39, we can compute the contribution of ground distortion to the induced voltage. Equation 6.9 gives the induced voltage without ground distortion, and Equation 6.39 gives the induced voltage with ground distortion. Hence, the difference between the two methods is

$$\Delta V = V_g - V_{wg} = \frac{q}{2\pi\varepsilon_0}\left(\ln\left(\frac{d'}{d}\right) - \ln\left(\frac{h}{d}\right)\right) = \frac{q}{2\pi\varepsilon_0}\ln\left(\frac{d'}{h}\right) \tag{6.40}$$

where
V_g is the induced voltage with ground effect
V_{wg} is the induced voltage without ground effect

The error can then computed as

$$\text{Error} = \frac{\Delta V}{V_g} = \frac{\ln(d') - \ln(h)}{\ln(d') - \ln(d)} \tag{6.41}$$

Figure 6.16 shows a sketch of Equation 6.41 with respect to the height of the tower. The figure suggests that the error due to ground distortion diminishes when the tower and point p are at high elevations.

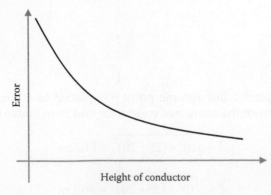

FIGURE 6.16
Effect of ground on induced voltage calculations.

Example 6.7

For the system in Example 6.5, compute the error if we ignore the ground effect.

Solution:

The error can be computed by direct substitution in Equation 6.41:

$$\text{Error} = \frac{\ln(d') - \ln(h)}{\ln(d') - \ln(d)} = \frac{\ln(45) - \ln(25)}{\ln(45) - \ln(5.1)} = 0.27$$

The error is very high and we cannot ignore the earth effect.

6.3.3 Bundled Conductors

As explained in Chapter 2, high-voltage lines above 340 kV are often arranged in bundle configuration to reduce the electric field strength at the surface of the conductor, thus reducing the corona discharges. A photo of a bundled conductor is shown in Figure 6.17.

Equation 6.36 can still be used for bundled conductor by assuming the radius of the bundle to be approximately equal to the centroid of the group. This is known as the *geometric mean radius* (GMR). Consider the general

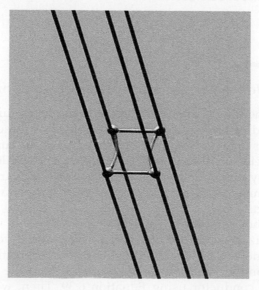

FIGURE 6.17
Bundled conductor with four subconductors.

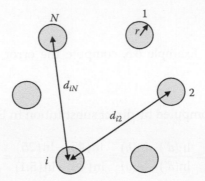

FIGURE 6.18
Bundled conductors.

arrangement in Figure 6.18 of a bundled conductor composed of N subconductors. The GMR of this bundle is

$$\text{GMR} = \sqrt[N^2]{\prod_{i=1}^{N}\prod_{j=1}^{N} d_{ij}} \qquad (6.42)$$

where
    d_{ij} is the distance between the center of subconductors i and j
    d_{ii} is the radius of subconductor i

6.4 Induced Voltage due to Multiple Energized Phases

Transmission line towers often carry more than one circuit, and each circuit is composed of three-phase wires. The effect of these multiple energized conductors on a de-energized object nearby is the cumulative effects of the electric fields of all energized conductors. Since multiple conductors in three-phase systems imply phase shifts between all variables, we must use complex number mathematics (phasor analysis) when we compute this cumulative effect.

Figure 6.19 shows multiple energized conductors at various heights and a de-energized conductor at point p. The induced voltage at p with respect to ground can be computed using the superposition theorem. In this case, we compute the voltages at point p due to the electric field of each energized conductor using Equation 6.39. Then, we add up these

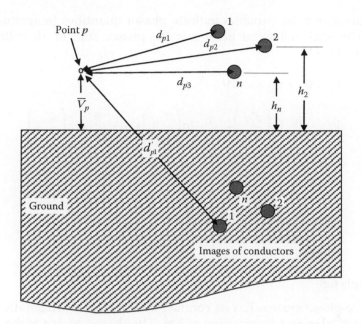

FIGURE 6.19
Induced voltage at a point above ground due to multiple energized conductors.

voltages using phasor analysis to obtain the cumulative effect of all fields at point p.

$$\bar{V}_{p1} = \frac{\bar{q}_1}{2\pi\varepsilon_0} \ln\left(\frac{d'_{p1}}{d_{p1}}\right)$$

$$\bar{V}_{p2} = \frac{\bar{q}_2}{2\pi\varepsilon_0} \ln\left(\frac{d'_{p2}}{d_{p2}}\right)$$

$$\vdots$$

$$\bar{V}_{pn} = \frac{\bar{q}_n}{2\pi\varepsilon_0} \ln\left(\frac{d'_{pn}}{d_{pn}}\right)$$

(6.43)

where
  V_{pi} is the voltage at p due to the charge of energized conductor i
  q_i is the charge of energized conductor i per unit length
  d_{pi} is the distance between point p and conductor i
  d'_{pi} is the distance between point p and the image of conductor i
  n is the total number of energized conductors

The bars above the variables indicate phasor quantities (magnitude and angle). The total voltage at point p is the phasor sum of all voltages in Equation 6.43.

$$\bar{V}_p = \bar{V}_{p1} + \bar{V}_{p2} + \cdots + \bar{V}_{pn}$$

$$\bar{V}_p = \frac{\bar{q}_1}{2\pi\varepsilon_0} \ln\left(\frac{d'_{p1}}{d_{p1}}\right) + \frac{\bar{q}_2}{2\pi\varepsilon_0} \ln\left(\frac{d'_{p2}}{d_{p2}}\right) + \cdots + \frac{\bar{q}_n}{2\pi\varepsilon_0} \ln\left(\frac{d'_{pn}}{d_{pn}}\right) \qquad (6.44)$$

or

$$\bar{V}_p = \frac{1}{2\pi\varepsilon_0} \sum_{i=1}^{n} \bar{q}_i \ln\left(\frac{d'_{pi}}{d_{pi}}\right) \qquad (6.45)$$

Example 6.8

A three-phase system has its conductors arranged horizontally at a height of 30 m and a separation of 5 m. The charge on one of the conductors is 1 μC/m. Assume that the charges on the three conductors are balanced. Compute the induced voltage at point p that is 30 m in height and at a horizontal distance of 6 m from the nearest conductor.

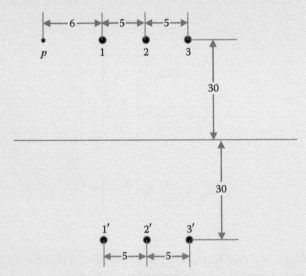

Solution:

The first step is to compute the distances between point p and all conductors including conductors' images.

$$d_{p1} = 6 \text{ m}$$

$$d_{p2} = 6 + 5 = 11 \text{ m}$$

$$d_{p3} = 6 + 10 = 16 \text{ m}$$

$$d'_{p1} = \sqrt{6^2 + (60)^2} = 60.3 \text{ m}$$

$$d'_{p2} = \sqrt{11^2 + (60)^2} = 61 \text{ m}$$

$$d'_{p3} = \sqrt{16^2 + (60)^2} = 62.1 \text{ m}$$

Use Equation 6.45 to compute the potential at point p due to all three phases:

$$\bar{V}_p = \frac{1}{2\pi\varepsilon_0} \sum_{i=1}^{n} \bar{q}_i \ln\left(\frac{d'_{pi}}{d_{pi}}\right)$$

$$\bar{V}_p = \frac{1}{2\pi\varepsilon_0} \left(\bar{q}_1 \ln\left(\frac{d'_{p1}}{d_{p1}}\right) + \bar{q}_2 \ln\left(\frac{d'_{p2}}{d_{p2}}\right) + \bar{q}_3 \ln\left(\frac{d'_{p3}}{d_{p3}}\right) \right)$$

Let us assume that conductor 1 is the reference (phase a) and conductor 2 is phase b, etc.:

$$\bar{V}_p = \frac{10^{-6}}{2\pi\,8.85\times10^{-12}} \left((1\angle 0)\ln\left(\frac{60.3}{6}\right) + (1\angle 120°)\ln\left(\frac{61}{11}\right) \right.$$

$$\left. + (1\angle -120°)\ln\left(\frac{62.1}{16}\right) \right)$$

$$\bar{V}_p = 14.97 \angle -21.8° \text{ kV}$$

6.4.1 Computation of Conductor Charge

As seen in Example 6.8, the induced voltage at any arbitrary point in space can be computed if we know the charges of the energized conductors. These charges, in turn, can be calculated if we know the phase voltages of the energized conductors (voltage with respect to ground).

Consider the system in Figure 6.19. If we move point p to the surface of conductor i, we obtain the phase voltage of conductor i:

$$\bar{V}_i = \frac{\bar{q}_1}{2\pi\varepsilon_0}\ln\left(\frac{d'_{i1}}{d_{i1}}\right) + \frac{\bar{q}_2}{2\pi\varepsilon_0}\ln\left(\frac{d'_{i2}}{d_{i2}}\right) + \cdots + \frac{\bar{q}_i}{2\pi\varepsilon_0}\ln\left(\frac{d'_{ii}}{d_{ii}}\right) + \cdots + \frac{\bar{q}_n}{2\pi\varepsilon_0}\ln\left(\frac{d'_{in}}{d_{in}}\right) \quad (6.46)$$

where
  d_{ij} is the distance between conductors i and j
  d'_{ij} is the distance between conductor i and the image of conductor j
  d_{ii} is the radius of conductor i
  d'_{ii} is the distance between conductor i and its image
  n is the number of energized conductors

$$d_{ii} = r_i$$

$$d'_{ii} = 2h_i$$

$$d_{ij} = d_{ji} \quad\quad\quad (6.47)$$

$$d'_{ij} = d'_{ji}$$

where r_i is the radius of conductor i. In bundled conductors, r_i is equal to the GMR of bundle i.

In general form, the phase voltage of conductor i is

$$\bar{V}_i = \sum_{j=1}^{n} \frac{\bar{q}_j}{2\pi\varepsilon_0}\ln\left(\frac{d'_{ij}}{d_{ij}}\right) \quad (6.48)$$

We can generalize Equation 6.48 to include all n energized conductors in matrix form:

$$
\begin{bmatrix} \bar{V}_1 \\ \bar{V}_2 \\ \vdots \\ \bar{V}_n \end{bmatrix} = \frac{1}{2\pi\varepsilon_0}
\begin{bmatrix}
\ln\left(\dfrac{2h_1}{r_1}\right) & \ln\left(\dfrac{d'_{12}}{d_{12}}\right) & \cdots & \ln\left(\dfrac{d'_{1n}}{d_{1n}}\right) \\
\ln\left(\dfrac{d'_{21}}{d_{21}}\right) & \ln\left(\dfrac{2h_2}{r_2}\right) & \cdots & \ln\left(\dfrac{d'_{2n}}{d_{2n}}\right) \\
\vdots & \vdots & \ddots & \vdots \\
\ln\left(\dfrac{d'_{n1}}{d_{n1}}\right) & \ln\left(\dfrac{d'_{n2}}{d_{n2}}\right) & \cdots & \ln\left(\dfrac{2h_n}{r_n}\right)
\end{bmatrix}
\begin{bmatrix} \bar{q}_1 \\ \bar{q}_2 \\ \vdots \\ \bar{q}_n \end{bmatrix} \quad (6.49)
$$

Hence, the charges of the energized conductor are

$$
\begin{bmatrix} \bar{q}_1 \\ \bar{q}_2 \\ \vdots \\ \bar{q}_n \end{bmatrix} = 2\pi\varepsilon_0 \begin{bmatrix} \ln\left(\dfrac{2h_1}{r_1}\right) & \ln\left(\dfrac{d'_{12}}{d_{12}}\right) & \cdots & \ln\left(\dfrac{d'_{1n}}{d_{1n}}\right) \\ \ln\left(\dfrac{d'_{21}}{d_{21}}\right) & \ln\left(\dfrac{2h_2}{r_2}\right) & \cdots & \ln\left(\dfrac{d'_{2n}}{d_{2n}}\right) \\ \vdots & \vdots & \ddots & \vdots \\ \ln\left(\dfrac{d'_{n1}}{d_{n1}}\right) & \ln\left(\dfrac{d'_{n2}}{d_{n2}}\right) & & \ln\left(\dfrac{2h_n}{r_n}\right) \end{bmatrix}^{-1} \begin{bmatrix} \bar{V}_1 \\ \bar{V}_2 \\ \vdots \\ \bar{V}_n \end{bmatrix} \tag{6.50}
$$

Since the ratio of the charge to the voltage is the capacitance, Equation 6.50 can be written as

$$
\mathbf{q} = \mathbf{C}\mathbf{V} \tag{6.51}
$$

where $\mathbf{C}$ is the system capacitance matrix. The unit of $\mathbf{C}$ is F/m:

$$
\mathbf{C} = 2\pi\varepsilon \begin{bmatrix} \ln\left(\dfrac{2h_1}{r_1}\right) & \ln\left(\dfrac{d'_{12}}{d_{12}}\right) & \cdots & \ln\left(\dfrac{d'_{1n}}{d_{1n}}\right) \\ \ln\left(\dfrac{d'_{21}}{d_{21}}\right) & \ln\left(\dfrac{2h_2}{r_2}\right) & \cdots & \ln\left(\dfrac{d'_{2n}}{d_{2n}}\right) \\ \vdots & \vdots & \ddots & \vdots \\ \ln\left(\dfrac{d'_{n1}}{d_{n1}}\right) & \ln\left(\dfrac{d'_{n2}}{d_{n2}}\right) & & \ln\left(\dfrac{2h_n}{r_n}\right) \end{bmatrix}^{-1} \tag{6.52}
$$

The elements in the matrix give the capacitances between the various conductors. They are shown in Figure 6.20 and are represented by

$$
\mathbf{C} = \begin{bmatrix} C_{11} & C_{12} & \cdots & C_{1n} \\ C_{21} & C_{22} & \cdots & C_{2n} \\ \vdots & \vdots & \ddots & \vdots \\ C_{n1} & C_{n2} & & C_{nn} \end{bmatrix} ; \quad \text{where} \quad C_{ij} = C_{ji} \tag{6.53}
$$

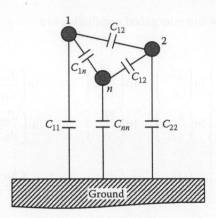

FIGURE 6.20
System capacitance.

Example 6.9

Assume a balanced three-phase system whose conductors are arranged as shown in Figure 6.21. The distances in the figure are in meters. Each conductor is made out of a bundle where the GMR is 0.25 m. The line-to-line voltage of the system is 500 kV and the system is balanced. Compute the charges of the conductors assuming that the height of conductor c is 15 m.

Solution:

To compute the distances between the conductors and the images of the conductors, we can use the Pythagorean theorem:

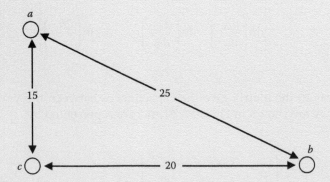

FIGURE 6.21
Three-phase conductors arranged as equilateral triangle.

$$d'_{ab} = \sqrt{(x_a - x'_b)^2 + (y_a - y'_b)^2} = \sqrt{(0-20)^2 + (30+15)^2} = 49.24 \text{ m}$$

$$d'_{ac} = \sqrt{(x_a - x'_c)^2 + (y_a - y'_c)^2} = \sqrt{(0-0)^2 + (30+15)^2} = 45 \text{ m}$$

$$d'_{bc} = \sqrt{(x_b - x'_c)^2 + (y_b - y'_c)^2} = \sqrt{(20-0)^2 + (15+15)^2} = 36 \text{ m}$$

$$
\begin{bmatrix}
\dfrac{500 \angle 0°}{\sqrt{3}} \\[2mm]
\dfrac{500 \angle -120°}{\sqrt{3}} \\[2mm]
\dfrac{500 \angle 120°}{\sqrt{3}}
\end{bmatrix}
=
\frac{1}{2\pi\varepsilon_0}
\begin{bmatrix}
\ln\left(\dfrac{60}{0.25}\right) & \ln\left(\dfrac{49.24}{25}\right) & \ln\left(\dfrac{45}{15}\right) \\[2mm]
\ln\left(\dfrac{49.24}{25}\right) & \ln\left(\dfrac{30}{0.25}\right) & \ln\left(\dfrac{36}{20}\right) \\[2mm]
\ln\left(\dfrac{45}{15}\right) & \ln\left(\dfrac{36}{20}\right) & \ln\left(\dfrac{30}{0.25}\right)
\end{bmatrix}
\begin{bmatrix}
q_a \\ q_b \\ q_c
\end{bmatrix}
$$

The solution of the above equation yields the charges of the conductors:

$$
\begin{bmatrix}
q_a \\ q_b \\ q_c
\end{bmatrix}
=
\begin{bmatrix}
3.63 \\ 3.8 \\ 4.07
\end{bmatrix}
\mu C/m
$$

6.4.2 Computational Steps for Induced Voltage on De-Energized Conductors

It is common for linemen to work on de-energized lines in the vicinity of energized circuits. Often the energized circuit is mounted on the same tower as the de-energized circuit. To assess the hazards, we need to compute the induced voltage on the de-energized circuit. This can be done in two steps:

1. Use the voltage of the energized circuit to compute the charges on its conductors (see Equation 6.50).
2. Place point p at the surface of the de-energized conductor and use the charges on the energized circuit to compute the induced voltage at p (see Equation 6.45).

Example 6.10

The tower of a double-circuit transmission line is shown in Figure 6.22. The dimensions in the figure are in meters. The GMR of each conductor is 30 cm. The voltage of the transmission line circuit at the left of the tower is 500 kV (line-to-line). The circuit on the right side of the tower is de-energized. Compute the induced voltage on conductor $a2$.

Solution:

The first step is to compute the charges of the energized conductors. To do this, we need to compute the distances between all energized conductors using the Pythagorean theorem:

$$d_{a1b1} = \sqrt{(0-3)^2 + (22-33)^2} = 11.4 \text{ m}$$

$$d_{a1c1} = \sqrt{(0-6)^2 + (22-22)^2} = 6 \text{ m}$$

$$d_{b1c1} = \sqrt{(3-6)^2 + (22-33)^2} = 11.4 \text{ m}$$

$$d'_{a1b1} = \sqrt{(0-3)^2 + (22+33)^2} = 55.08 \text{ m}$$

$$d'_{a1c1} = \sqrt{(0-6)^2 + (22+22)^2} = 44.41 \text{ m}$$

$$d'_{b1c1} = \sqrt{(3-6)^2 + (22+33)^2} = 55.08 \text{ m}$$

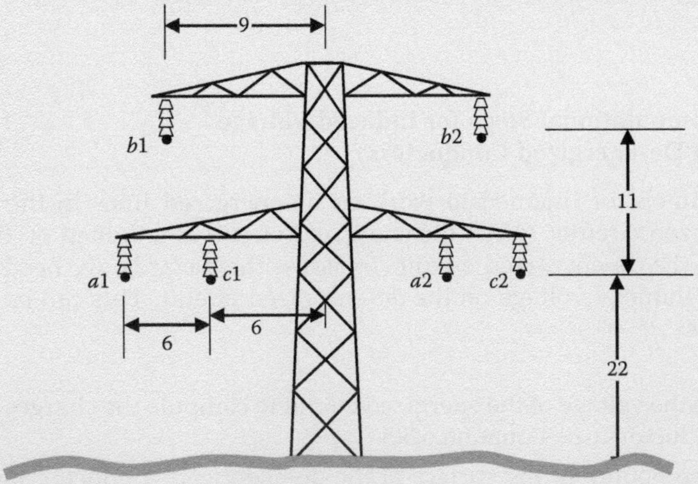

FIGURE 6.22
Configuration of a double-circuit transmission line.

Now use Equation 6.50 to compute the charges of the energized conductors:

$$\begin{bmatrix} \bar{q}_{a1} \\ \bar{q}_{b1} \\ \bar{q}_{c1} \end{bmatrix} = 2\pi\varepsilon_0 \begin{bmatrix} \ln\left(\dfrac{2h_{a1}}{r_{a1}}\right) & \ln\left(\dfrac{d'_{a1b1}}{d_{a1b1}}\right) & \ln\left(\dfrac{d'_{a1c1}}{d_{a1c1}}\right) \\ \ln\left(\dfrac{d'_{a1b1}}{d_{a1b1}}\right) & \ln\left(\dfrac{2h_{b1}}{r_{b1}}\right) & \ln\left(\dfrac{d'_{b1c1}}{d_{b1c1}}\right) \\ \ln\left(\dfrac{d'_{a1c1}}{d_{a1c1}}\right) & \ln\left(\dfrac{d'_{b1c1}}{d_{b1c1}}\right) & \ln\left(\dfrac{2h_{c1}}{r_{c1}}\right) \end{bmatrix}^{-1} \begin{bmatrix} \bar{V}_{a1} \\ \bar{V}_{b1} \\ \bar{V}_{c1} \end{bmatrix}$$

$$\begin{bmatrix} \bar{q}_{a1} \\ \bar{q}_{b1} \\ \bar{q}_{c1} \end{bmatrix} = 2\pi\times8.85\times10^{-12} \begin{bmatrix} \ln\left(\dfrac{44}{0.3}\right) & \ln\left(\dfrac{55.08}{11.4}\right) & \ln\left(\dfrac{44.41}{6}\right) \\ \ln\left(\dfrac{55.08}{11.4}\right) & \ln\left(\dfrac{66}{0.3}\right) & \ln\left(\dfrac{55.08}{11.4}\right) \\ \ln\left(\dfrac{44.41}{6}\right) & \ln\left(\dfrac{55.08}{11.4}\right) & \ln\left(\dfrac{44}{0.3}\right) \end{bmatrix}^{-1}$$

$$\times \begin{bmatrix} \dfrac{500,000\,\angle 0}{\sqrt{3}} \\ \dfrac{500,000\,\angle -120°}{\sqrt{3}} \\ \dfrac{500,000\,\angle 120°}{\sqrt{3}} \end{bmatrix}$$

The solution of the above equation yields the charges of the lines:

$$\begin{bmatrix} \bar{q}_{a1} \\ \bar{q}_{b1} \\ \bar{q}_{c1} \end{bmatrix} = \begin{bmatrix} 5.1\,\angle 5.77° \\ 4.2\,\angle -120° \\ 5.1\,\angle 125.77° \end{bmatrix} \mu C/m$$

The second step is to place point p on the de-energized conductor $a2$. Use Equation 6.45 to compute the induced voltage on $a2$:

$$\bar{V}_{a2} = \frac{\bar{q}_{a1}}{2\pi\varepsilon_0}\ln\left(\frac{d'_{a1a2}}{d_{a1a2}}\right) + \frac{\bar{q}_{b1}}{2\pi\varepsilon_0}\ln\left(\frac{d'_{b1a2}}{d_{b1a2}}\right) + \frac{\bar{q}_c}{2\pi\varepsilon_0}\ln\left(\frac{d'_{c1u2}}{d_{c1a2}}\right)$$

where

$$d_{a1a2} = 18 \text{ m}$$

$$d_{b1a2} = \sqrt{(15)^2 + (11)^2} = 18.6 \text{ m}$$

$$d_{c1a2} = 12 \text{ m}$$

$$d'_{a1a2} = \sqrt{(0-18)^2 + (22+22)^2} = 47.54 \text{ m}$$

$$d'_{b1a2} = \sqrt{(3-18)^2 + (33+22)^2} = 57 \text{ m}$$

$$d'_{c1a2} = \sqrt{(6-18)^2 + (22+22)^2} = 45.6 \text{ m}$$

Hence,

$$V_{a2} = 30.55 \ \angle 145.7° \text{ kV}$$

6.5 Effect of Line Configuration on Induced Voltage

The configuration of the energized circuit can change the induced voltage on nearby de-energized objects. Take for example the configurations in Figures 6.23 and 6.24. In Figure 6.23, the energized conductors are arranged linearly; in Figure 6.24, they are arranged as an equilateral triangle. If the distance between point p and conductor a (d_{pa}) is larger than the separation between the conductors (d), the triangle configuration will yield a lower induced voltage on point p (see Example 6.11).

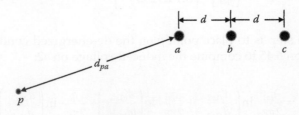

FIGURE 6.23
Energized conductors in linear configuration.

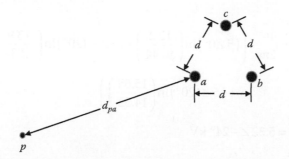

FIGURE 6.24
Energized conductors in triangle configuration.

Example 6.11

Assume a balanced three-phase circuit with each conductor charged with 5 μC/m. In the linear and triangular configurations in Figures 6.23 and 6.24, the coordination of point p is (0,2) and that for phase a is (10,5). The distance d between the conductors is 2 m. Compute the induced voltage at point p in both cases.

Solution:

For the linear configuration, the coordinates of conductors b and c are (12,5) and (14,5), respectively. The coordinates of the images of a, b, and c are (10,−5), (12,−5), and (14,−5), respectively.

The distances between all energized conductors using the Pythagorean theorem are as follows:

$$d_{pa} = \sqrt{(0-10)^2 + (2-5)^2} = 10.44 \text{ m}$$

$$d_{pb} = \sqrt{(0-12)^2 + (2-5)^2} = 12.37 \text{ m}$$

$$d_{pc} = \sqrt{(0-14)^2 + (2-5)^2} = 14.32 \text{ m}$$

$$d'_{pa} = \sqrt{(0-10)^2 + (2+5)^2} = 12.2 \text{ m}$$

$$d'_{pb} = \sqrt{(0-12)^2 + (2+5)^2} = 13.9 \text{ m}$$

$$d'_{pc} = \sqrt{(0-14)^2 + (2+5)^2} = 15.65 \text{ m}$$

Using Equation 6.45, we can compute the induced voltage at point p:

$$\bar{V}_p = \frac{1}{2\pi\varepsilon_0} \sum_{i=1}^{n} \bar{q}_i \ln\left(\frac{d'_{pi}}{d_{pi}}\right) = \frac{1}{2\pi\varepsilon_0}\left(\bar{q}_a \ln\left(\frac{d'_{pa}}{d_{pb}}\right) + \bar{q}_b \ln\left(\frac{d'_{ph}}{d_{pb}}\right) + \bar{q}_c \ln\left(\frac{d'_{pc}}{d_{pc}}\right)\right)$$

$$\bar{V}_p = \frac{10^{-6}}{2\pi\varepsilon_0}\left((5\angle 0°)\ln\left(\frac{12.2}{10.44}\right)+(5\angle -120°)\ln\left(\frac{13.9}{12.37}\right)\right.$$

$$\left.+(5\angle 120°)\ln\left(\frac{15.65}{14.32}\right)\right)$$

$$\bar{V}_p = 5.32\angle -24° \text{ kV}$$

For the triangular configuration, the coordinates of conductors b and c are (12,5) and (11,6.73), respectively. The coordinates of the images of a, b, and c are (10,−5), (12,−5), and (11,−6.73), respectively.

The distances between all energized conductors using the Pythagorean theorem are as follows:

$$d_{pa} = \sqrt{(0-10)^2+(2-5)^2} = 10.44 \text{ m}$$

$$d_{pb} = \sqrt{(0-12)^2+(2-5)^2} = 12.37 \text{ m}$$

$$d_{pc} = \sqrt{(0-11)^2+(2-6.73)^2} = 11.97 \text{ m}$$

$$d'_{pa} = \sqrt{(0-10)^2+(2+5)^2} = 12.2 \text{ m}$$

$$d'_{pb} = \sqrt{(0-12)^2+(2+5)^2} = 13.9 \text{ m}$$

$$d'_{pc} = \sqrt{(0-11)^2+(2+6.73)^2} = 14.04 \text{ m}$$

Using Equation 6.45, we can compute the induced voltage at point p:

$$\bar{V}_p = \frac{1}{2\pi\varepsilon_0}\sum_{i=1}^{n}\bar{q}_i\ln\left(\frac{d'_{pi}}{d_{pi}}\right)=\frac{1}{2\pi\varepsilon_0}\left(\bar{q}_a\ln\left(\frac{d'_{pa}}{d_{pb}}\right)+\bar{q}_b\ln\left(\frac{d'_{pb}}{d_{pb}}\right)+\bar{q}_c\ln\left(\frac{d'_{pc}}{d_{pc}}\right)\right)$$

$$\bar{V}_p = \frac{10^{-6}}{2\pi\varepsilon_0}\left((5\angle 0°)\ln\left(\frac{12.2}{10.44}\right)+(5\angle -120°)\ln\left(\frac{13.9}{12.37}\right)\right.$$

$$\left.+(5\angle 120°)\ln\left(\frac{14.04}{11.97}\right)\right)$$

$$\bar{V}_p = 3.75\angle 64.5° \text{ kV}$$

By arranging the three-phase conductors in triangle configuration, the induced voltage at p is reduced by

$$\Delta V = \frac{5.32 - 3.75}{5.32} = 29.5\%$$

6.6 Induced Voltage due to Energized Cables

Figure 6.25 shows a typical three-phase cable. It consists of an outer layer made of material that can withstand the abrasive conditions of the surrounding environment. The armored steel wire layer provides the cable with flexibility and strength. This layer is grounded. The cushion and conductors are placed inside the armored layer.

The three-phase cables produce less induced voltage on objects placed outside the cable because of three reasons:

1. The three phases are arranged in triangular formation (see Example 6.11).
2. The cables are surrounded by a conductive armored layer that is grounded. This would confine most of the electric field inside the cables.
3. If the cable is buried, any leaked electric field is trapped inside the soil.

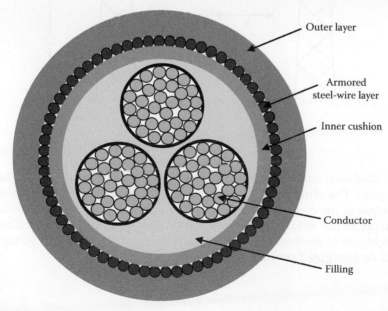

Outer layer

Armored
steel-wire layer

Inner cushion

Conductor

Filling

FIGURE 6.25
Main components of a three-phase cable.

Exercises

6.1 A balanced three-phase system has its conductors arranged as equilateral triangle. The distance between any two conductors is 10 m and the magnitude of the charge in each conductor is 3 μC/m. Compute the electric field at point p located at the orthocenter of the triangle. If one of the conductors is disconnected, compute the electric field strength at point p. Can you draw a conclusion?

6.2 A transmission line tower is shown in the following figure. The dimensions in the figure are in meters. The radius of the conductors is 6 cm. The voltage of the transmission line is 230 kV (line-to-line). Compute the induced voltage at point p located 10 m under phase c.

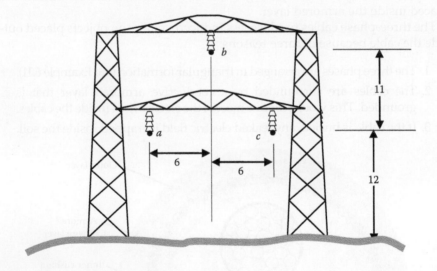

6.3 A double-circuit transmission line tower is shown in the following figure. The dimensions in the figure are in meters. The equivalent radius of the conductors is 30 cm. The voltage of the transmission line circuit at the left of the tower is 340 kV (line-to-line). The circuit on the right side of the tower is de-energized. Compute the induced voltage on all de-energized conductors.

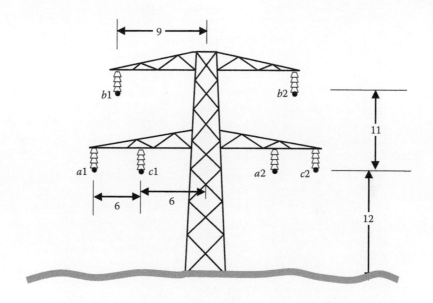

7

Induced Voltage due to Magnetic Field

A magnetic field can induce voltage on nearby de-energized and isolated metallic objects. For transmission lines, the induced voltage due to magnetic field
is usually less than that induced by the electric field. Thus, it is sometimes
ignored for transmission lines operating under normal conditions. However,
the effect of a magnetic field cannot be ignored for the following cases:

- When the energized conductor is under transient or fault conditions—this is because the surge in current is high enough to produce
  substantial induced voltage on de-energized conductors
- In distribution networks where the current is high, the voltage is
  low, and towers are short
- When industrial equipment is carrying high currents
- When the energized and de-energized lines are running in parallel
  on the same tower for a long distance

IEEE Standard 524a-1993 describes the hazards of induced voltages due to
magnetic fields. The code states

> Magnetic fields can induce hazardous open circuit voltages in partially
> grounded loops of de-energized lines adjacent to energized lines. This
> voltage can be as high as 300 V/mile under normal load conditions or
> as high as 5000 V/mile under short circuit conditions. In addition, haz-
> ardous current levels can be induced in grounded loops to levels that
> create hazardous step or touch voltages at or near ground electrodes.

7.1 Flux and Flux Linkage

An energized transmission line forms a single loop where the current passes
inside the conductor and returns through the ground. The current inside
an energized conductor produces a magnetic flux Φ that surrounds the conductor, as shown in Figure 7.1. Ideally, these flux contours are circular and
concentric with the conductor.

There are two important terms for flux that need to be explained: *flux* Φ
and *flux linkage* Λ. A flux is generated by a current passing through a conductor. The flux linkage is the amount of flux that surrounds an object located
within the flux contours. Flux linkage is determined by the location of the

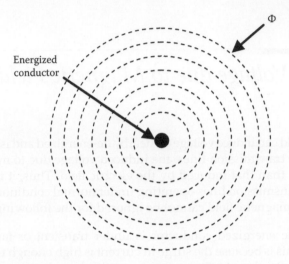

FIGURE 7.1
Flux generated by energized conductor.

object inside the flux contours. Therefore, we cannot say "flux linkage" without identifying the object it is linked to.

Since a transmission line conductor is a single loop, the flux linkage of the energized conductor in Figure 7.1, Λ due to the current inside the conductor is the same as the total flux Φ produced by the current:

$$\Lambda = \Phi \tag{7.1}$$

The unit of Λ is Weber (Wb or V s). The flux linkage of the energized conductor is often normalized per unit length of the conductor; hence,

$$\lambda = \frac{\Lambda}{l} \tag{7.2}$$

where
l is the length of the conductor
λ is the flux linkage of the energized conductor per meter; the unit of λ is Wb/m

Assume an energized conductor is in space. If we place a de-energized conductor near the energized conductor as shown in Figure 7.2, some of the total flux will surround the de-energized conductor (dashed lines in the figure). These are the *flux linking the de-energized conductor*. The flux that links the de-energized conductor λ_p is all the flux contours that pass through the de-energized conductor from its location to infinity.

For the purpose of computing the induced voltage on the de-energized conductor, we use the ground as a reference frame. In this case, the flux linking the de-energized conductor and induces voltage above the ground level is all the flux that passes between the de-energized conductor and point n on the ground,

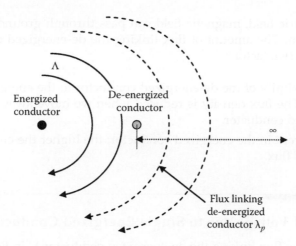

FIGURE 7.2
Flux and flux linkage in space.

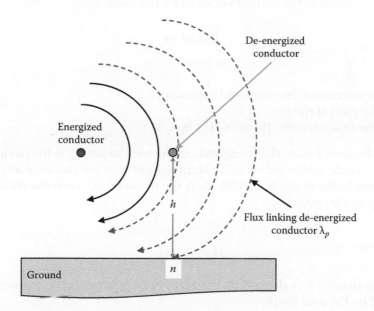

FIGURE 7.3
Flux and flux linkage for induced voltage calculations.

as shown in Figure 7.3. This is represented by a rectangular plane whose dimensions are the length of the de-energized line and the height of the de-energized conductor above ground. The flux that induced voltage on the de-energized conductor, with respect to point n, is the perpendicular flux to the plane. In our calculations, we shall assume the worst condition by assuming all the flux linkage of the de-energized conductor are perpendicular to the plane.

Unlike electric field, magnetic field can pass through ground with no or little distortion. The amount of flux linking the de-energized conductor is dependent on two factors:

1. The proximity of the de-energized conductor to the energized conductor. The flux density is reduced when we move away from the energized conductor.
2. The current in the energized conductor; the higher the current, the higher is flux.

7.2 Induced Voltage due to Single Energized Conductor

To compute the flux linking the de-energized conductor λ_p in Figure 7.3, we need to use Ampere's law, which states that the flux intensity H enclosed by a path S is equal to the current enclosed by the path:

$$\oint H \, dS = i \tag{7.3}$$

where
i is the current of the energized conductor
S is the path of the flux
H is the flux intensity; the unit of H is A/m

At a distance x from the energized conductor, the path S is the circumference of a circle with a radius x. We shall assume that the radius of any energized conductor is much smaller than the distance between the conductor and any nearby object.

Hence

$$H_x(2\pi x) = i \tag{7.4}$$

The flux density B is defined as the amount of flux falling on a given area divided by the area itself:

$$B = \frac{d\Phi}{dA} \tag{7.5}$$

where
Φ is the total flux produced by the energized conductor in Wb or V s
$d\Phi$ is the flux falling in a small area dA
dA is the area receiving $d\Phi$
B is the flux density in Wb/m^2

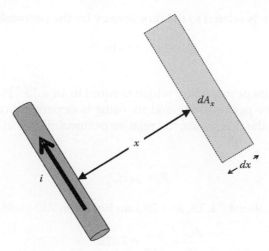

FIGURE 7.4
Conductor carrying current and adjacent area.

Consider Figure 7.4 where a conductor carries a current i. The nearby small area dA_x is at a distance x from the conductor. The flux density B_x in this area is

$$B_x = \frac{d\Phi_x}{dA_x} = \frac{d\Phi_x}{dl\, dx} \qquad (7.6)$$

where
$d\Phi_x$ is the flux falling on the area dA_x
dx is the width of the area
dl is the length of the area

Because a transmission conductor forms a single loop, the flux linkage term is used instead of just the flux. The flux linkage $d\lambda_x$ at a distance x per unit length of the conductor is

$$d\lambda_x = \frac{d\Lambda_x}{dl} \qquad (7.7)$$

Hence, the flux density at a distance x is

$$B_x = \frac{d\lambda_x}{dx} \qquad (7.8)$$

The flux intensity is related to the flux density by the permeability μ

$$B_x = \mu H_x = \mu_0 \mu_r H_x \tag{7.9}$$

where
    μ_0 is the absolute permeability which is equal to $4\pi \times 10^{-7}$ H/m
    μ_r is the relative permeability and its value is dependent on the medium
    that houses the flux; in air, the relative permeability is 1

Hence

$$B_x = \mu_0 H_x \tag{7.10}$$

Substituting Equations 7.4, 7.8, and 7.9 into Equation 7.10 yields

$$\frac{d\lambda_x}{dx} = \frac{\mu_0 i}{2\pi x} = 2 \times 10^{-7} \frac{i}{x} \tag{7.11}$$

Equation 7.11 is general and can be used to compute the flux linking an object near power lines. In Figure 7.5, for example, we can compute the flux linking the de-energized conductor λ_p by integrating Equation 7.11 with respect to distance x. The induced voltage of the de-energized conductor is computed with respect to the nearest ground potential. In this case, the limits of x are from d to D. Hence,

$$\lambda_p = 2 \times 10^{-7} i \int_d^D \frac{1}{x} dx \tag{7.12}$$

$$\lambda_p = 2 \times 10^{-7} i \ln\left(\frac{D}{d}\right) \tag{7.13}$$

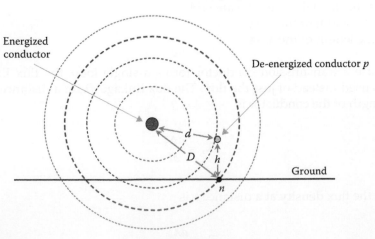

FIGURE 7.5
De-energized conductor in the vicinity of an energized conductor.

where

d is the distance between the energized conductor and the de-energized conductor

D is the distance between the energized conductor and the ground point below the de-energized conductor (n)

The induced voltage e_p on the de-energized conductor per unit length is proportional to the rate of change of flux linkage λ_p, with respect to time.

$$e_p = \frac{d\lambda_d}{dt} = 2\times10^{-7}\ln\left(\frac{D}{d}\right)\frac{di}{dt} \qquad (7.14)$$

Since λ_p is the flux linkage per meter, e_p is the induced voltage on the de-energized conductor per meter.

If you assume that the current of the energized line is sinusoidal

$$i = I_{max}\sin(\omega t) \qquad (7.15)$$

then

$$\frac{di}{dt} = \omega I_{max}\cos(\omega t) = \sqrt{2}\,\omega I\cos(\omega t) \qquad (7.16)$$

where I is the rms value of the current. Now, we can rewrite Equation 7.14 in rms quantities as

$$E_p = 2\times10^{-7}\omega I\ln\left(\frac{D}{d}\right) \qquad (7.17)$$

E_p is the induced voltage in rms on the de-energized conductor per unit length.

Example 7.1

A de-energized conductor is placed at $d = 0.5$ m from an energized conductor. The distance $D = 10$ m. The two conductors are in parallel for 10 km. If the current of the energized conductor is 500 A, compute the induced voltage on the de-energized conductor.

Solution:

Use Equation 7.17:

$$E_p = 2\times10^{-7}\omega I\ln\left(\frac{D}{d}\right)$$

$$E_p = 2 \times 10^{-7} \times 377 \times 500 \times \ln\left(\frac{10}{0.5}\right) = 112.94 \text{ mV/m}$$

Since the two lines are in parallel for 10 km, the total induced voltage V_p on the de-energized conductor is

$$V_p = E_p \, l = 112.94 \times 10,000 = 1.1294 \text{ kV}$$

7.3 Induced Voltage due to Multiple Energized Conductors

For multiconductor systems, shown in Figure 7.6, each one of the conductors contributes to the induced voltage on the de-energized conductor. If the currents of the energized conductors are not all in phase, phasor analysis must be used to compute the total induced voltage at p.

To compute the total flux linking the de-energized conductor at p due to all energized conductors, we can use the superposition theorem

$$\bar{\lambda}_p = \bar{\lambda}_1 + \cdots + \bar{\lambda}_n = \sum_{k=1}^{n} \bar{\lambda}_k \tag{7.18}$$

where
λ_k is the flux linking p due to the current of energized conductor k
n is the total number of energized conductors

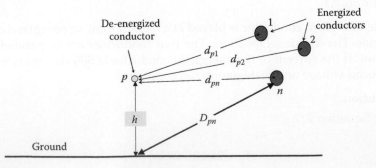

FIGURE 7.6
De-energized conductor in the vicinity of several energized conductors.

Note that the variables in Equation 7.18 are all phasor quantities to account for the effect of the phase shifts between the currents of the energized conductors. Substituting Equation 7.13 into Equation 7.18 yields

$$\bar{\lambda}_p = \sum_k \bar{\lambda}_k = 2\times10^{-7} \sum_k \bar{i}_k \ln\left(\frac{D_{pk}}{d_{pk}}\right) \tag{7.19}$$

where
i_k is the current inside the energized conductor k
d_{pk} is the distance between the de-energized conductor p and the energized conductor k
D_{pk} is the distance between the energized conductor k and the projection point of the de-energized conductor on ground
λ_p is the total flux linking the de-energized conductor p

Since the induced voltage on the de-energized conductor is the derivative of all flux linking the de-energized conductor, hence

$$e_p = \frac{d\lambda_p}{dt} = 2\times10^{-7} \sum_k \ln\left(\frac{D_{pk}}{d_{pk}}\right)\frac{d\bar{i}_k}{dt} \tag{7.20}$$

7.3.1 Induced Voltage due to Steady-State Current

For a three-phase system under steady-state conditions, the currents of the energized conductors can be represented by

$$i_a = I_{a\max} \sin(\omega t - \theta_a) = I_a \angle -\theta_a$$

$$i_b = I_{b\max} \sin(\omega t - \theta_b) = I_b \angle -\theta_b \tag{7.21}$$

$$i_c = I_{c\max} \sin(\omega t - \theta_c) = I_c \angle -\theta_c$$

where
$I_{a\,max}$ is the maximum (peak) current of phase a
θ_a is the phase angle of phase a

The derivatives of these currents are

$$\frac{di_a}{dt} = \omega I_{a\max} \cos(\omega t - \theta_a)$$

$$\frac{di_b}{dt} = \omega I_{b\max} \cos(\omega t - \theta_b) \tag{7.22}$$

$$\frac{di_c}{dt} = \omega I_{c\max} \cos(\omega t - \theta_c)$$

In rms quantities,

$$\frac{di_a}{dt} \Rightarrow \omega I_a \angle\left(90° - \theta_a\right)$$

$$\frac{di_b}{dt} \Rightarrow \omega I_b \angle\left(90° - \theta_b\right) \qquad (7.23)$$

$$\frac{di_c}{dt} \Rightarrow \omega I_c \angle\left(90° - \theta_c\right)$$

Equation 7.20 can now be written in rms quantities for the three-phase system as

$$E_p = 2\times10^{-7}\,\omega\left[\ln\left(\frac{D_{pa}}{d_{pa}}\right) \quad \ln\left(\frac{D_{pb}}{d_{pb}}\right) \quad \ln\left(\frac{D_{pc}}{d_{pc}}\right)\right]\begin{bmatrix} I_a \angle\left(90° - \theta_a\right) \\ I_b \angle\left(90° - \theta_b\right) \\ I_c \angle\left(90° - \theta_c\right) \end{bmatrix} \qquad (7.24)$$

Example 7.2

For the system in Figure 7.7, circuit 1 on the left of the tower is energized and circuit 2 is de-energized. All dimensions are in meters. The current in each of the energized conductors is 1 kA, and the two circuits run in parallel for 10 km. Compute the induced voltage on conductor $a2$ due to the magnetic field.

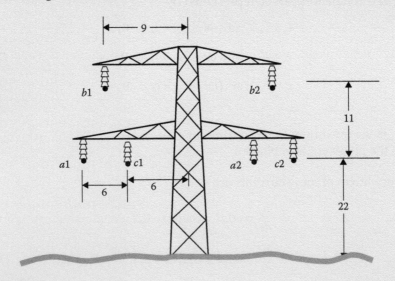

FIGURE 7.7
Double-circuit transmission lines.

Solution:

The first step is to compute all distances:

$$d_{a2-a1} = 18 \text{ m}$$

$$d_{a2-b1} = \sqrt{15^2 + 11^2} = 18.6 \text{ m}$$

$$d_{a2-c1} = 12 \text{ m}$$

$$D_{a2-a1} = \sqrt{18^2 + 22^2} = 28.43 \text{ m}$$

$$D_{a2-b1} = \sqrt{15^2 + 33^2} = 36.25 \text{ m}$$

$$D_{a2-c1} = \sqrt{12^2 + 22^2} = 25.06 \text{ m}$$

Assume that the currents in the energized conductors are balanced, that is,

$$\theta_a = 0°$$

$$\theta_b = -120°$$

$$\theta_c = 120°$$

Using Equation 7.24, the induced voltage per unit length on conductor E_{a2} is

$$E_{a2} = 2\times10^{-7}\, \omega\left[\ln\left(\frac{28.43}{18}\right)\times10^3 \angle 0 + \ln\left(\frac{36.25}{18.6}\right)\times10^3 \angle -120° \right.$$

$$\left. + \ln\left(\frac{25.06}{12}\right)\times10^3 \angle 120° \right)$$

$$E_{a2} = 0.019 \text{ V/m}$$

For a 10 km line, the total induced voltage on conductor $a2$ is

$$V_{a2} = E_{a2}\, l = 190 \text{ V}$$

In Chapter 6, we calculated the induced voltage on conductor $a2$ due to the electric field. The result was over 30 kV, which is much higher than the induced voltage due to the magnetic field. This is why the calculation of the induced voltage due to the magnetic field is often ignored unless the line experiences high current transients due to lightning, faults, or switching action. Hence, unless the two lines run in parallel for a long time, the magnetic field effect can be ignored.

Example 7.3

The conductors of a three-phase circuit are arranged at the same height of 10 m above ground. The separation between any two adjacent conductors is 4 m. The conductors carry a balanced three-phase current of 1.0 kA. A de-energized conductor is located 3 m above the middle conductor. All conductors run in parallel for 10 km. Compute the induced voltage on the de-energized conductor.

Solution:

The first step is to calculate all relevant distances between the de-energized conductor and the three energized conductors:

$$d_{pa} = \sqrt{3^2 + 4^2} = 5 \text{ m}$$

$$d_{pb} = 3 \text{ m}$$

$$d_{pc} = 5 \text{ m}$$

$$D_{pa} = \sqrt{4^2 + 10^2} = 10.77 \text{ m}$$

$$D_{pb} = 10 \text{ m}$$

$$D_{pc} = \sqrt{4^2 + 10^2} = 10.77 \text{ m}$$

$$E_p = 2 \times 10^{-7} \, \omega \left[\ln\left(\frac{D_{pa}}{d_{pa}}\right) \quad \ln\left(\frac{D_{pb}}{d_{pb}}\right) \quad \ln\left(\frac{D_{pc}}{d_{pc}}\right) \right] \begin{bmatrix} I \angle 0° \\ I \angle -120° \\ I \angle 120° \end{bmatrix}$$

$$= 2 \times 10^{-7} \times 377 \left[\ln\left(\frac{10.77}{5}\right) \quad \ln\left(\frac{10}{3}\right) \quad \ln\left(\frac{10.77}{5}\right) \right]$$

$$\times \begin{bmatrix} 1000 \angle 0° \\ 1000 \angle -120° \\ 1000 \angle 120° \end{bmatrix} \approx 0.038 \text{ V/m}$$

If the conductors run in parallel for 10 km, the total induced voltage is

$$V = E_p \times l = 0.038 \times 10 = 380 \text{ V}$$

Note that if the de-energized conductor is far away from the three energized conductors and the currents are balanced, the ration (D/d) is almost the same for all conductors. Hence, no appreciable voltage is induced on the de-energized conductor. Also, if the conductors are close to each other (as with the case of three-phase cables), the induced voltage on any de-enegized object nearby is small.

7.3.2 Induced Voltage due to Transient Current

Transient conditions occur in power systems due to several reasons, including

- Lightning striking directly or indirectly the transmission lines
- Faults on power lines due to falling trees, earthquakes, animal intrusion, faulty equipment, etc.
- Switching transients due to circuit breakers or recloser operations
- Inrush currents due to sudden change in large loads

In all these cases, the magnitude and rate of change of the current di/dt are often very high. Consider the typical fault current characteristic in Figure 7.8 without the dc offset. When a fault occurs, the rate of change of the current is very high during the initial transient period. After the transients damp out, the current reaches a steady state, but its magnitude is still above the normal value.

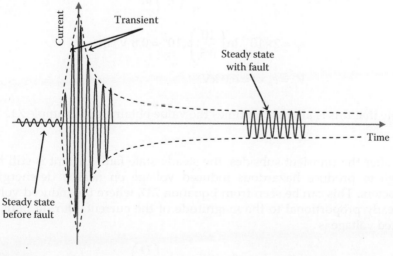

FIGURE 7.8
Typical fault current characteristics.

During the transient period, the high rate of change of current could produce excessive induced voltage on nearby de-energized conductor as given by Equation 7.14. The equation indicates that the induced voltage in this period is linearly proportional to the rate of change of current

$$e_{pf} = 2 \times 10^{-7} \ln\left(\frac{D}{d}\right)\frac{di_f}{dt} \tag{7.25}$$

where

e_{pf} is the induced voltage per unit length at conductor p due to transient or fault current

$\frac{di_f}{dt}$ is rate of change of current during transient or fault

Example 7.4

A de-energized conductor is placed 0.5 m away from an energized conductor. The distance D is 10 m. The two conductors are in parallel for 10 km. Assuming that a fault on the energized line produced a current surge of 1 kA/ms, compute the induced voltage on the de-energized conductor.

Solution:

Use Equation 7.25:

$$e_{pf} = 2 \times 10^{-7} \ln\left(\frac{D}{d}\right)\frac{di_f}{dt}$$

$$e_p = 2 \times 10^{-7} \ln\left(\frac{10}{0.5}\right) \times 10^6 = 0.6 \text{ V/m}$$

$$V_p = e_p \times l = 6.0 \text{ kV}$$

Note that this is almost five times the value obtained in Example 7.1.

Even after the transient subsides, the steady-state fault current is still high enough to produce hazardous induced voltage on nearby de-energized conductors. This can be seen from Equation 7.17, where the induced voltage is linearly proportional to the magnitude of the current. During fault, the induced voltage is

$$E_{pf} = 2 \times 10^{-7} \omega I_f \ln\left(\frac{D}{d}\right) \tag{7.26}$$

where

E_{pf} is the rms-induced voltage per unit length due to the steady-state fault current

I_f is the rms steady-state fault current, which is much higher than the operating current of the line

7.4 Induced Voltage due to Electric and Magnetic Fields

The total induced voltage due to electric and magnetic fields can be computed using the superposition theorem. The effect of each field is computed independently, and then the total effect is the phasor sum of the two.

Example 7.5

The system in Figure 7.9 consists of an energized conductor and a de-energized conductor. The voltage of the energized conductor with respect to ground is 140 kV and carries 1 kA at a power factor of 0.9 lagging. The radius of the energized conductor is 1 cm and that of the de-energized conductor is 0.5 cm. All conductors run in parallel for 30 km. Compute the induced voltage on the de-energized conductor due to both the electric and magnetic fields.

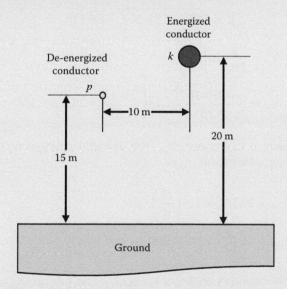

FIGURE 7.9
Study system.

Solution:

The first step is to calculate the induced voltage due to the electric field. To do this, we need to compute the charge of the energized conductor. As given in Chapter 6, the relationship between the phase voltage of the energized conductor and its charge is

$$V_{kn} = \frac{q}{2\pi\varepsilon_0} \ln\left(\frac{2h}{r}\right)$$

$$140,000 = \frac{q}{2\pi\varepsilon_0} \ln\left(\frac{40}{0.01}\right)$$

$$q = 0.94 \ \mu C$$

The induced voltage on the de-energized conductor due to the electric field V_{pe} can be computed as given in Chapter 6. Assume that the voltage of the energized conductor is our reference. Hence, the angle of the charge is zero.

$$V_{pe} = \frac{q}{2\pi\varepsilon_0} \ln\left(\frac{d'}{d}\right)$$

where

$$d = \sqrt{10^2 + 5^2} = 11.18 \ m$$

$$d' = \sqrt{10^2 + (15 + 20)^2} = 36.4 \ m$$

$$V_{pe} = \frac{q}{2\pi\varepsilon_0} \ln\left(\frac{d'}{d}\right)$$

$$V_{pe} = \frac{0.98 \times 10^{-6}}{2\pi \times 8.85 \times 10^{-12}} \ln\left(\frac{36.4}{11.18}\right) = 20.8 \angle 0° \ kV$$

The second step is to compute V_{pm}; the induced voltage on conductor p due to the magnetic field is

$$\bar{E}_{pm} = 2 \times 10^{-7} \ \omega \ \bar{I} \ln\left(\frac{D}{d}\right)$$

$$\bar{E}_{pm} = 2 \times 10^{-7} \times 377 \left(1000\angle - \cos^{-1}(0.9)\right)$$

$$\times \ln\left(\frac{\sqrt{10^2 + 20^2}}{11.18}\right) = 52.26 \angle - 25.84° \ mV/m$$

The conductors run in parallel for 30 km. Hence, the total induced voltage due to magnetic field is

$$V_{pm} = E_{pm} \times l = 52.26 \angle -25.8° \times 30 = 1.568 \angle -25.84° \text{ kV}$$

The total induced voltage V_p is the phasor sum of V_{pe} and V_{pm}:

$$\overline{V}_p = \overline{V}_{pe} + \overline{V}_{pm} = 20.8 \angle 0° + 1.568 \angle -25.84° = 22.22 \angle -1.76° \text{ kV}$$

Exercises

7.1 A balanced three-phase system has its conductors arranged as an equilateral triangle. The distance between any two conductors is 10 m, and conductor c is 20 m above ground. Each conductor carries 10 kA, and the three conductors form a balanced three-phase system. Compute the induced voltage per meter at point p located at the orthocenter of the triangle. Repeat the solution assuming the conductor of phase a is disconnected.

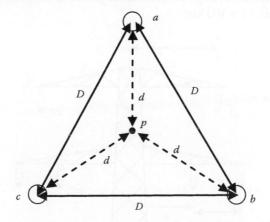

7.2 A transmission line tower is shown in the following figure. The dimensions in the figure arc in meters. The current in the three-phase balanced circuit is 1 kA. All lines run in parallel for 1 km. Compute the induced voltage on the de-energized conductor.

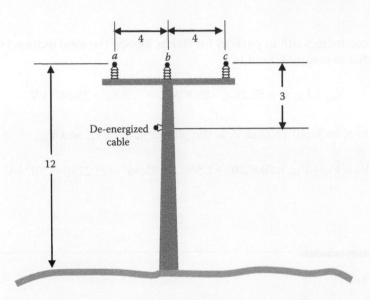

7.3 Assume a fault occurred on conductor *b* of the system in the previous problem. The fault resulted in a current surge of 1 MA/s in phase *b*. Compute the induced voltage on the de-energized conductor.

7.4 A double-circuit transmission line tower is shown in the following figure. The dimensions in the figure are in meters. The current of the transmission line circuit on the left is balanced at 500 A. The circuit on the right side of the tower is de-energized. Compute the induced voltage on the de-energized conductor phase *a2* assuming the two lines run in parallel for 100 km.

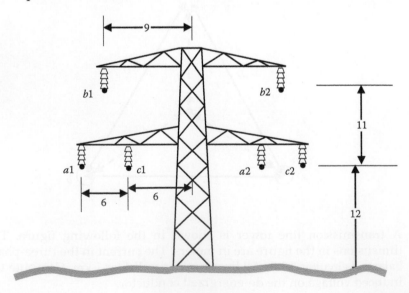

8

De-Energized Line Work

Working on power lines is a daring job that demands full knowledge of site hazards, safe working techniques, full attention to the performed tasks, and continuous monitoring of the work by others to continually correct any unsafe practice at the site. Fatality and severe injuries occur when workers use wrong equipment, do not follow the established rules, or work without proper protective gear. A large number of injuries are attributed to workers ignoring the safety rules due to time pressure, lack of knowledge, or poor supervision. The United States through the Occupational Safety and Health Administration (OSHA) and the Institute of Electrical and Electronics Engineers (IEEE) has established safety codes and standards for power line work that are widely used worldwide.

Power line work can be divided into two categories: live work and de-energized work. Each of these two types requires different sets of safety procedures. This chapter deals with de-energized work and the next one covers energized work. One of the great sources for de-energized line work is IEEE Standard 1048.

8.1 Definition of a De-Energized Conductor

There is a great deal of confusion about the definition of a de-energized conductor. Here are a few of them:

- A conductor that is disconnected from all energy sources by opening all circuit breakers or switches feeding the line
- A conductor that is disconnected from any energy source and free from induced voltage due to electromagnetic coupling
- A conductor that is disconnected from both of its ends and intentionally grounded at the line sides of the circuit breakers
- A conductor that is disconnected from both of its ends and grounded at the worksite

The definition of the de-energized conductor used in this book is the first one, "Conductor that is disconnected from all energy sources by opening all circuit breakers or switches feeding the line." This is the definition used by OSHA and IEEE. By this definition, there is no assumption for any kind of grounding.

Power lines always share the right-of-way (ROW) and run adjacent to each other. If anywhere in the system the de-energized line is in the vicinity of an energized line, the electromagnetic field (electric field and magnetic field) produced by the energized line induces voltage on the de-energized line, as depicted in Figure 8.1. In fact, the de-energized conductor can attain enough charges due to the electromagnetic coupling to elevate the voltage to hazardous levels. Hence, the de-energized conductor is a hot wire. It is a fatal mistake to assume that the de-energized conductors are safe to touch (dead); they are not.

The induced voltage on a de-energized line can be extremely high, especially when it shares a tower with high-voltage energized circuits.

The severity of the induced voltage depends on several factors, including the following:

- Voltage and current of the energized line
- Proximity of the two lines

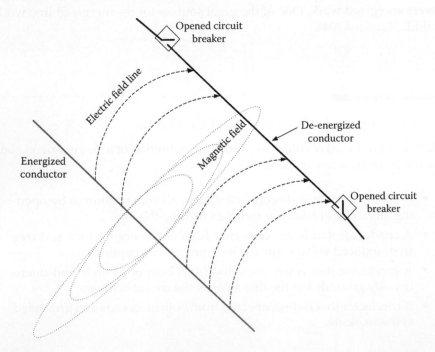

FIGURE 8.1
Electromagnetic coupling between energized and de-energized lines.

- Length by which the two lines run adjacent to each other
- Fault-current on adjacent lines

Besides induced voltage, the de-energized lines could become energized due to other events such as

- Lightning strikes
- Switching equipment malfunction
- Human error

8.2 Methods of Detecting Induced Voltage

Linemen use several methods to detect the presence of voltage on de-energized lines. Although it is good to know if the line has induced voltage, the information cannot be used to determine safety procedure. The code states that the safety procedure must be based on the worst voltage at the site. This could mean the highest nominal (rated) voltage of the line being worked on or the highest voltage of nearby circuit sharing the same towers. IEEE Standard 1048-2003 states

> Voltage detection is the process of sensing voltage on a line to determine whether the line voltage is present and is used only for confirmation of isolation and only after standard clearance procedures are complete.

In addition, IEEE Standard 524a states

> If the line under construction comes in contact with an energized line, the de-energized line will be energized by the same voltage as the energized line. The de-energized line should be adequately grounded to protect the workers in the event of this condition.

Among the common detection methods are the following:

- *Buzzing (fuzzing)*: An old method accomplished by touching the conductor with the metal cap at the end of a live line tool. If a buzzing sound is produced, it indicates the presence of elevated voltage. This method is not used anymore by most utilities as it is ineffective at low, yet hazardous, voltages.
- *Glowing neon or light-emitting diode*: The neon lamps typically require about 90 V to glow. The light-emitting diodes can be activated at even lower voltage. These are good methods to detect low voltages, but the light could be hard to see in bright and sunny conditions. Also, the device may light up due to induced voltage from nearby circuits.

- *Noise-producing devices*: In high-voltage circuits, sharp edges have high electric field strength that could ionize surrounding air. The ionization creates leakage current in air known as corona. The corona emits light and the ionization produces a distinct high-frequency noise. The method is not accurate for detecting hazardous low voltages.

- *Remote voltage detection*: The method uses the presence of electromagnetic field surrounding the de-energized line. It produces audible beeping as well as visual light. The detection is from a safe distance and there is no need to contact the surface being tested. For a 50 KV system, it can detect the presence of voltage from as far away as 150 m (500 ft). The method, however, cannot identify the voltage of a specific line among a group of lines.

- *Direct voltage detectors*: This is the most accurate method. A voltmeter fitted into a hot stick touches the line. The method is highly accurate and can cover a wide range of voltage.

8.3 Main Protection Techniques

There are basically three general methods to protect workers from electric shocks:

1. Isolate: Construct barriers to prevent people from entering work areas

2. Insulate: Create an extremely high resistance path between the person and the ground

3. Grounding: Bring the potential of the de-energized object to the local ground potential

These general principals, although simple, require a great deal of careful assessment and design.

8.3.1 Isolation

The main function of isolation is to create physical barriers between people and any de-energized conductor or object that could carry charges while performing the work, as shown in Figure 8.2. The barrier should have warning signs to keep people away from the work area. If during the work a current is injected into the ground, the barricaded area must expand to reduce the step potential outside the barriers.

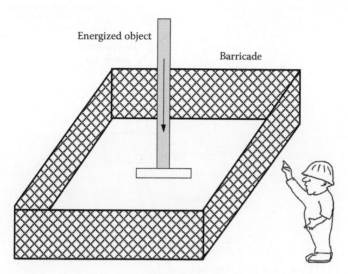

Energized object

Barricade

FIGURE 8.2
Isolating de-energized object that could carry a charge.

IEEE Standard 1048 states

Isolation may be provided by physical restraints such as barricades or barriers. No one should be inside the isolating perimeter unless protected by method (a) or method (b). Isolation, if properly used, provides a positive means of protecting the public. With respect to the step voltage hazard, the isolation distance may vary from a few meters to 9.5 meters or more depending on the available fault current and voltage.

Method (a) in the standard is grounding and method (b) is insulation, which are discussed next.

8.3.2 Insulation

Insulation is a method used to create high-enough resistance between the person touching an object and the ground. This can be achieved by any combination of the following methods:

- Place workers on insulated platform away from any grounded object. Figure 8.3 shows a worker on an insulated boom truck.
- Use insulated gloves and sleeves designed for the *available* voltage at the site. The available voltage is often interpreted as the nominal (rated) voltage of the de-energized line. In a multicircuit tower, the available voltage is the highest voltage among all circuits regardless of which one is being worked on.
- Cover all energized object within direct or indirect reach by insulating equipment (blankets, hoses, etc.).

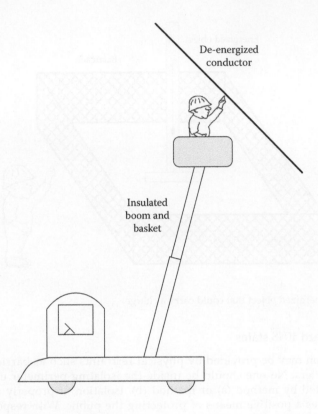

De-energized
conductor

Insulated
boom and
basket

FIGURE 8.3
Insulation.

IEEE Standard 1048 states

> Workers are insulated by gloves, footwear, mats, platforms, insulated
> booms, etc., suitable for the voltage resulting from maximum fault cur-
> rents and the voltage available at the worksite.

8.3.3 Grounding

The basic principle behind grounding is to ensure that the voltage difference
between any two points of contact is not hazardous, and the potential of
all nearby objects is equal to the potential of the local ground. Assume that
a person touches a de-energized line while his feet are on the ground (or
grounded object), as shown on the left side of Figure 8.4. If for some reason a
voltage is induced on this line (or the line is accidentally energized), the person provides a path for the current to ground. This could be lethal because
the hands and feet of the person are touching two different potential points.
However, if before the person touches the line a temporary grounding wire

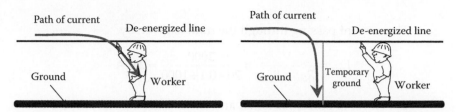

FIGURE 8.4
Worker without and with temporary ground.

with very low resistance is installed between the line and the ground, as shown on the right side of Figure 8.4, almost all of the current flow through the ground wire, and the person is safe to work on the de-energized line. In this case, the hands and feet of the person are at equal potential points (equipotential).

Example 8.1

Assume that a 20 km de-energized line is accidentally energized by a 2 kV source while a lineman is working at the far end of the line. The line resistance is 0.1 Ω/km, and the worker's body resistance plus his ground resistance is 1000 Ω.

1. Compute the current passing through the lineman and the voltage across his body.
2. Assume a 0.001 Ω temporary grounding conductor is installed between the line and the ground in parallel with the worker. Compute the current passing through the lineman and the voltage across his body.

Solution:

1. The equivalent circuit of the system without the temporary ground is shown in Figure 8.5.

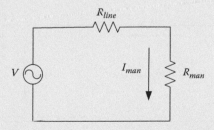

FIGURE 8.5
Equivalent circuit of part 1.

The current passing through the man is given by

$$I_{man} = \frac{V}{R_{line} + R_{man}} = \frac{2000}{20 \times 0.1 + 1000} \approx 2.0 \text{ A}$$

This level of current is lethal and the worker cannot survive this electric shock. What makes the situation even worse is that the circuit breaker cannot clear this fault as the current is too low.

2. Now assume that a grounding wire is installed. The equivalent circuit in this case is given in Figure 8.6.

The parallel resistance of the man and grounding wire is

$$R = \frac{R_{man} \, R_{ground \, wire}}{R_{lman} + R_{ground \, wire}} = \frac{1000 \times 0.001}{1000.001} \approx 0.001 \, \Omega$$

The current in the line is

$$I_{line} = \frac{V}{R_{line} + R} = \frac{2000}{2 + 0.001} \approx 999.5 \text{ A}$$

The current is very high and will likely trip the circuit breaker. If not, the current passing through the man is

$$I_{man} = I_{line} \frac{R_{ground \, wire}}{R_{ground \, wire} + R_{man}} = 999.5 \frac{0.001}{0.001 + 1000} \approx 1.0 \text{ mA}$$

This level of current is safe.

Note that with the grounding wire, the current through the lineman is small. However, the current passing through the grounding conductor is very large. The worker is only protected if the grounding conductor is not damaged by this high current.

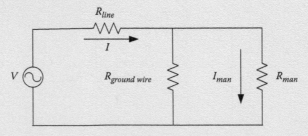

FIGURE 8.6
Equivalent circuit of part 2.

As seen in the previous example, the temporary ground wire provides two main advantages:

1. It protects the worker from electric shock.
2. It causes the fault current to be very high. Without the temporary ground, the current is just 2 A, and with the ground it is almost 1000 A. When the breakers of the line sense the high fault current, they interrupt the circuit.

8.4 Grounding System

Because worksites have various equipment and structures in contact with earth, an effective grounding system must be designed to bring the potential of all metallic objects at the worksite to the potential of the local ground. This is essential to limit the amount of currents though all workers under the worst scenarios at the worksite. Effective grounding is achieved by using a group of *temporary grounds* arranged at the worksite in a specific manner. OSHA 1910-269 (n.3) states

> Temporary grounds shall be placed at such locations and arranged in such a manner as to prevent each employee from being exposed to hazardous difference in potential.

Besides limiting the current passing through all workers, the protective grounding system must achieve two additional objectives:

- Provide a very low impedance path to ground. If a fault occurs, the fault current is high enough to trip the circuit breakers feeding the fault.
- Prevent people in the vicinity of the worksite from being exposed to electric shocks due to excessive touch and step potentials.

The two objectives would look contradictory. This is because high fault current to ground through the temporary grounding system increases the ground potential rise GPR as well as the step potential. However, careful design of the protective ground system and using additional equipment can achieve all these objectives. But before we explain the grounding design in details, let us discuss the shapes and features of temporary grounds.

8.4.1 Protective Grounds

Protective grounds are also called *temporary grounds, jumpers, working grounds, traveler grounds* (for stringing operations), and *personal grounds*. Some utilities use the term "personal ground" to identify a very specific grounding conductor that is placed across the worker's body.

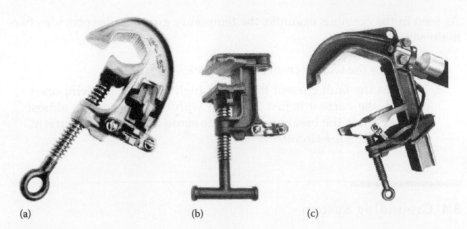

(a)   (b)   (c)

FIGURE 8.7
Protective ground clamps: (a) serrated conductor, (b) serrated flat tower clamp, and (c) busbar clamp.

The protective ground consists of a conductor with clamps at both ends. The clamps come in various shapes depending on the surface they are connected to. Some of the clamps are shown in Figure 8.7.

A schematic of a protective ground is shown in Figure 8.8. The total resistance of the protective ground is the aggregated resistance of its conductor, clamps, and ferrules, as well as the resistance of the surface of the objects on both ends.

Each component of the grounding system must have the following features:

- Sufficiently low resistance
- Withstand the worst fault conditions (current magnitude and duration) without being damaged or fused
- Do not deteriorate mechanically under the most adverse combination of fault magnitude and duration

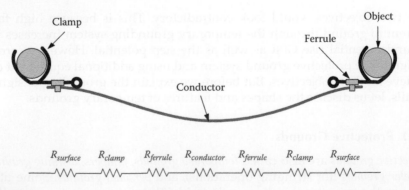

Clamp

Object

Ferrule

Conductor

$R_{surface}$ R_{clamp} $R_{ferrule}$ $R_{conductor}$ $R_{ferrule}$ R_{clamp} $R_{surface}$

FIGURE 8.8
Resistance of protective ground.

These conditions can be achieved by the following:

1. *Low conductor resistance*: In case of accidental energization, low resistance accomplishes two objectives: it reduce a the current passing through workers and causes immediate operation of the protective devices (circuit breaker, reclosers, etc.)

   The generic expression of a conductor resistance is

$$R = \rho \frac{l}{A} \tag{8.1}$$

   where
   R is the resistance of the conductor
   ρ is the resistivity of the conductor's material
   A is the cross-sectional area of the conductor
   l is the length of the conductor

   To reduce the resistance of the conductor, the following must be implemented:
   - The length of the conductor should not be longer than needed. In Equation 8.1, the resistance increases linearly with respect to length. Also, long conductors pose hazards during faults as fault current develops enough force to swing the conductors and hit nearby people.
   - The cross section of the conductor must be large enough. In Equation 8.1, the resistance is inversely proportional to the cross section.
   - The conductor must not be coiled. Coiling a conductor increases its inductive reactance substantially, especially during high-frequency events such as switching transients or faults.

2. *Electrical rating*: Grounding conductors must be sized to withstand the maximum fault current at the site for the time it takes to clear the fault by the protection devices. A fault current of 20–30 kA is common, but it can be as high as 100 kA. The next subsection discusses this point further.

3. *Clamps*: Proper types of clamps must be used to match the surface they are bonded to. This way, the contact area is maximized and the bonding is strong mechanically. The surface resistance is reduced and the protective grounds are securely attached to the surface. A clamp may have either smooth or serrated jaws. Smooth-jawed clamps used to minimize conductor damage. Serrated-jawed clamps are used to break through the normal corrosion or oxides that could have been accumulated on the surface.

4. *Securing clamps*: Clamps must be mechanically secured and must not dislodge even due to severe force of the fault current. They must be tightened to clean surfaces to reduce surface resistance and ensure that the developed force on the conductor during fault does not dislodge the protective ground. IEEE Standard 524a states

   It is very important to give attention to restraint of grounding jumpers to minimize possible severe mechanical movement should a fault occur.

   The mechanical force is especially severe during faults. IEEE Standard 1048 provides the formula to estimate this force:

   $$F = 2 \times 10^{-7} \frac{l i^2}{d} \qquad (8.2)$$

   where
    F is the force between two parallel conductors carrying the same current (N)
    l is the length of the conductor (m)
    i is the instantaneous current (A)
    d is the separation between the conductors (m)

   For 100 kA fault and 1 m separation, the force is 2000 N/m. This force is deadly if the conductor hits a person.

5. *Clean surface*: Clamps must not be installed over galvanized, painted, or rusted frames. A brittle, corrosive layer could cause the clamps to loosen. In addition, it increases the surface resistance. IEEE Standard 1048 states

   Instruct workers in proper surface preparation at the connection points of grounding cable clamps to ensure low contact resistance. Low contact resistance will prevent the clamps from being "blown off" by mechanical forces. Failure to remove the high-resistance oxide layer at the connection point can lead to excessive resistance heating and consequent melting at the connection. The heating and melting will result in loosening and dislodging of the clamp. A brittle, corrosive layer could also cause the clamp to loosen.

6. *Location*: Protective grounds must be utilized at the immediate site. If placed away from the site, the resistance between the person and the grounding conductor increases the current passing through the worker.

7. *Parallel protective grounds*: Some utilities allow the workers to parallel protective grounds. The reason for this is to increase the

current-carrying capability of the grounding system. The division of current among parallel grounds depends on the total impedance of each ground set, including connection impedance. When two protective grounds are placed in parallel, they must be identical in material, cable length, cable size, and clamp type and size.

8. *Installation and removal of grounds*: Installation and removal of protective grounds must be made in a specific sequence. The simple rule for installing protective grounds is to have the first clamp of each grounding conductor attached to the grounded side. When dismantling the system, the sequence is reversed; that is, the ground side of the protective ground is removed last. This way, the loose end of the grounding conductor is at the ground potential. Keep in mind that the installation and removal of the protective grounds must be done by live-line tools and the minimum approach distance must be maintained.

8.4.2 Sizing of Temporary Ground Conductors

Table 8.1 shows the resistance of copper conductors with various sizes. This is the steady-state resistance that is computed based on Equation 8.1.

The selection of the temporary ground conductor is not only based on its resistance but also on its ampacity and fusion temperature. A typical ampacity table for copper conductors is given in Table 8.2.

TABLE 8.1

Resistance of Copper Conductors

AWG	Diameter		Copper Resistance	
	(in.)	(mm)	(mΩ/m)	(mΩ/FT)
0000 (4/0)	0.4600	11.684	0.1608	0.04901
000 (3/0)	0.4096	10.404	0.2028	0.06180
00 (2/0)	0.3648	9.266	0.2557	0.07793
0 (1/0)	0.3249	8.252	0.3224	0.09827
1	0.2893	7.348	0.4066	0.1239
2	0.2576	6.544	0.5127	0.1563
3	0.2294	5.827	0.6465	0.1970
4	0.2043	5.189	0.8152	0.2485
5	0.1819	4.621	1.028	0.3133
6	0.1620	4.115	1.296	0.3951
7	0.1443	3.665	1.634	0.4982
8	0.1285	3.264	2.061	0.6282
9	0.1144	2.906	2.599	0.7921
10	0.1019	2.588	3.277	0.9989

TABLE 8.2

Withstand Ampacity of Copper Conductors in KA

AWG	15 Cycles	30 Cycles	45 Cycles	60 Cycles
0000 (4/0)	430	30	25	22
000 (3/0)	34	24	20	17
00 (2/0)	27	19	16	14
0 (1/0)	21	15	12	11
1	16	12	9	8
2	14	9	7	7

To consider the fusion temperature of a conductor as well as the duration of the current, we can use the formula in IEEE Standard 80:

$$A_{kcmil} = I_f \, K \sqrt{t_c} \qquad (8.3)$$

where

A_{kcmil} is the area of the conductor in kcmil

I_f is the maximum fault current in kiloamperes

t_c is the current duration in seconds (clearing time of the circuit breaker)

K is a constant that is dependent on the fusing temperature (T_m) of the conductor's material and the ambient temperature (T_a); at ambient temperature of 40°C, K can be obtained from Table 8.3

Keep in mind that the area in kcmil (A_{kcmil}) is equal to the area in square millimeter (A_{mm}) multiplied by 1.974:

$$A_{kcmil} = 1.974 \times A_{mm} \qquad (8.4)$$

TABLE 8.3

Material Constants

Material	T_m (°C)	K
Copper, annealed soft-drawn	1083	7.00
Copper, commercial hard-drawn	1084	7.06
Copper, commercial hard-drawn	250	11.78
Copper-clad steel wire	1084	10.45
Copper-clad steel wire	1084	12.06
Copper-clad steel rod	1084	14.64
Aluminum EC grade	657	12.12
Aluminum 5005 alloy	652	12.41
Aluminum 6201 alloy	654	12.47
Aluminum-clad steel wire	657	17.20
Steel 1020	1510	15.95
Stainless clad steel rod	1400	14.72
Zinc-coated steel rod	419	28.96
Stainless steel 304	1400	30.05

Source: IEEE standard 80, *Guide for Safety in AC Substation Grounding.*

Example 8.2

Compute the size of an aluminum-clad steel grounding conductor for a circuit with 10 kA fault current. The breaker of the circuit has a clearing time of 200 ms.

Solution:

Using Equation 8.3 and Table 8.3, the cross section of the grounding conductor is

$$A_{kcmil} = I_f \ K \sqrt{t_c} = 10 \times 17.2 \times \sqrt{0.2} = 77 \text{ kcmil}$$

The size of the conductor would be the higher value from 77 kcmil.

8.5 Grounding Methods

A very simple scenario where a single grounding conductor is sufficient to protect a worker is given in Example 8.1. However, worksites have more complex arrangement such as the following:

- Several workers could be at the worksite performing different tasks.
- Power line towers could carry several circuits; each one is composed of three conductors forming a three-phase circuit. All conductors can be at different voltage levels or have different phase shifts.
- Conductive equipment could be within reach.
- Grounded equipment could be within reach.
- Some of the work could be aerial and others could be at the ground level.

Because of the complexity of worksites, the grounding methods for worksites are more complex. They require the full knowledge of the site configuration and how the tasks are to be performed. The grounding method must create an *equipotential zone* that protects all workers from hazardous touch and step potentials.

8.5.1 Equipotential Zone

According to OSHA and IEEE, people at any worksite are protected if they are placed inside an equipotential zone. If establishing the equipotential zone is not desired, the de-energized work can still be carried out but with only live-line tools and by maintaining the minimum approach distance based on the maximum nominal voltage at the site.

The equipotential zone is a physical area encompassing the worksite where any two points within direct or indirect reach have the same potential. The indirect reach is the maximum distance that could be reached when workers use conductive tools.

The equipotential zone must protect all workers at the site. It can be grounded or floating. If the worksite includes grounded objects or equipment on or near earth, the equipotential zone must be grounded. The floating equipotential zone is mainly for live-line work or any work that does not involve people or equipment at low potential or ground potentials.

The equipotential zone is established by installing several protective grounds at the worksite in a way that guarantees the safety of all workers. This includes, but is not limited to, the following:

- Workers on aerial booms
- Workers standing on ground level touching trucks or lift equipment
- Workers on ground touching towers
- People walking near the site

All people at the site must be protected against the worst possible scenario that could occur at the site. If anyone at the worksite is not protected, we do not have equipotential zone, and the worksite is hazardous.

IEEE Standard 1048 states that the worksite has grounded equipotential zone only when

> *All* equipment is electrically interconnected by bonding cables and grounded to a system neutral, system ground, and/or grids that provide negligible potential difference across the zone.

OSHA 1910-269 (n.3) describes the way to establish an equipotential as follows

> Temporary grounds shall be placed at such locations and arranged in such a manner as to prevent *each* employee from being exposed to hazardous difference in potential.

OSHA leaves the design of the equipotential zone to companies' experts as it could be different depending on the work to be performed, the hazards at the site, and site configuration. The underlying condition is to "prevent *each* employee from being exposed to hazardous difference in potential."

Grounding at the worksite must have key features to be categorized as equipotential zone, including

- Use proper protective grounds, as discussed in Section 8.2
- Form single-point grounding
- Bond the system to the best ground at the site
- Ground all metallic objects within reach
- Ground all phases at the site

8.5.1.1 Single-Point Grounding

All protective grounds and grounded equipment at the worksite must be bonded to a single-point ground. Multiple-point ground systems can lead to hazardous conditions due to variations in the GPR. Take, for example, the case in Figure 8.9. The conductor at the worksite is grounded through

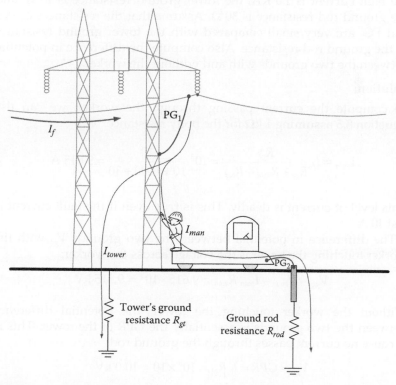

FIGURE 8.9
Hazards due to multiple grounds.

the tower using protective ground PG_1. The truck is grounded through PG_2 and the ground rod. Since the truck ground and tower grounds are not bonded, any worker standing on the truck and touching the tower is exposed to the potential difference between the two grounds. If the line being worked on is accidentally energized and a fault current flows to the site, the current through the worker could be lethal. Assuming the resistances of PG_1 and PG_2 are very small, the current passing through the worker is

$$I_{man} = I_f \frac{R_{gt}}{R_{gt} + R_{man} + R_{rod}} \tag{8.5}$$

Example 8.3

Compute the current passing through the worker in Figure 8.9. Assume the fault current is 1.0 KA, the tower ground resistance is 10 Ω, and the ground rod resistance is 30 Ω. Assume that the resistances of PG_1 and PG_2 are very small compared with the tower ground resistance or the ground rod resistance. Also compute the difference in potential between the two grounds with and without the worker.

Solution:

To compute the current passing through the worker, we can use Equation 8.5 assuming 1 kΩ for the body resistance:

$$I_{man} = I_f \frac{R_{gt}}{R_{gt} + R_{man} + R_{rod}} = 10^3 \frac{10}{10 + 1000 + 30} = 9.615 \text{ A}$$

This level of current is deadly. This is true even if the fault current is just 10 A.

The difference in potential between the two grounds V_{gg} with the worker touching the tower is the voltage across the worker:

$$V_{gg} = V_{man} = I_{man} R_{man} = 9.615 \times 10^3 = 9.615 \text{ kV}$$

Without the worker touching the tower, the potential difference between the two grounds is essentially the GPR of the tower. This is because no current passes through the ground rod:

$$V_{gg} = GPR = I_f R_{gt} = 10^3 \times 10 = 10.0 \text{ kV}$$

As seen in the previous example, with separated grounds, the worker touching the tower and the truck is exposed to a lethal level of current. To remove this hazard, we can use additional protective ground (PG_3) between the truck

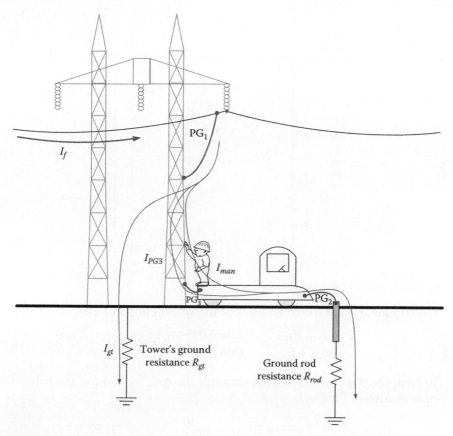

FIGURE 8.10
Single ground system.

and the tower, as shown in Figure 8.10. This way, the fault current is branched into three paths: one through the tower ground (I_{gt}), the second through PG$_3$ (I_{PG3}), and the third through the man (I_{man}). To have the current passing through the man within the safe limit, the resistance of PG$_3$ must be very low.

Example 8.4

Repeat Example 8.3, but assume a 0.005 Ω protective ground (PG$_3$) is installed between the tower and the truck.

Solution:

PG$_3$ is in parallel with the worker. The equivalent circuit of the system in Figure 8.10 is shown in Figure 8.11.

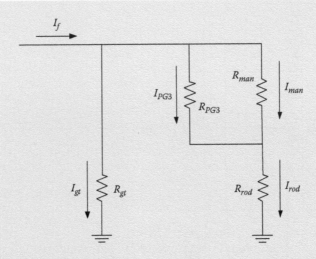

FIGURE 8.11
Equivalent circuit of the system in Figure 8.10.

The combination of the resistances of the worker and PG$_3$ is

$$R = \frac{R_{man}\,R_{PG3}}{R_{man} + R_{PG3}} = \frac{1000 \times 0.005}{1000 + 0.005} = 0.004999\ \Omega$$

To compute the current passing through the rod, we can use the principle of current divider:

$$I_{rod} = I_f \frac{R_{gt}}{R_{gt} + R + R_{rod}} = 10^3 \frac{10}{10 + 0.004999 + 30} = 249.97\ \text{A}$$

The current passing through the man is

$$I_{man} = I_{rod} \frac{R_{PG3}}{R_{PG3} + R_{man}} = 249.97 \frac{0.005}{0.005 + 1000} = 1.25\ \text{mA}$$

This is a low current and the person is safe even during the fault.

From the previous two examples, it would seem that by not using the ground rod, the worker is safe as there will be no current passing through him. This is a wrong assumption because of two main reasons:

1. The tires of the truck provide some conductivity, especially when moist, and will provide a path to ground. Also, truck outriggers, or leaning equipment on the truck, provide some conductivity to the earth.
2. If the worksite involves people on the ground touching the truck, they will not be protected as explained later in this chapter.

8.5.1.2 Bonding to Best Ground at the Site

The equipotential zone must be bonded to the best ground at the site, the one with the lowest ground resistance and adequate ampacity. If overhead ground wire (OHGW; also called static or shield wire) is present, the equipotential zone must be bonded to it. This is because the static wire is grounded at all towers as well as at the substations. The two main reasons for using the static wire are

1. Its ground resistance is very low. Most of the fault current will pass through the static wire; thus, reducing the currents in the protective grounds at the worksite.

2. Small portion of the fault current reaches the local ground. Thus, the step potential at the worksite is substantially reduced.

Figure 8.12 shows a case where a de-energized line is bonded to the structure by a temporary ground, and the structure is bonded to the static wire by

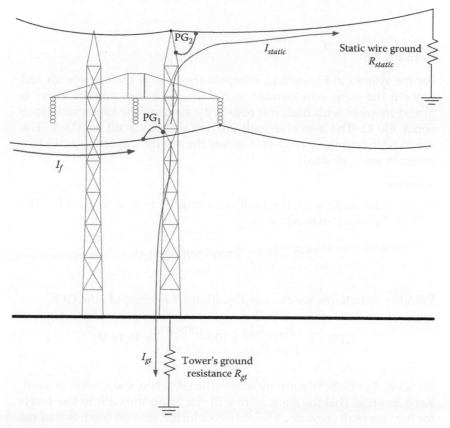

FIGURE 8.12
Bonding to the shield wire.

another temporary ground at the top of the tower. If a fault occurs, the fault current I_f is divided into two components: one goes through the structure to the static wire I_{static}, and the other goes through the structure to the local ground of the tower I_{gt}:

$$I_{gt} = I_f \frac{R_{static}}{R_{gt} + R_{static}} \qquad (8.6)$$

The ground potential rise of the tower is

$$GPR = I_{gt} R_{gt} = I_f \frac{R_{static} R_{gt}}{R_{gt} + R_{static}} \qquad (8.7)$$

Hence, the ground potential rise of the tower depends on the parallel combination of R_{static} and R_{gt}. Since often $R_{static} < R_{gt}$, the GPR is reduced when the static wire is bonded to the structure.

Example 8.5

For the system in Figure 8.12, compute the GPR of the tower with and without the static wire bonded to the structure. Assume the tower is placed in areas with high soil resistivity, making the tower resistance about 100 Ω. The available fault current of the circuit is 100 A. The shield wire resistance is 0.5 Ω. Assume the resistances of the protective grounds are very small.

Solution:

Without the static wire, the fault current goes only through PG_1. The GPR of the tower in this case is

$$GPR = I_f R_{gt} = 100 \times 100 = 10 \text{ kV}$$

With the static wire, we can use Equation 8.7 to compute the GPR:

$$GPR = I_f \frac{R_{static} R_{gt}}{R_{gt} + R_{static}} = 100 \frac{100 \times 0.5}{100 + 0.5} = 49.75 \text{ V}$$

As seen, the GPR is substantially reduced when static wire is used. Keep in mind that the static wire will not bring the GPR to low levels for heavier fault currents. The only conclusion we can reach is that the static wire will reduce the GPR.

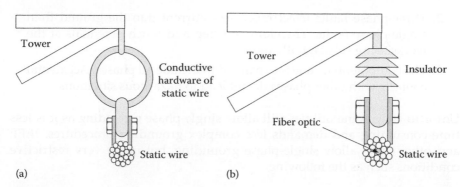

FIGURE 8.13
Static wire assembly: (a) static wire conductive hardware and (b) static wire with insulator.

The static wire can either be bonded to the tower through its hardware or isolated from the tower. Figure 8.13 shows schematics of these two types. On the left side of the figure, the static wire is bonded to the tower through its conductive hardware. In most cases, it is also bonded to the structure by a conductor. On the right side, the static wire is isolated from the tower. The isolation is needed to provide low-voltage power for obstruction lights near airports or to energize repeaters for communication signals. This type of static wire often includes fiber-optic cables inside them.

Because the static wire is exposed to harsh lightning strikes, and its hardware is exposed to the environment, it is wise to assume that the resistance between the static wire and the tower is high. Even if a conductor bonds the static wire to the structure, it could be blown off by earlier events such as lightning strikes or high fault currents. Therefore, temporary protective ground must be used to establish a low-resistance bonding between the static wire and the tower structure. This is PG_2 in Figure 8.12.

For the isolated static wire, its power must be disconnected and must be grounded at the substation and bonded to the tower structure.

8.5.1.3 Single versus Three-Phase Grounding

For three-phase de-energized circuits, all phases should be grounded even if the work involves one phase only. This is because of three main reasons:

1. The magnitude of three-phase short-circuit current is often higher than that for single-phase faults, especially for systems with high ground resistance. Therefore, grounding the three phases is a sure way to trip the circuit breaker due to accidental energization.

2. Three-phase faults inject much less current into the ground than single-phase faults. Therefore, the step and touch potentials at the worksite are substantially reduced.

3. The equipotential zone is expanded to include all phases. Accidental contacts with other phases do not lead to hazardous situations.

Unfortunately, some utilities still allow single-phase grounding as it is less time-consuming and demands less complex grounding procedures. IEEE and other codes allow single-phase grounding, but under very restrictive conditions such as the following:

- Ground resistance at the site is low enough to ensure consistent fault clearing for single-line-to-ground shorts
- Step and touch voltages are within acceptable limits or can be guarded against
- Working clearances (minimum approach distances) are maintained for the ungrounded phase conductors

IEEE Standard 1048 states

> The magnitude of three-phase short-circuit currents may be higher than the magnitude of a single-phase short, especially when the ground resistance is high. Single-phase fault current of a three-phase distribution line, grounded only on one phase through a high-resistance ground, may be insufficient to cause the line circuit breaker to open. Protective grounds applied to all phases will provide more certain, and generally more rapid, circuit breaker operation when ground resistance is high.
>
> Three-phase grounding means that only a small part of the fault current of a three-phase fault would flow to ground at the structure, thereby reducing the step and touch voltages at the base of the structure. If ground resistances are low enough to ensure consistent fault clearing, and the step and touch voltages are within acceptable limits or can be guarded against, then grounding of only one phase of a three-phase line might be permitted. However, working clearances shall always be maintained for the ungrounded phase conductors.

8.5.2 Examples of Equipotential Zone Design

The equipotential zone can be established by installing protective grounds in a configuration that protects all workers at the site. The configuration depends on a number of factors including the nature of work, location and type of aerial and ground equipment, nearby grounded objects, and access to worksite. Three worksite scenarios are discussed in this section.

8.5.2.1 Worksite 1: Aerial Work Away from Towers with Isolated Ground Equipment

Figure 8.14 shows a worksite away from any tower. The site includes a three-phase circuit, a grounded static wire, and a conductive boom and truck. The workers can ground the system by the following steps:

1. Selecting the static wire to be the reference ground since it has the lowest ground resistance at the site.

2. Bonding the truck to a ground rod through protective ground PG_1. This way the touch potential of the truck can be reduced, but not eliminated. The effectiveness of the ground rod is substantially

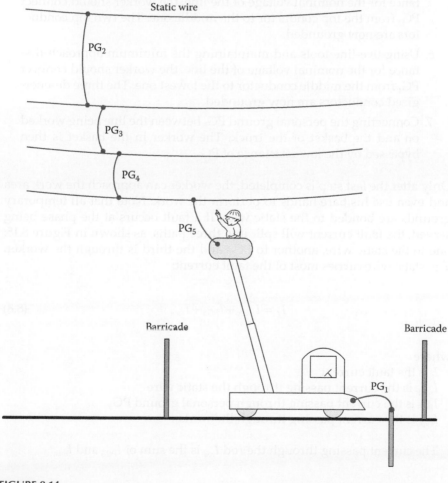

FIGURE 8.14
Worksite away from towers.

reduced if the resistivity of the soil is high or the rod is not driven deep enough into the soil.

3. Isolating the truck and any other attached equipment using barricades to prevent ground personnel from entering the work area. This is because the GPR of the truck will be lethal if a fault occurs at the site. The barricaded area should be large enough to limit the step potential outside it even during faults.

4. Using live-line tools and maintaining the minimum approach distance for the nominal voltage of the line, the worker should connect protective ground PG_2 to the static wire first and then to the nearest conductor. The top wire is now grounded.

5. Using live-line tools and maintaining the minimum approach distance for the nominal voltage of the line, the worker should connect PG_3 from the top conductor to the middle one. The two top conductors are now grounded.

6. Using live-line tools and maintaining the minimum approach distance for the nominal voltage of the line, the worker should connect PG_4 from the middle conductor to the lowest one. The three de-energized conductors are now grounded.

7. Connecting the personal ground PG_5 between the line being worked on and the basket of the truck. The worker in the basket is then bypassed by the low resistance of PG_5.

Only after the last step is completed, the worker can approach the work area and even use his bare hands to perform the work. Note that all temporary grounds are bonded to the static wire. If a fault occurs at the phase being served, the fault current will split into three paths, as shown in Figure 8.15: one to the static wire, another to PG_5, and the third is through the worker. The static wire carries most of the fault current:

$$I_f = I_{static} + I_{PG5} + I_{man} \tag{8.8}$$

where
  I_f is the fault current
  I_{static} is the current passing through the static wire
  I_{PG5} is the current passing through personal ground PG_5
  I_{man} is the current passing through the worker

The current passing through the rod I_{rod} is the sum of I_{PG5} and I_{man}

$$I_{rod} = I_{PG5} + I_{man} \tag{8.9}$$

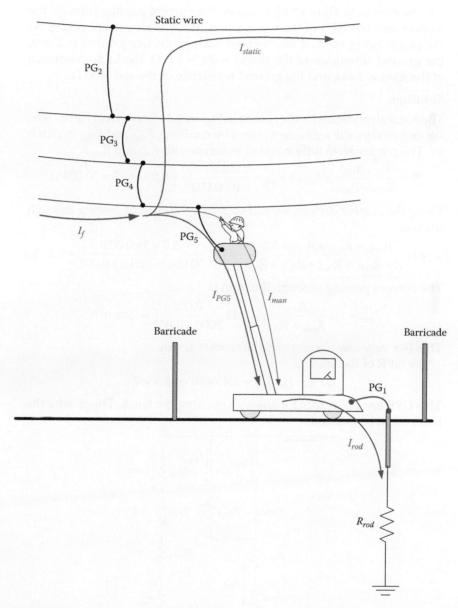

FIGURE 8.15
Fault current flow for the system in Figure 8.14.

Example 8.6

For the system in Figure 8.15, compute the current passing through the worker and the GPR of the truck when 10 kA fault current occurs on the phase being worked on. Assume each protective ground is 2 mΩ, the ground resistance of the shield wire is 5.0 Ω, the body resistance of the man is 1 kΩ, and the ground resistance of the rod is 30 Ω.

Solution:

The equivalent circuit for the system in Figure 8.15 is shown in Figure 8.16. We can modify the equivalent circuit by combing R_{man} and R_{PG5} in parallel. The combination is then added in series with R_{rod} and R_{PG1}.

$$R = \frac{R_{PG5} R_{man}}{R_{PG5} + R_{man}} + R_{PG1} + R_{rod} = \frac{0.002 \times 1000}{1000.002} + 0.002 + 30 = 30.004 \ \Omega$$

Using the current divider, we can compute the current passing through the rod:

$$I_{rod} = I_f \frac{R_{static} + R_{PG2} + R_{PG3} + R_{PG4}}{R + R_{static} + R_{PG2} + R_{PG3} + R_{PG4}} = 10^4 \frac{5.0 + 3 \times 0.002}{30.004 + 5.0 + 3 \times 0.002} = 1.43 \ \text{kA}$$

The current passing through the man is

$$I_{man} = I_{rod} \frac{R_{PG5}}{R_{man} + R_{PG5}} = 1.43 \frac{0.002}{1000 + 0.002} = 2.86 \ \text{mA}$$

This is a very low current and the worker is safe.

The GPR of the truck is

$$\text{GPR} = I_{rod} R_{rod} = 1.43 \times 30 = 42.9 \ \text{kV}$$

This GPR level is deadly for anyone touching the truck. This is why the truck must be isolated.

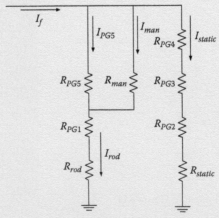

FIGURE 8.16
Equivalent circuit for the system in Figure 8.15.

8.5.2.2 Worksite 2: Aerial Work near Towers with Isolated Ground Equipment

Figure 8.17 shows a worksite that includes a three-phase circuit, metallic tower, static wire, and conductive boom and truck. Because of the proximity to the tower structure, the worker could contact the tower while working on the line. The workers can establish the grounding system by the following steps:

1. Selecting the static wire to be the reference ground since it has the lowest ground resistance at the site.

2. Bonding the truck to a ground rod through protective ground PG_1. This way, the touch potential of the truck can be reduced, but not eliminated. The effectiveness of the ground rod is substantially reduced if the resistivity of the soil is high or the rod is not driven deep enough into the soil.

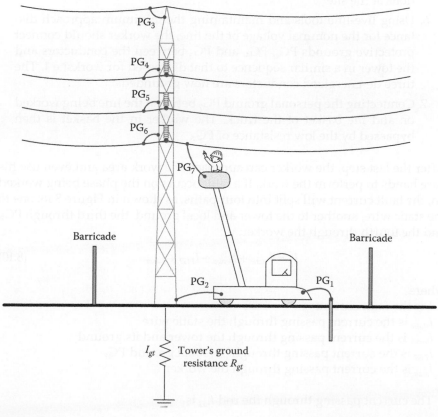

FIGURE 8.17
Worksite near towers.

3. Isolating the truck, tower, and any other attached equipment using barricades to prevent ground personnel from entering the work area. This is because the GPR of the truck will be lethal if a fault occurs at the site. The barricaded area should be large enough to limit the step potential outside it even in the case of faults.

4. Bonding the truck to the base of the tower through PG_2. This ties the grounds of the truck and tower to a single point. If the ground resistance of the tower is tested and verified to be low and the truck is close to the tower, the ground rod may not be needed.

5. Using live-line tools and maintaining the minimum approach distance for the nominal voltage of the line, the worker should connect protective ground PG_3 between the tower and the static wire. This can also be done by climbing the tower to make the connection. No one should assume that the static wire is bonded to the tower before installing PG_3. Rust and corrosion of the hardware can add a few ohms to the static wire path. This is enough to cause lethal conditions at the site.

6. Using live-line tools and maintaining the minimum approach distance for the nominal voltage of the line, the worker should connect protective grounds PG_4, PG_5, and PG_6 between the conductors and the tower in a similar sequence to that discussed for worksite 1. The three de-energized conductors are now grounded.

7. Connecting the personal ground PG_7 between the line being worked on and the basket of the truck. The worker in the basket is then bypassed by the low resistance of PG_7.

After the last step, the worker can approach the work area and even use his bare hands to perform the work. If a fault occurs on the phase being worked on, the fault current will split into four paths, as shown in Figure 8.18: one to the static wire, another to the tower and local ground, the third through PG_7, and the fourth through the worker:

$$I_f = I_{static} + I_{tower} + I_{PG7} + I_{man} \tag{8.10}$$

where
I_f is the fault current
I_{static} is the current passing through the static wire
I_{tower} is the current passing through the tower and its ground
I_{PG7} is the current passing through personal ground PG_7
I_{man} is the current passing through the worker

The current passing through the rod I_{rod} is

$$I_{rod} = I_{PG7} + I_{man} - I_{PG2} \tag{8.11}$$

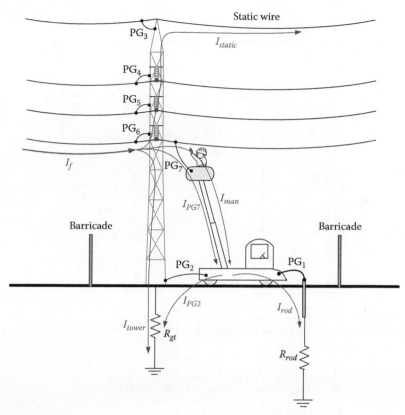

FIGURE 8.18
Fault current flow for the system in Figure 8.17.

Example 8.7

For the system in Figure 8.18, compute the current passing through the man and the GPR of the truck when a 10 kA fault current occurs on the phase being worked on. Assume each protective ground is 2 mΩ, the ground resistance of the shield wire is 5.0 Ω, the body resistance of the man is 1 kΩ, the ground resistance of the rod is 30 Ω, and the ground resistance of the tower is 20 Ω.

Solution:

The equivalent circuit for the system in Figure 8.18 is shown in Figure 8.19. Using the circuit theory, we can solve for the current anywhere in the circuit. We can also simplify the circuit by assuming R_{PG2} is almost zero. In this case, R_{man}, R_{PG7}, and R_{PG6} are in parallel. Also $R_{static} + R_{PG3}$, R_{gt}, and $R_{PG1} + R_{rod}$ are all in parallel.

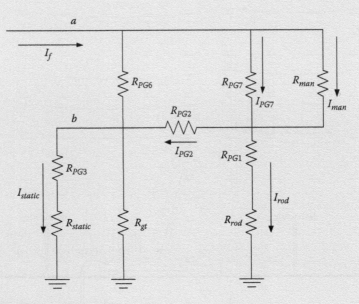

FIGURE 8.19
Equivalent circuit for the system in Figure 8.18.

$$\frac{1}{R_1} = \frac{1}{R_{man}} + \frac{1}{R_{PG6}} + \frac{1}{R_{PG7}} = \frac{1}{1000} + \frac{1}{0.002} + \frac{1}{0.002}$$

$$R_1 = 0.001\,\Omega$$

$$\frac{1}{R_2} = \frac{1}{R_{gt}} + \frac{1}{R_{PG3} + R_{static}} + \frac{1}{R_{PG1} + R_{rod}} = \frac{1}{20} + \frac{1}{5.002} + \frac{1}{30.002}$$

$$R_2 = 3.53\,\Omega$$

Using the voltage divider, we can compute the voltages V_{ab} and V_b:

$$V_{ab} = I_f R_1 = 10^4 \times 10^{-3} = 10.0\text{ V}$$

The voltage between point b and the ground is

$$V_b = I_f R_2 = 10^4 \times 3.53 = 35.3\text{ kV}$$

The current passing through the man is

$$I_{man} = \frac{V_{ab}}{R_{man}} = \frac{10.0}{1000} = 10.0\text{ mA}$$

This is a low current and the worker is safe. For higher fault currents, lower protective ground resistance should be used.

The GPR of the truck and tower is the same as V_b:

$$GPR = V_b = 35.3 \, kV$$

This level of GPR is deadly. This is why the truck, tower, and all attached equipment must be isolated.

8.5.2.3 Worksite 3: Aerial and Ground Work

The arrangements in Figures 8.14 and 8.17 cannot be classified as equipotential zone if ground equipment and towers are not isolated. The reason for this is the elevated GPR during fault conditions that can be deadly to workers on the ground.

If the work requires ground personnel to contact ground equipment (truck, tower, etc.), the ground workers must be protected as well. If not fully protected, the site is not inside an equipotential zone, and touching grounded equipment can be hazardous, especially during faults.

To protect ground personnel, the worker must stand on a ground mat bonded to the equipment (truck, tower, etc.), as shown in Figure 8.20. A ground mat is shown in Figure 8.21. If the work involves a number of ground personnel, the truck should be parked over a ground mat large enough to accommodate the truck and any person working in its immediate vicinity.

During faults, the current distribution, as shown in Figure 8.22, can be expressed by

$$I_f = I_{static} + I_{PG5} + I_{man1} \tag{8.12}$$

$$I_{PG5} + I_{man1} = I_{rod} + I_{man2} + I_{PG6} \tag{8.13}$$

where
I_f is the fault current
I_{static} is the current passing through the static wire
I_{man1} is the current passing through the aerial worker
I_{PG5} is the current passing through personal ground PG5
I_{PG6} is the current passing through personal ground PG6
I_{man2} is the current passing through the ground worker
I_{rod} is the current passing through the ground rod

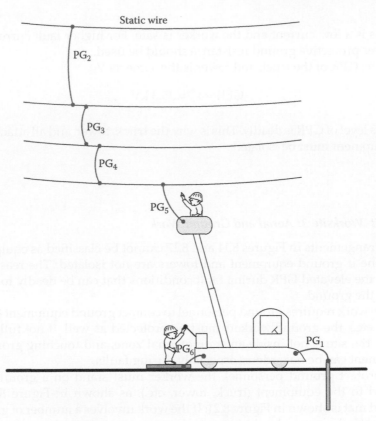

FIGURE 8.20
Equipotential zone to protect aerial and ground workers.

FIGURE 8.21
Worker standing on ground mat attached to truck. (Chance® tools and grounding equipment depicted courtesy of Hubbell Power Systems, Inc.)

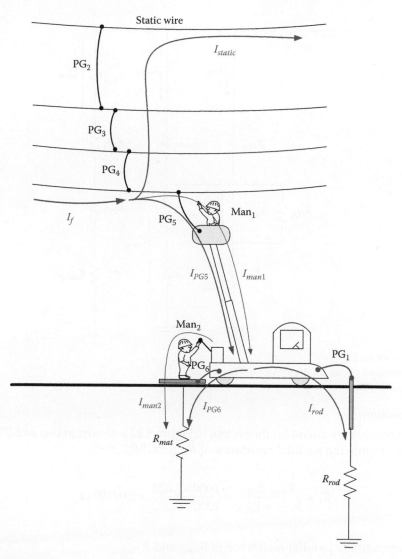

FIGURE 8.22
Fault current flow for the system in Figure 8.20.

Example 8.8

For the system in Figure 8.22, compute the current passing through both men when a 10 kA fault current occurs on the phase being worked on. Assume each protective ground is 2 mΩ, the ground resistance of the shield wire is 5.0 Ω, the body resistance of each man is 1 kΩ, the ground resistance of the rod is 30 Ω, and the ground resistance of the mat is 50 Ω.

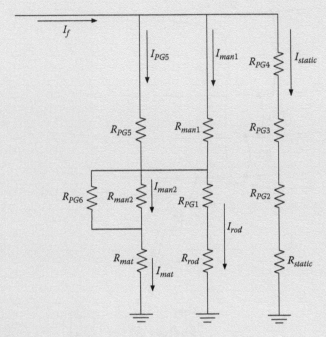

FIGURE 8.23
Equivalent circuit for the system in Figure 8.22.

Solution:

The equivalent circuit for the system in Figure 8.22 is shown in Figure 8.23.
Compute the parallel resistance of R_{man1} and R_{PG5}:

$$R_1 = \frac{R_{man1} R_{PG5}}{R_{man1} + R_{PG5}} = \frac{1000 \times 0.002}{1000 + 0.002} = 0.00199 \; \Omega$$

Compute the parallel resistance of R_{man2} and R_{PG6}:

$$R_2 = \frac{R_{man2} R_{PG6}}{R_{man2} + R_{PG6}} = \frac{1000 \times 0.002}{1000 + 0.002} = 0.00199 \; \Omega$$

Compute the parallel combination of $R_2 + R_{mat}$ and $R_{PG1} + R_{rod}$:

$$R_3 = \frac{(R_2 + R_{mat})(R_{PG1} + R_{rod})}{(R_2 + R_{mat}) + (R_{PG1} + R_{rod})} = \frac{50.00199 \times 30.002}{50.00199 + 30.002} = 18.75 \; \Omega$$

R_1 and R_3 are in series:

$$R_4 = R_1 + R_3 = 0.00199 + 18.75 = 18.75199 \; \Omega$$

The current $I_{man1} + I_{PG5}$ is

$$I_{man1} + I_{PG5} = I_f \frac{R_{static} + R_{PG2} + R_{PG3} + R_{PG4}}{R_4 + R_{static} + R_{PG2} + R_{PG3} + R_{PG4}}$$

$$= 10^4 \frac{5.006}{18.75199 + 5.006} = 2.107 \; \text{kA}$$

The current passing through the aerial worker I_{man1} is

$$I_{man1} = 2107 \frac{R_{PG5}}{R_{man1} + R_{PG5}} = 2107 \frac{0.002}{1000 + 0.002} = 4.21 \; \text{mA}$$

Now, the aerial worker is protected.

To compute the current passing through the ground worker I_{man2}, we need to compute the current passing through the mat first:

$$I_{mat} = 2107 \frac{R_{PG1} + R_{rod}}{R_{PG1} + R_{rod} + R_2 + R_{mat}} = 2107 \frac{30.002}{30.002 + 0.00199 + 50} = 790.14 \; \text{A}$$

The current passing through ground worker is

$$I_{man2} = 790.14 \frac{R_{PG6}}{R_{PG6} + R_{man2}} = 790.14 \frac{0.002}{0.002 + 1000} = 1.58 \; \text{mA}$$

The ground worker is also safe.

For low-voltage towers, the equipotential zone can be established without using boom trucks. The lineman just climbs the tower and uses a hot stick to establish the equipotential zone, as shown in Figure 8.24.

8.5.3 Bracketed Grounds

Some utilities, unfortunately, allow bracketed grounding without protective grounds at the immediate worksite. A design of bracketed grounding is

FIGURE 8.24
Lineman establishing the equipotential zone using a hot stick. (Courtesy of Salisbury by Honeywell, Bolingbrok, IL.)

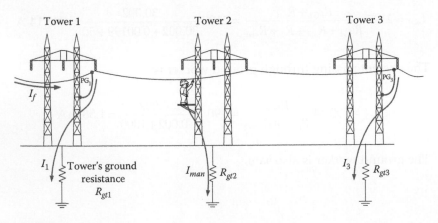

FIGURE 8.25
Bracketed grounding.

shown in Figure 8.25. The protective grounds are installed at towers located one or more spans away from the worksite on both sides. The figure shows three towers with protective grounds installed at towers 1 and 3. The worksite is tower 2 where the lineman is working on the conductor while standing on a platform on tower 2 (or using conductive aerial equipment).

Bracketed grounding is used because of two alleged advantages:

- Work can be done quickly because it is installed for a wide work area. There is no need to keep installing and dismantling the grounding system as long as the work is performed between the brackets.
- Fault current is distributed throughout the two towers; thus reducing the step potential.

However, the bracketed grounds method in Figure 8.25 does not protect the worker. If a fault occurs on the conductor being worked on, the lineman will receive a deadly shock, as shown in the next two examples.

Example 8.9

For the system in Figure 8.25, assume a fault current of 10 kA. Assume the ground resistance of each tower is just 10 Ω. Estimate the current passing through the worker on tower 2.

Solution:

The fault current is branching into three paths. One is through the protective ground (PG_1) of tower 1 and then to the ground. The second is through the man touching the line and to the ground of tower 2. The third is through the protective ground PG_3 of tower 3 and then to the ground.

The equivalent circuit of the bracketed ground is shown in Figure 8.26. The grounds of towers 1 and 3 can be combined in parallel.

$$\frac{1}{R} = \frac{1}{R_{gt1} + R_{PG1}} + \frac{1}{R_{gt3} + R_{PG3}} = \frac{1}{10.002} + \frac{1}{10.002}$$

$$R \approx 5\,\Omega$$

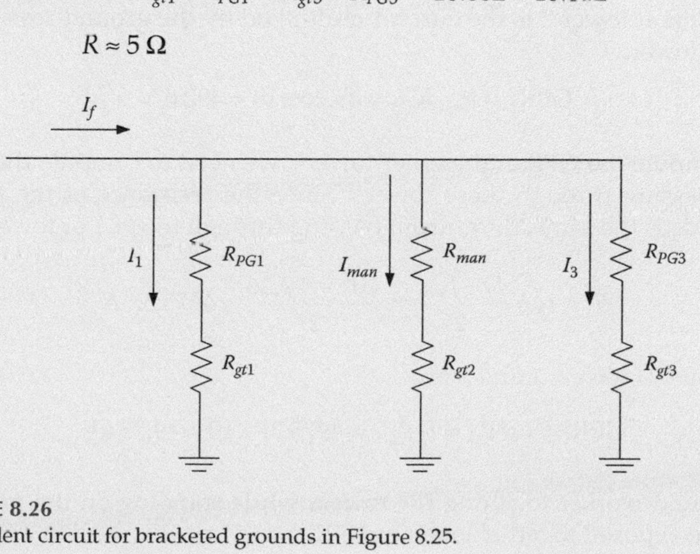

FIGURE 8.26
Equivalent circuit for bracketed grounds in Figure 8.25.

Assuming 1 kΩ for the body resistance, the current passing through the man is then

$$I_{man} = I_f \frac{R}{R + R_{gt2} + R_{man}} = 10^4 \frac{5}{5 + 10 + 1000} = 49.26 \text{ A}$$

The worker on tower 2 cannot survive this level of current. Note that even when the ground resistance of all towers is very low, the current passing through the man is deadly. Also, if you assume just 1000 A for fault current, the current is still deadly.

As seen in the previous example, the bracketed grounding does not protect workers at tower 2. A lethal voltage appears between tower 2 and the conductor during faults. So why is it used sometimes? Because it is fast and requires less work, it is convenient for large work areas. It is also used because of the tendency of the workers to dismiss the possibility of faults while working at the site.

Another problem of bracketed grounding is the high GPR at towers 1 and 3. This can be lethal for people on the ground touching either tower, as shown in the next example.

Example 8.10

For the case in Example 8.9, compute the ground potential rise at all towers.

Solution:

The GPR at tower 2 is the current multiplied by the ground resistance of the tower:

$$GPR_2 = I_{man} R_{gt2} = 49.26 \times 10 = 492.6 \text{ V}$$

To compute the GPR at the other towers, we need to compute the current passing through these towers. Since the resistance of the tower grounds is the same, the current passing through tower 1 or tower 3 is

$$I_1 = I_3 = \frac{I_f - I_{man}}{2} = \frac{10^4 - 49.26}{2} = 4975.37 \text{ A}$$

The GPA of towers 1 and 3 is

$$GPR_1 = GPR_3 = I_1 R_{gt1} = 4975.37 \times 10 = 49.75 \text{ kV}$$

A ground worker touching the towers while standing on the ground will be exposed to lethal levels of GPR.

8.5.4 Circulating Current

When work is performed on a de-energized line, the circuit breakers of the line are opened and their line sides are sometimes grounded. This is done to ensure that accidental closure of any of the breakers will result in a bolted fault that would quickly open the circuit breakers. If both circuit breakers are grounded, a circulating current could exist at the worksite if ground current is present. Figure 8.27 shows two circuit breakers serving a line. At each breaker, a temporary ground wire R_{PG} is used to connect the line conductor to the local ground R_g.

Since points a and b are at some distance from each other, any ground current causes a potential difference between these two points. If we assume constant current density for the ground current, we can compute the voltage between the two points, as discussed in Equation 4.5.

$$V_{ab} = \frac{\rho I_g}{2\pi} \left[\frac{1}{r_a} - \frac{1}{r_b} \right] \tag{8.14}$$

where

V_{ab} is the voltage between points a and b
ρ is soil resistivity
I_g is the ground current
r_a is the distance from where the current is injected into the ground and point a
r_b is the distance from where the current is injected into the ground and point b

The circulating current can be computed by Thevenin's theorem, where Thevenin's voltage is V_{ab} without the presence on the line and temporary grounds. Thevenin's resistance R_{th} is

$$R_{th} = R_{g1} + R_{g2} \tag{8.15}$$

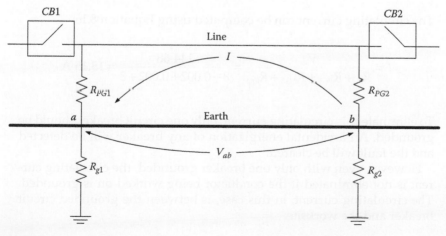

FIGURE 8.27
Grounding circuit breakers.

Hence, the circulating current I is

$$I = \frac{V_{ab}}{R_{th} + R_{PG1} + R_{PG2} + R_{line}} \tag{8.16}$$

where
   I is the circulating current
   R_{PG} is the resistance of the protective ground
   R_{line} is the resistance of the line

Example 8.11

A fault current of 10 kA occurs 1.0 km away from point a in Figure 8.27. The distance between the two circuit breakers is 10 km. Assume the ground grid resistance at either circuit breakers is 4 Ω, the resistance of the temporary ground is 2 mΩ, the line resistance is 3 Ω, and the soil resistivity is 100 Ω-m. Estimate the circulating current.

Solution:

Equation 8.14 can be used to estimate Thevenin's voltage:

$$V_{ab} = \frac{\rho I_g}{2\pi}\left[\frac{1}{r_a} - \frac{1}{r_b}\right] = \frac{100 \times 10^4}{2\pi}\left[\frac{1}{10^3} - \frac{1}{11 \times 10^3}\right] = 144.68 \text{ V}$$

The circulating current can be computed using Equation 8.16:

$$I = \frac{V_{ab}}{R_{th} + R_{PG1} + R_{PG2} + R_{line}} = \frac{144.68}{8 + 0.002 + 0.002 + 3} = 13.15 \text{ A}$$

To eliminate this circulating current, only one circuit breaker should be grounded. An accidental energization of any breaker will be detected and the fault will be cleared.

However, even with only one breaker grounded, the circulating current is not eliminated if the conductor being worked on is grounded. The circulating current, in this case, is between the grounded circuit breaker and the worksite.

8.6 Case Studies

Several case studies are presented in this section. All are real cases that resulted in injuries or fatalities. These could have been avoided if the correct procedures, which often require more time at the site, were implemented. In some of the cases, the victims were aware of the inadequate protection during faults, but dismissed the likelihood of fault happening while working at the site. In other cases, wrong equipments were used or incorrectly installed.

8.6.1 Case Study 1

The worksite is shown in Figure 8.28. It was away from any tower and was de-energized. The worker used two ground rods with two protective grounds: PG_1 and PG_2. The circuit breakers feeding the worksite were open and one of them was grounded at the substation. The truck, boom, and basket were conductive. The protective ground was AWG 04, which is a high-resistance conductor. In addition, PG_2 was long and coiled.

The worksite had several violations, including:

- Inadequate size of protective grounds was used.
- Although PG_2 was coiled just two turns, the impedance of the protective ground is high during high-frequency transients.

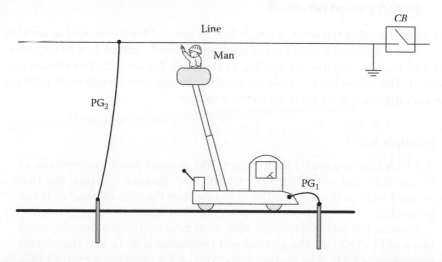

FIGURE 8.28
Worksite with violations for case study 1.

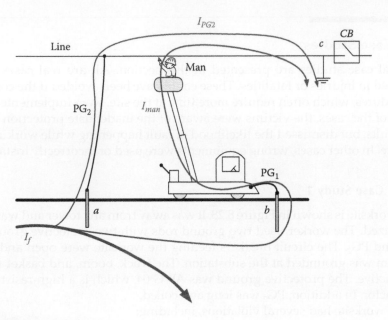

FIGURE 8.29
Case study 1 during nearby fault.

- Worksite had multiple ground points.
- No personal ground was used between the basket and the conductor.
- No ground mat was provided and no isolation was constructed to protect ground personnel.

While performing the work, a single line to ground fault occurred at another circuit on a different tower. Part of the high ground current flew to the worksite and went through several paths, as shown in Figure 8.29. The worker was injured. The following example shows that the site was unsafe even with the conservative numbers we used in the example.

Example 8.12

A 10 kA line-to-ground fault occurs 500 m away from the worksite in Figure 8.29 and on a different circuit. The distance between the two ground rods at the site is 50 m. Assume that the circuit breaker is not grounded.

Assume the soil is dry with 1000 Ω-m resistivity. Assume the resistance of PG_1 is 2 mΩ, the ground rod resistance is 30 Ω, and the worker resistance is 1 kΩ. The worker uses AWG 4 for protective ground PG_2. The length of PG_2 is 50 m. Estimate the current passing through the worker if the circuit breaker is not grounded.

Solution:

The first step is to compute the potential V_{ab} due to the fault currents without the presence of worksite resistances:

$$V_{ab} = \frac{\rho I_f}{2\pi}\left[\frac{1}{r_a} - \frac{1}{r_b}\right] = \frac{10^3 \times 10^4}{2\pi}\left[\frac{1}{500} - \frac{1}{550}\right] = 289.37\ V$$

This is Thevenin's voltage between the two rods. The resistance of PG_2, for AWG 4, is 0.8152 mΩ/m. The total resistance of PG_2 is

$$R_{PG2} = 0.8152 \times 50 = 40.76\ m\Omega$$

Thevenin's resistance is the series combination of R_{rod} and $R_{PG1} + R_{rod}$ (Figure 8.30):

$$R_{th} = 2R_{rod} + R_{PG1} = 60 + 0.002 = 60.002\ \Omega$$

The current passing through the worker is

$$I_{man} = \frac{V_{th}}{R_{th} + R_{PG2} + R_{man}} = \frac{289.37}{60.002 + 40.76 + 1000} = 262.88\ mA$$

The current passing through the worker is hazardous, even when the fault is on another circuit.

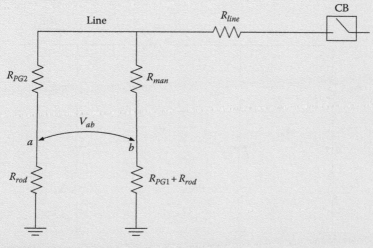

FIGURE 8.30
Equivalent circuit for the system in Figure 8.29 without grounding CB.

Example 8.13

Repeat the previous example, but assume that the circuit breaker is grounded, as shown in Figure 8.29. The distance between the site and the circuit breaker is 2 km. Assume the ground grid resistance at the circuit breaker is 5 Ω, R_{PGg} is 2 mΩ, and the resistance of the transmission line between the site and the circuit breaker is 1 Ω. Estimate the current passing through the worker.

Solution:

The first step is to compute the potential V_{bc} due to fault currents:

$$V_{bc} = \frac{\rho I_f}{2\pi}\left[\frac{1}{r_b} - \frac{1}{r_c}\right] = \frac{10^3 \times 10^4}{2\pi}\left[\frac{1}{550} - \frac{1}{2000}\right] = 2.098 \text{ kV}$$

The equivalent circuit of the system is shown in Figure 8.31. Using the superposition theorem, we can compute the current passing through the worker. By superposition, you assume one voltage source at the time and compute its contribution to the worker's current. The total current passing through the worker is the sum of the contributions of all sources. However, because $V_{bc} \gg V_{ab}$, we can assume that the current due to V_{bc} is much higher than that due to V_{ab}.

Thevenin's resistance is the series combination of R_{rod}, R_{PG1}, and R_g:

$$R_{th} = R_{rod} + R_{PG1} + R_g = 30 + 0.002 + 5 = 35.002 \ \Omega$$

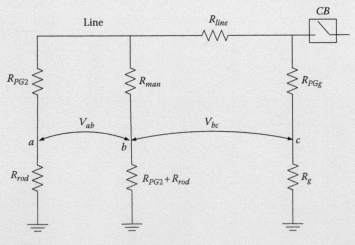

FIGURE 8.31
Equivalent circuit for the system in Figure 8.29.

The current passing through the worker is

$$I_{man} = \frac{V_{th}}{R_{th} + R_{PGg} + R_{line} + R_{man}} = \frac{2098}{35.002 + 0.002 + 1.0 + 1000} = 2.025 \text{ A}$$

Note that the current passing through the worker is much more hazardous when the circuit breaker is grounded.

Based on the previous examples, the worksite is hazardous whether the CB is grounded or not. Of course, if the soil is wet organic, the current passing through the worker is reduced, but not eliminated.

The worksite can be made safe by implementing the following procedures:

- Barricading the worksite to prevent anyone from entering the area, as shown in Figure 8.32.
- Using a short conductor with adequate cross section for PG_2.
- Establishing a single-point ground, as shown in Figure 8.32.
- Installing a personal ground between the basket and the conductor, as shown in Figure 8.32.

The flow of currents during fault is shown in Figure 8.33. The current passing through the worker will be very low if the suggested corrections are implemented as shown in the Example 8.14.

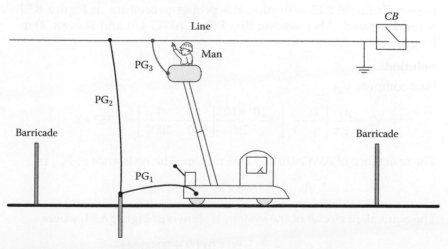

FIGURE 8.32
Correcting worksite violations in Figure 8.28.

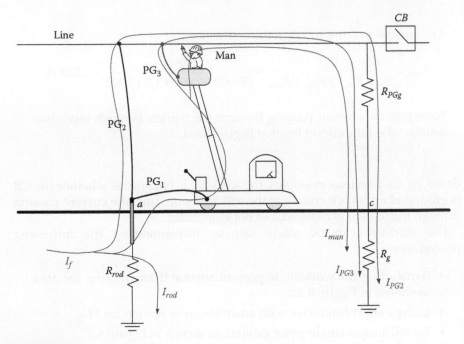

FIGURE 8.33
Worksite in Figure 8.32 during faults.

Example 8.14

Repeat Example 8.13 assuming the proper procedure in Figure 8.32 is implemented. Also assume that PG_2 is AWG 4/0 and is even 30 m long.

Solution:

First compute V_{ac}:

$$V_{ac} = \frac{\rho I_f}{2\pi} \left[\frac{1}{r_b} - \frac{1}{r_c} \right] = \frac{10^3 \times 10^4}{2\pi} \left[\frac{1}{500} - \frac{1}{2000} \right] = 2.387 \text{ kV}$$

The resistance of AWG 4/0 is 0.1608 mΩ/m. The resistance of PG_2 is

$$R_{PG2} = 0.1608 \times 30 = 4.824 \text{ m}\Omega$$

The equivalent circuit of the system is shown in Figure 8.34, where

$$R_{th} = R_{rod} + R_g = 30 + 0.002 = 30.002 \ \Omega$$

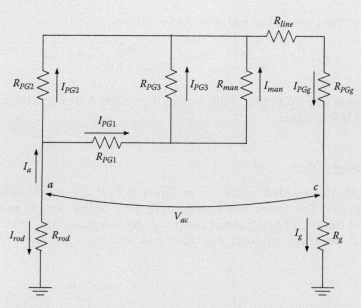

FIGURE 8.34
Equivalent circuit of system in Figure 8.33.

To compute the current passing through the worker, we need to compute the equivalent resistance of R_{PG1}, R_{PG3}, and R_{man}:

$$R_1 = \frac{R_{man} R_{PG3}}{R_{man} + R_{PG3}} + R_{PG1} = \frac{1000 \times 0.002}{1000.002} + 0.002 = 0.004 \; \Omega$$

The combination of R_{PG2} and R_1 is

$$R_2 = \frac{R_1 R_{PG2}}{R_1 + R_{PG2}} = \frac{0.004 \times 0.00767}{0.004 + 0.00767} = 0.00263 \; \Omega$$

The current I_a is

$$I_a = \frac{V_{ac}}{R_{th} + R_2 + R_{line} + R_{PGg}} = \frac{2387}{30.002 + 0.00263 + 1 + 0.002} = 77 \; \text{A}$$

I_{PG1} is

$$I_{PG1} = I_a \frac{R_{PG2}}{R_{PG2} + R_1} = 77 \frac{0.00767}{0.00767 + 0.004} = 50.6 \; \text{A}$$

The current passing through the worker is then

$$I_{man} = I_{PG1} \frac{R_{PG3}}{R_{PG3} + R_{man}} = 50.6 \frac{0.002}{0.002 + 1000} = 0.1\,\text{mA}$$

The site is safe. Note that if the worker uses AWG 3/0 or 2/0, he or she would still be safe.

8.6.2 Case Study 2

The site for this incident is shown in Figure 8.35. It consists of two three-phase circuits. The tower has a static wire. One of the circuits was energized and the circuit being maintained was de-energized. The field worker grounded the truck through a ground rod and installed one protective

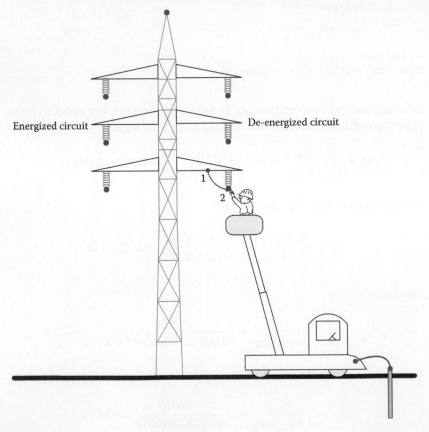

FIGURE 8.35
Worksite with violations for case study 2.

ground between the conductor being worked on (lowest conductor) and the tower. The clamps of the protective ground were class C, which is suitable for conductors. Therefore, terminal 1 of the protective ground was not tightly connected to the flat structure of the tower. Before we go any further, let us list the violations:

- Worksite has multiple ground points.
- Only one phase was grounded.
- Connections between the OHGW and tower were not verified.
- No personal ground was used between the basket and the conductor.
- Wrong types of clamps were used.
- Site was not barricaded.

During the work, the conductor vibrated severely enough to cause terminal 1 of the protective ground to become loose and fall while the worker was still in contact with the conductor, as shown in Figure 8.36. Because of the proximity of the energized circuit, the induced voltage on the touched conductor was high enough to cause the death of the worker.

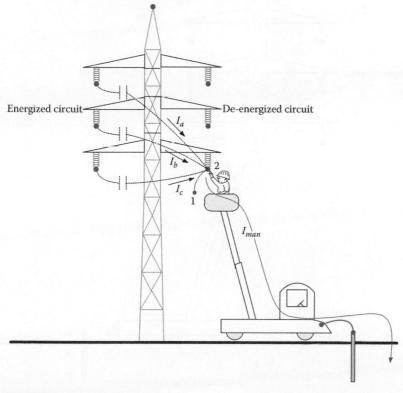

FIGURE 8.36
Fatality for the worksite in Figure 8.35.

8.6.3 Case Study 3

In this case, shown in Figure 8.37, the worker installed a ground rod using PG_1, and then installed one protective ground (PG_2) between the conductor and the tower. He then moved his basket near the tower where he touched the structure. The person received a primary shock. Upon examination of the site, the following were found:

- External bonding of the static wire was damaged due to earlier events. The static wire was bonded to the tower through the hardware only (shoe) that was galvanized. The resistance between the tower and the static wire was therefore high.
- The ground resistance of the tower was elevated due to corrosions.

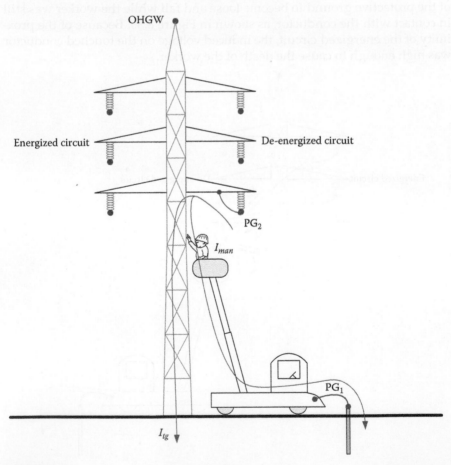

FIGURE 8.37
Worksite with violations for case study 3.

The site has several violations, including

- The static wire was not bonded to the tower by a protective ground.
- The ground resistance of the tower was high, which increased the GPR.
- The worksite has multiple ground points. The grounds of the tower and truck were separated.
- No personal ground was used between the basket and the tower.
- The site was not isolated.
- Only one phase was grounded.

The injury happened when the induced voltage on the conductor being worked on caused a current flow to the structure, and then to the worker on its way to ground, as shown in Figure 8.37. There was no fault in the circuit and the induced voltage was enough to cause the injury.

Example 8.15

For the system in Figure 8.37, assume that the induced voltage leaks just 10 A through the tower. If the ground rod resistance is 30 Ω, and the tower ground resistance is 15 Ω, estimate the current passing through the worker.

Solution:

The equivalent circuit for the case in this example is shown in Figure 8.38. Use the current divider to compute the current passing through the worker:

$$I_{man} = I \frac{R_{tg}}{R_{tg} + R_{man} + R_{PG1} + R_{rod}} = 10 \frac{15}{15 + 1000 + 0.002 + 30} = 143.54 \text{ mA}$$

This level of current is hazardous. The worker was not able to let go, and the shock time was therefore long enough to cause cardiac arrest.

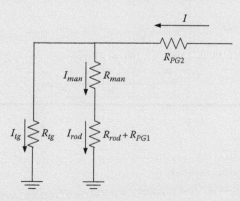

FIGURE 8.38
Equivalent circuit for the system in Figure 8.37.

To fix the violations at this site, the following should be implemented:

- The static wire should be bonded to the tower by a protective ground.
- Single-point ground should be established by adding protective ground PG_3, as shown in Figure 8.39.
- Personal ground should be added between the basket and the tower, as shown in Figure 8.39.
- The site should be isolated.
- All three phases should be grounded.

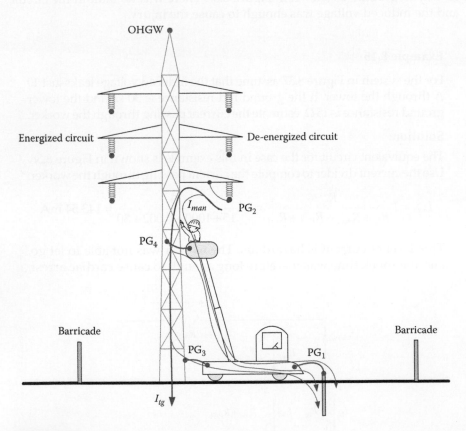

FIGURE 8.39
Correction of violations for case study 3.

Example 8.16

Repeat Example 8.15 but assume PG$_3$ and PG$_4$ are in place, as shown in Figure 8.39.

Solution:

The equivalent circuit for the case in Figure 8.39 is shown in Figure 8.40. R_{PG} is the parallel equivalent of R_{PG3} and R_{PG4}. For identical protective ground, R_{PG} is

$$R_{PG} = 0.5R_{PG3} = 0.5R_{PG3} = 0.001 \ \Omega$$

The parallel combination of R_{man} and R_{PG} is

$$R = \frac{R_{man} R_{PG}}{R_{man} + R_{PG}} = \frac{1000 \times 0.001}{1000.001} \approx 0.001 \ \Omega$$

The current passing through the rod is

$$I_{rod} = I \frac{R_{tg}}{R_{tg} + R + R_{PG1} + R_{rod}} = 10 \frac{15}{15 + 0.001 + 0.002 + 30} = 3.33 \ \text{A}$$

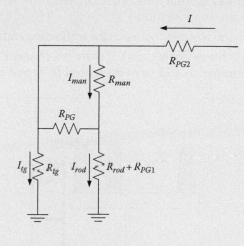

FIGURE 8.40
Equivalent circuit for the system in Figure 8.39.

The current passing through the man is

$$I_{man} = I_{rod} \frac{R_{PG}}{R_{PG} + R_{man}} = 3.33 \frac{0.001}{0.001 + 1000} = 3.33 \, \mu A$$

Because of the single-point ground, the current passing through the worker is very low.

8.6.4 Case Study 4

In the case shown in Figure 8.41, the work was to maintain a disconnect switch in a substation. The circuit was de-energized and the worker installed three-phase grounding system on the right side of the disconnect switch. The grounding system was bonded to the substation grid. The worker was standing on conductive scaffolding that was in contact with the ground. After finishing with the right side of the disconnect switch, the worker moved the scaffolding to the left side of the switch to complete the rest of the maintenance work. While working on the left side of the switch, the worker used some force to loosen a bolt; in the process, he unintentionally opened the switch, as shown in Figure 8.42. The worker received a shock.

The violations of the site include the following:

- Grounding system was not installed at the immediate site. The fact that there is a device that can open between the worker and the grounding system is a violation.
- No personal ground was used between the scaffolding and the conductor.
- Disconnect switch was not secured. If there is a chance that the switch could open, a jumper should have been used to bypass the switch.

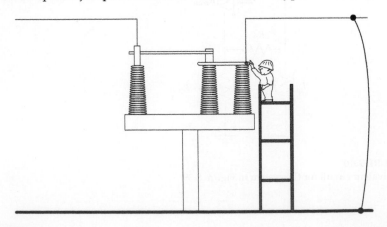

FIGURE 8.41
Worksite with violations for case study 4.

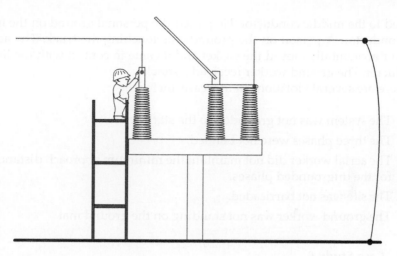

FIGURE 8.42
Injury due to open disconnect switch.

8.6.5 Case Study 5

The site description for this case is shown in Figure 8.43. The work was performed in midspan. The tower of the three-phase circuit has static wires. The truck was grounded through grounding rod and the aerial worker used personal ground. The worker finished working on the lower conductor and

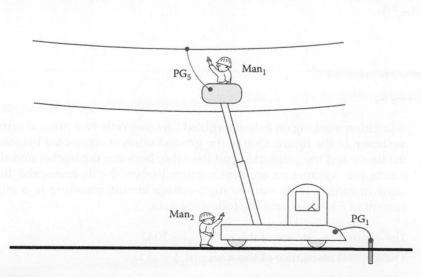

FIGURE 8.43
Worksite with violations for case study 5.

moved to the middle conductor. He placed the personal ground on the middle conductor. A person on the ground was touching the truck. The aerial worker accidentally moved the basket and it came in contact with the lower conductor. The ground worker received a shock.

There are several violations in this case, including

- The system was not grounded to the static wire.
- The three phases were not bonded.
- The aerial worker did not maintain the minimum approach distance for the ungrounded phases.
- The site was not barricaded.
- The ground worker was not standing on the ground mat.

8.6.6 Case Study 6

In this case study, a lineman was working on an overhead line that had no other circuit sharing the right of way at the worksite. The two circuit breakers of the line were opened, locked, and tagged. Their high-voltage sides were grounded. Regrettably, the worker had determined that there was no chance for induced voltage on the line from another circuit. Therefore, he attempted to work without installing protective ground. Unfortunately, the worker died.

The investigation showed that the line being worked on shared towers with another circuit over 20 miles away from the site. The sharing was just for three spans, which was enough to induce lethal voltage on the line being worked on.

Exercises

8.1 A lineman working on a de-energized line uses only two ground wires as shown in the figure. One of the ground wires is connected between the tower and the conductor and the other between the bucket and the conductor. Assume an accident occurred when the de-energized line came in contact with another high-voltage circuit, resulting in a fault current of 5 kA. Assume the following data:

The ground resistance of the rod (R_{rod}) = 30 Ω

The ground resistance of the tower (R_{gt}) = 15 Ω

The ground resistance of the static wire including the wire resistance (R_{static}) = 0.01 Ω

The resistance of any protective ground assembly (R_{gw}) = 0.002 Ω

The body resistance of either man (R_{man}) = 1000 Ω

The ground resistance (foot resistance) of the man on the ground (R_f) = 1000 Ω

Assume that the lineman fails to remove the high-resistance oxide layer at the connection point where ground wire 2 is attached to the tower. This results in a 1 Ω contact resistance. Ignore the resistance of the tower structure and evaluate the grounding system.

8.2 Repeat the previous problem assuming that the man on the ground is standing on a ground mat attached to the truck through a ground cable of 0.02 Ω.

8.3 What is the definition of de-energized conductor?

8.4 What are the factors that determine the severity of induced voltage on de-energized conductors?

8.5 Name the methods used to detect voltage on de-energized conductors.

8.6 What are the main protection techniques for working on de-energized lines?

8.7 What are the main features of protective grounds?

8.8 What is the equipotential zone?

8.9 What are the key features of equipotential zone?

8.10   What are the advantages of grounding the three-phases versus single-phase grounding?

8.11   How will you ground a site with aerial work and away from towers?

8.12   How will you ground a site with aerial work near towers?

8.13   What is the importance of ground mats?

9

Live-Line Work

To avoid the loss of power to customers or void network vulnerability, live-line work is sometimes the only method available to perform maintenance tasks. Because the equipment is energized, live-line work comes with considerable hazards. Therefore, strict procedures and rules have been established by various government organizations to ensure the safety of all workers at the site. Some of these procedures were actually developed back in the early twentieth century for relatively low-voltage works. During the second half of the twentieth century, several methods were developed for almost all voltage levels. Generally, live-line work can be done by one of three methods:

Hot stick: Where the worker remains at a distance from the live parts and carries out the work by hot stick or shotgun.

Insulate and isolate: The worker is in direct contact with the live parts but only after wearing insulated gloves, insulated sleeves, and other insulating personal protective equipment (PPE). The worker is placed on insulated platform and is isolated from any other object with different potential. This technique is limited to distribution system voltages up the to 36 kV class.

Bare hand: The potentials of the worker and the energized equipment are equalized. Then, the work is carried out on the live parts by bare hand. At all times, the worker must be isolated from the surroundings objects that are at different potentials.

9.1 Hot Stick Method

Hot sticks were initially made out of baked wood to remove internal moist. They were handy tools for simple jobs such as replacing fuses and insulators. Because wood can crack and attract moisture, its use was limited to lower voltages. In the middle of the twentieth century, fiberglass replaced wood because of its high dielectric strength and light weight. Nowadays, hot sticks can have mechanical remote control to allow the worker to perform more complex tasks. These types of hot sticks are known as shotguns. The length of the rod is dependent on the voltage of the line and can be as long as 20 ft (about 6 m). A shotgun is shown in Figures 9.1 and 9.2. The tip of the shotgun in Figure 9.2 fits inside the round eye and

FIGURE 9.1
Hot stick shotgun.

FIGURE 9.2
End of shotgun.

allows the worker to tighten or loosen clamps from a distance. Figure 9.3a shows one of the tools that are used to pull fuses. Figure 9.3b shows round eye screws of temporary hardware. More sophisticated hot sticks can operate power tools. These tools must be pneumatically or hydraulically controlled.

(a)   (b)

FIGURE 9.3
Various tools used on hot stick: (a) fuse puller tool and (b) round eye screws.

While using the hot stick, the worker cannot get closer than an established minimum distance to the energized objects. This is known as the *minimum approach distance* (MAD), which is given in Table 9.1. The MAD is a function of the voltage of the line. The *phase-to-ground exposure* means that the worker is exposed to only one line. The *phase-to-phase exposure* is for the work on one line while the other is nearby.

The established rule is that the worker and all conductive equipment in contact with the worker must stay outside the MAD; the entire body of the worker must be outside the corresponding MAD, as shown in Figure 9.4. If any part of his body at any moment is at a distance less than the MAD, the

TABLE 9.1

Minimum Approach Distance for Live-Line Work

	MAD	
Line Voltage (kV)	Phase-to-Ground Exposure (m)	Phase-to-Phase Exposure (m)
1.1–15	0.64	0.66
15.1–36	0.72	0.77
36.1–46	0.77	0.85
46.1–72.5	0.90	1.05
72.6–121	0.95	1.29
138–145	1.09	1.50
161–169	1.22	1.71
230–242	1.59	2.27
345–362	2.59	3.8
500–550	3.42	5.5
765–800	4.53	7.91

Source: OSHA Regulation.

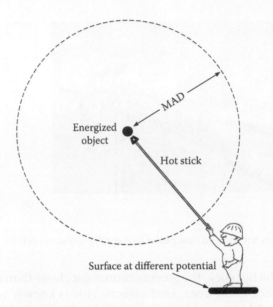

FIGURE 9.4
Working outside the MAD zone.

FIGURE 9.5
Using hot stick in insulated aerial baskets. (Courtesy of Seattle City Light.)

FIGURE 9.6
Using shotgun at ground level. (Courtesy of Seattle City Light.)

person is not protected. Figure 9.5 shows a crew performing maintenance work using hot stick while standing on aerial equipment. Figure 9.6 shows work at the ground level using hot stick.

9.2 Insulate and Isolate Method

If the employees performing the work must enter the MAD, they can do so if

- PPE is worn to insulate the worker from the energized objects
- Isolating protective equipment (IPE) is used to isolate the workers from nearby energized objects

PPEs are items worn by workers. They include insulating rubber gloves with leather protectors, insulating sleeves, safety glasses, face shields, hard hats, safety shoes, and flame-resistant (FR) clothing. Figure 9.7 shows a glove and sleeve. PPEs come in various classes depending on the nominal voltage of the energized objects. These classes are shown in Table 9.2. Although class 4 is for 36 kV, it is not common to use this technique at these high-voltage levels without having the workers on an insulated platform.

IPE includes rubber line hose, blankets, hoods, isolating barriers made of fiberglass or phenolic resin, and plastic or fiberglass hardcover. They are installed using hot sticks, and the worker must maintain the MAD during the installation process. IPE alone without PPE cannot protect the workers. This is because some IPEs are designed to protect workers from brush contact only. Figure 9.8 shows workers using PPEs and IPEs.

FIGURE 9.7
Rubber glove and sleeve.

TABLE 9.2

Maximum Voltage for Rubber Insulating
Equipment

Class of Equipment	Maximum Line-to-Line Voltage (kV)
00	0.5
0	1.0
1	7.5
2	17.0
3	26.5
4	36.0

Source: National Electric Safety Code (NESC), Table 441-6.

Besides the PPEs and IPEs, utilities require linemen to perform the work from an insulating platform. The workers can be placed inside an insulated aerial lift or can simply be placed on wooden pole.

9.3 Bare-Hand Method

With the bare-hand technique, the qualified worker should be in direct contact with the energized object, but must be isolated from any other object with different potential. The common methods for bare-hand techniques are as follows:

FIGURE 9.8
Workers with PPE and IPE. (Courtesy of Salisbury by Honeywell, Bolingbrok, IL.)

- *Insulated aerial lift*: This is the most common method. Linemen can access the work area by insulating aerial device (IAD), which is a boom truck with conductive basket and insulating boom. The boom must be a good insulator with clean surface to protect workers. It must not allow a leakage current above just a few microamperes even when the basket is in direct contact with the energized object. Table 9.3 shows the allowable leakage current for various voltage levels. These IADs must be tested according to ANSI Standard A92.2, where the aerial lift intentionally contacts the energized object for at least 3 min and the leakage current must be constant and within the allowable levels during the entire test period. If the current exceeds the allowable value, the work must be terminated. The basket of the IAD is made of conductive material with the purpose of providing the required bonding with the energized object being worked on.

TABLE 9.3

Allowable Leakage Current of Insulated Aerial Boom

Phase-to-Phase Voltage (kV)	Phase-to-Ground Voltage (kV)	Acceptable Dielectric Current for Insulated Booms (µA)
72.6–121	67	Less than 70
138–145	80	Less than 80
161–169	93	Less than 100
230–242	133	Less than 140
345–362	200	Less than 200
500–550	318	Less than 320

FIGURE 9.9
Workers with PPE bonded to energized conductor for the bare-hand technique. (Images courtesy of KT Power Inc., Waddington, NY.)

This way, the potential of the worker matches that of the energized object. OSHA 1910.269 (q) (3) requires that a conductive bucket liner, conductive mesh, or other conductive device be provided for bonding the basket to the energized line or equipment (see Figure 9.9). The industry practice is to use one lead per person to bond the basket to the energized line. As long as the workers are bonded to the energized object, they are safe. This is known as "touch only that you are clipped to rule." To limit the touch potential on ground, OSHA demands that before the boom of an aerial lift is elevated, the body of the truck be grounded, or barricaded and treated as energized.

- *Insulated ladders or platforms*: Instead of the IAD, the worker can stand on an insulating ladder that moves toward the energized object by means of nonconductive rope or equipment.

- *Helicopters*: Helicopters are the fastest way to perform live-line work. They can be used to lower the linemen into the work area using non-conductive line as shown in Figure 9.10. They can also be used to bring the lineman alongside the energized object and allow him to work from a platform attached to the helicopter, as shown in Figure 9.11. For the second scenario, the maintenance work is performed after the helicopter is connected electrically to the object being worked on. This method cannot be performed during windy conditions.

- *Traveling baskets*: The worker is placed on a traveling basket that resembles a ski cable car, as shown in Figure 9.12. The basket moves along the line, and the worker can perform the tasks along the line quite fast.

FIGURE 9.10
Helicopter transporting linemen to working site.

FIGURE 9.11
Working on a power line using a helicopter. (Image courtesy of USA Airmobile, Inc., Davie, FL.)

- *On conductors*: The worker is placed directly on the energized conductor, as shown in Figure 9.13. The photo on the right side shows
  a worker performing maintenance work while surrounded by high-
  voltage wires. The four subconductors in the figure are actually bundled together to form one phase of a transmission line; thus all of the
  four subconductors have the same potential. The worker inside the
  bundle is safe even when he contacts any two of these subconductors.

FIGURE 9.12
Workers moving into the traveling basket for the bare-hand technique. (Image courtesy of KT Power Inc., Waddington, NY.)

FIGURE 9.13
Workers on energized conductors. (Right-side image courtesy of USA Airmobile, Inc., Davie, FL.)

Because the workers are very close to the energized equipment, they are exposed to a strong electric field; for every 1 kV/m, about 15 µA passes through a worker's body. To protect themselves from the high-electric fields, the workers wear Faradaysuits. This conductive PPE includes conductive hooded coverall, gloves, and socks, as shown in Figures 9.9 through 9.13. The suite is bonded to the energized object through a spring ankle clamp or by wearing shoes with conductive soles.

As the lineman approaches the energized object, an arc is formed between the object and the lineman. The arc charges the worker's body and his suit to bring their voltage to that for the energized object. This arc is often just a few microamperes. However, it could be annoying. Therefore, the worker must rapidly bond himself to the energized object or use a conducting wand to make the first contact. Because arcs could be extremely bright and could have high ultraviolet component, workers use tinted goggles.

OSHA demands that the MAD be maintained between the workers and any nearby grounds or other circuits with different potentials. This includes the approaching and leaving of the energized object. The MAD is not required if all grounded objects and other lines are covered by insulating guards.

9.4 Case Study

The case study is shown in Figure 9.14. The task was to replace the arm of a disconnect switch without interrupting the service. Two workers were assigned to the task. The process calls for installing a jumper across the terminals of the switch, removing the existing arm, installing a new arm, and lastly, removing the jumper. The workers used insulated scaffolding made of insulated rods with assembly joints made of aluminum. One worker bonded himself to the energized line. The other worker bonded himself to the first worker.

Because of the bonding with the energized object, the hot stick and isolate and insulate techniques were not used. Instead, they used the bare-hand technique from an aerial platform with several violations; the most important ones are as follows:

- The platform was not made of conductive material that can be bonded to the energized object.
- Workers were bonded to each other.

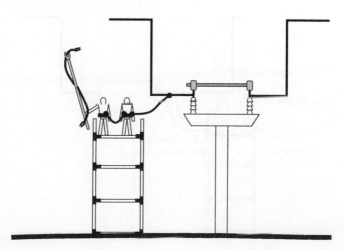

FIGURE 9.14
Case study.

The workers used a hot stick to install the jumper; one end of the jumper was attached to the tip of the hot stick. For an unknown reason, the stick fell to the ground with one end of the jumper on the ground and the other end jammed in one of the platform conductive joints. At the same time, one of the workers was also in contact with the same joint. This caused a fatal current to pass through both workers, as shown in Figure 9.15.

This accident could have been avoided if the procedure in Figure 9.16 were implemented:

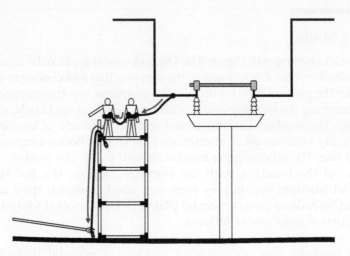

FIGURE 9.15
The accident.

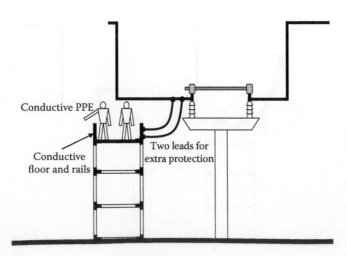

FIGURE 9.16
Proper procedure.

- Platform of the scaffolding should have been made of conductive material that is bonded to the energized conductor by two leads.
- Workers should have worn conductive PPE.
- Workers should not have been bonded to each other.

Exercises

9.1 Why do we need to perform maintenance on energized equipment?

9.2 A lineman uses a 4′ 8″ (1.32 m) long hot stick to work on a 230 kV system. Is he in violation of any rule?

9.3 While using the hot stick technique, can the worker stand on grounded object?

9.4 While using the insulate and isolate technique, can the worker stand on grounded object?

9.5 Which class of PPE must be used for 24 kV work?

9.6 Is MAD applicable for the insulate and isolate technique?

9.7 While covering energized nearby object by IPEs for the insulate and isolate technique, is MAD applicable?

9.8 For the insulate and isolate technique, are IPEs enough to protect the worker without insulating PPE?

9.9 What are the main bare-hand technique methods?

9.10 While testing an insulated aerial lift for bare-hand technique on a 140 kV system, the measured leakage current was 50 μA. Can the aerial lift be used?

9.11 Must we use conductive basket for the bare-hand technique from insulated aerial lift? Why?

9.12 How many leads should be used between the conductive basket and the energized conductor?

9.13 Can conductive suit be used for the hot stick or isolate and insulate techniques?

9.14 What is the purpose of the conductive suit?

9.15 When a helicopter approaches an energized line, a wand can be used to touch the conductor. Why is there an arc between the wand and the energized conductor when there is no electric circuit?

9.16 Is MAD applicable for the bare-hand technique?

9.17 If two workers are on the same basket during the bare-hand technique, can they be directly bonded to each other?

9.18 Can a bare-hand technique be performed from insulated scaffolding?

10

Arc Flash

Arc flash is a form of hazard that does not involve current passing through a person. This hazard occurs when air is ionized between two points that have enough potential difference to cause an arc between them. This often happens accidentally during maintenance work when tools make contacts with live parts of difference potentials. It can also be caused by dust, water vapor, corrosion, overvoltage, dropping foreign object, misalignment of moving contacts, and intrusion of animals. If the energy in the arc is high, the arc can be deadly as the temperature and pressure caused by it can rapidly increase to excessive values near the work area. In the United States alone, 2000 persons are admitted to hospitals due to arc flash every year.

Arc flash problems have existed since the inception of electricity. However, only in the 1980s, serious developments were made to protect people from the hazards of arc flash. Today, there are agencies with standards to assess and prevent arc flash. Among the standards in the United States are

- NFPA 70-2002, *National Electrical Code*
- NFPA 70E-2000, *Standard for Electrical Safety Requirements for Employee Workplaces*
- IEEE Standard 1584-2002, *Guide for Performing Arc Flash Hazard Calculations*
- OSHA 29, *Code of Federal Regulations (CFR) Part 1910 Subpart S*

10.1 Arc Flash Phases

Arcs can occur due to human error or equipment failure. Based on its energy, the arc phenomenon can develop into three phases:

1. *Arc fault*: The initiating event that causes an arc that ionizes air.
2. *Arc flash*: It is an escalating event where the arc possesses enough energy to elevate the temperature of air to levels that could be higher than the temperature on the sun surface.
3. *Arc blast*: This is the most dangerous phase when the energy is high enough to cause an explosion.

10.1.1 Arc Fault

Arc currents exhibit different characteristics than bolted fault currents. On the one hand, a bolted short circuit path is often low impedance and the fault current is therefore high. This high current is cleared quickly by circuit breakers or fuses. Any arc produced by the switching action of protection devices is contained within the device and is extinguished rapidly before releasing a destructive amount of heat.

On the other hand, arcing fault path is high impedance as it involves air. Therefore, the current is often lower than the tripping level of the protection device. Hence, the arc fault lasts longer than the bolted fault, which creates tremendous amount of destructive heat.

An arc fault is often initiated by actions that create a path of conduction or by a failure of insulation. During arc fault, air is ionized and turned into highly conductive plasma. This occurs for voltages exceeding 120 V. Lower voltages do not normally sustain the arc. The arc fault is sustained by the plasma that conducts as much energy as is available from the source and is only limited by the impedance of the arc.

The arc flash can be difficult to interrupt by circuit breakers because the arc fault current could be less than the available bolted fault current. Often, the arc fault current could be as low as 20% of the bolted fault current.

Arc fault can turn into a serious situation causing severe arc flash and arc blast. This occurs if the energy of the arc is high enough to cause these problems. Several factors determine the energy level of the arc; the most important ones are as follows:

- *Available fault current*: If the circuit involving the arc has high available fault current, the arc current (energy) is expected to be high as well.
- *Circuit impedance*: Impendence in series with the arc determines the magnitude of the arc current.

10.1.2 Arc Flash

If the arc fault has enough potential energy, a flash will occur. The flash is a rapid release of energy that superheats the air, turning it into plasma with temperature that can reach 35,000°F (19,427°C). This is almost 3.5 times the temperature of the sun. This amazing heat in the air can cause third-degree burns to the unprotected human body. In addition to the excessive heat, the arc flash creates intense light that can damage the cornea and retina.

10.1.3 Arc Blast

After the arc flash, a pressure wave is developed with increasing intensity. For high-energy arc flash, the excessive temperature can cause metals to instantly vaporize and expand tens of thousands times their original volumes. Air also expands rapidly. These rapid expansions create an immense pressure wave with very high velocity. The pressure waves, called arc blasts, can cause lungs to collapse and can easily damage eardrums. The pressure wave can also force a victim to inhale high-temperature metal vapor and toxic substances found in the fumes. The blast can toss a person several meters away and perhaps into other hazards. If the blast is inside an electric panel, the panel could disintegrate sending debris all over the work area. The sound of the arc blast can reach 160 dB, which is well above the threshold of pain and may cause permanent damage to hearing.

10.2 Assessment of Arc Flash

The key data in arc flash assessment is the incident energy of the arc. Once computed, protection boundaries can be identified to keep personnel outside the hazardous zones. IEEE Standard 1584–2002, *Guide for Performing Arc Flash Calculations*, provides numerical methods to obtain three key quantities:

1. *Arc flash current*: The current of the arc, which is often less than the bolted fault current.
2. *Incident energy*: It is a measure of the thermal energy at a distance from an arc fault.
3. *Flash boundary*: A minimum distance from the energized object where the person cannot encroach.

To obtain these quantities, three key information need to be known:

1. Available bolted fault current of the circuit being worked on. This is done by normal system analysis.
2. Clearing time of the protection device serving the circuit.
3. Expected distance between the worker and the energized object.

10.2.1 Calculation of Arc Flash Current

The current of the arc is often less than the bolted fault current or the available fault current at the site. This is one of the dangers of arc flash as it may not be detected and cleared by the protection devices.

According to IEEE, an arc current can be computed using one of two empirical formulas: the first is for circuits with nominal voltages below 1 kV and the second is for higher voltages. For voltages less than 1000 V, the following equation is used:

$$\log I_a = K + (0.662 + 0.5588\,V - 0.00304\,G)\log I_{bf} + 0.0966\,V + 0.000526\,G \quad (10.1)$$

where

I_a is arc current in kA

K is a constant equal to -0.153 for open air and -0.097 for enclosures

I_{bf} is the three-phase bolted fault current in kA

V is the system nominal voltage (line-to-line) in kV

G is the gap between arcing points in mm (for some known gaps, check Table 10.1)

For voltages higher than 1 kV, we can use the following formula:

$$\log I_a = 0.00402 + 0.983 \log I_{bf} \qquad (10.2)$$

Example 10.1

A 480 V motor control center (MCC) has a bolted fault current of 10 kA. Compute the arc current.

Solution:

Using the typical gap in Table 10.1, we can directly substitute in Equation 10.1

$$\log I_a = K + (0.662 + 0.5588V - 0.00304G)\log I_{bf} + 0.0966V + 0.000526G$$

$$\log I_a = -0.097 + (0.662 + 0.5588 \times 0.480 - 0.00304 \times 25) \times \log 10$$

$$+ 0.0966 \times 0.480 + 0.000526 \times 25$$

$$\log I_a = 0.816742$$

Hence

$$I_a = 10^{0.816742} = 6.5576 \text{ kA}$$

TABLE 10.1

Typical Values of G in Known Systems

Voltage (kV)	Equipment Type	Typical Gap between Conductors (mm)
0.208–1	Open air	10–40
	Switchgear	32
	Motor control center and panels	25
	Cable	13
>1–5	Open air	102
	Switchgear	13–102
	Cables	13
>5–15	Open air	13–153
	Switchgear	153
	Cables	13

Source: Table 4 in IEEE Standard 1584, 2002.

10.2.2 Calculation of Incident Energy

The incident energy is the amount of energy impressed on a surface at a certain distance from an arcing source. This incident energy can be calculated using IEEE Standard 1584. The calculations require the knowledge of the normalized incident energy E_n, which can be computed as follows:

$$\log E_n = K_1 + K_2 + 1.081 \times \log I_a + 0.0011G \tag{10.3}$$

where

E_n is the incident energy normalized in time and distance in cal/cm²
K_1 is a constant equal to -0.792 for open air and -0.555 for enclosures
K_2 is a constant equal to 0 for ungrounded system (or system with high ground resistance) and -0.113 for grounded system
G is the gap between arcing points in mm (for some known gaps, check Table 10.1)

The incident energy of the arc is determined by the following equation for system voltages up to 15 kV:

$$E = C_f E_n \left(\frac{t}{0.2} \right) \left(\frac{610}{D} \right)^x \tag{10.4}$$

where

E is the incident energy of the arc in cal/cm²
C_f is a calculation factor; it is 1.5 for system voltages at 1 kV or lower, and 1.0 for system voltages higher than 1 kV
t is the arcing time in seconds; it is the time it takes for the protection device to clear the fault
D is the distance between the arcing point and the person in mm; the typical value used is 460 mm (~18 in.)
x is the distance factor; for known systems, x can be obtained from Table 10.2

For system voltages above 15 kV, the following equation can be used:

$$E = 5.12 \times 10^5 V I_{bf} \left(\frac{t}{D^2} \right) \tag{10.5}$$

The incident energy determines the amount of heat at the calculated distance. Depending on its magnitude, the incident energy causes various degrees of burns to the human body. The threshold for second-degree burns is 1.2 cal/cm². For values higher than this, third- and fourth-degree burns can occur. Table 10.3 shows the classifications of burns and their impacts.

TABLE 10.2

Typical Values of x in Known Systems

Voltage (kV)	Equipment Type	Distance Factor, x
0.208–1	Open air	2.0
	Switchgear	1.473
	Motor control center and panels	1.641
	Cable	2.0
>1–5	Open air	2.0
	Switchgear	0.973
	Cables	2.0
>5–15	Open air	2.0
	Switchgear	0.973
	Cables	2.0

Source: Table 4 in IEEE Standard 1584, 2002.

TABLE 10.3

Classification of Burns Based on Incident Energy

Burns	Body Layers Involved	Time to Healing
First degree	Epidermis	<1 week
Second degree	Extends into superficial dermis	2–3 weeks
Third degree	Extends through entire dermis	Requires excision
Fourth degree	Extends through skin, subcutaneous tissue, and into underlying muscles and bones	Requires excision

Example 10.2

For the case in Example 10.1, compute the incident energy at 18 in. (457.2 mm) away from the arcing point. Assume that the protection device clears the arc in 50 ms and the MCC system is ungrounded.

Solution:

Using the values of I_a and G in Example 10.1 to compute the normalized incident energy by Equation 10.3

$$\log E_n = K_1 + K_2 + 1.081 \times \log I_a + 0.0011 G$$

$$\log E_n = -0.555 + 0 + 1.081 \times \log(6.5576) + 0.0011 \times 25 = 0.3554$$

$$E_n = 10^{0.3554} = 2.2667 \text{ cal}/\text{cm}^2$$

Use Equation 10.4 to compute the incident energy of the arc at 18 in. distance. The distance factor x in the Equation is obtained from Table 10.2. For 480 V MCC, it is 1.641.

$$E = C_f \, E_n \left(\frac{t}{0.2} \right) \left(\frac{610}{D} \right)^x = 1.5 \times 2.2667 \times \frac{0.05}{0.2} \times \frac{610^{1.641}}{457.2^{1.641}} = 1.364 \text{ cal}/\text{cm}^2$$

NFPA 70E states that second-degree burns occur when heat energy at the skin level is 1.2 cal/cm². Without protection, the incident energy in this example can cause third-degree burns.

10.3 Calculation of Arc Flash Protection Boundary

IEEE Standard 1584 defines the flash protection boundary as *"An approach limit at a distance from live parts that are uninsulated or exposed within which a person could receive a second-degree burn."*

Second-degree burns occur at an incident energy level of 1.2 cal/cm². If we can set E in Equation 10.4 or 10.5 to 1.2 cal/cm², D is the flash protection boundary and is denoted by D_b.

For system voltage up to 15 kV, the flash protection boundary D_b is

$$D_b = 610 \sqrt[x]{C_f \frac{E_n}{E} \left(\frac{t}{0.2} \right)} = 610 \sqrt[x]{4.167 C_f \, E_n \, t} \qquad (10.6)$$

For voltages above 15 kV, D_b is

$$D_b = \sqrt{5.12 \times 10^5 V I_{bf} \left(\frac{t}{E} \right)} = 653.22 \sqrt{V I_{bf} \, t} \qquad (10.7)$$

The arc flash protection boundary is shown in Figure 10.1. You can imagine it as a sphere where the arcing object is in its center. If a person is inside the boundary, he can receive a second-degree burn.

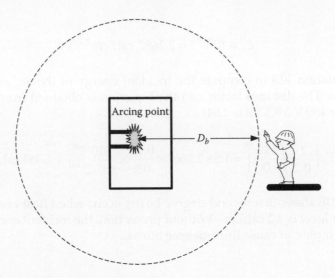

FIGURE 10.1
Arc flash protection boundary.

Example 10.3

For the case in Example 10.2, compute the arc flash boundary.

Solution:

Using Equation 10.6, for a distance factor of 1.641, the minimum distance is

$$D_b = 610\sqrt[x]{4.167C_f E_n t} = 610(4.167 \times 1.5 \times 2.2667 \times 0.05)^{\frac{1}{1.641}} = 494.41 \text{ mm}$$

10.4 Personal Protection Equipment

What if the work must be performed at a shorter distance than D_b? In this case, suitable personal protection equipment (PPE) must be used. The type of PPE is determined by the calculated incident energy at the work distance. PPE includes cloth, face and eye protection, head protection, foot protection, hand protection, protective ointments, shields, barriers, and restraints. Table 10.4 shows the clothing of the PPE from NPFA 70E. The table shows the five categories identified by NPFA based on the intensity of the incident energy.

Most of the recorded incident energy levels at a distance of 18 in. from arc flashes are less than 2.5 cal/cm², which requires PPE0 or PPE1. About 25%

TABLE 10.4

Protective Clothing Characteristics

PPE Category	E (cal/cm²)	Clothing
PPE0	0–1.2	Untreated cotton
PPE1	1.2–5	Flame retardant (FR) shirt and FR pants
PPE2	5–8	Cotton underwear FR shirt and FR pants
PPE3	8–25	Cotton underwear FR shirt, FR pants, and FR coveralls
PPE4	25–40	Cotton underwear FR shirt, FR pants, and double-layer switching coat and pants

Source: NPFA 70E.

FIGURE 10.2
Forty calories arc flash jacket and bib overall. (Courtesy of Salisbury by Honeywell, Bolingbrook, IL.)

of arc flashes have incident energy higher than 8 cal/cm², which requires PPE3. About 10% have 40 cal/cm², which requires PPE4. Unfortunately, 5% of arc flashes have incident energy higher than 80 cal/cm². At this excessive level, there is no full protection, and de-energizing the equipment may be the only option to perform the work. Figure 10.2 shows the PPE used for 40 cal/cm² arc flash incident energy.

10.5 Approach Boundaries

Besides arc flash, NFPA 70E creates three additional approach boundaries based on the voltage level of the circuit being serviced: (1) limited approach boundary (LAB), (2) restricted approach boundary (RAB), and (3) prohibited approach boundary (PAB). In each of these boundaries, the proper PPE must be used.

Limited approach boundary: LAB is a distance from an exposed live part within which a shock hazard exists. LAB can be crossed only by qualified persons using the *appropriate PPE*. Unqualified personnel with proper PPE must be accompanied by qualified personnel.

Restricted approach boundary: RAB is a distance from an exposed live part within which there is an increased risk of shock due to inadvertent movement made by personnel working in close proximity to live parts. The RAB can be crossed only by qualified persons using *shock protection techniques* and wearing appropriate PPE.

Prohibited approach boundary: PAB is a shock protection boundary that can be crossed only by qualified persons with proper PPE. The persons must use the same procedures and protection for workers who directly contact live parts.

The various protection boundaries are depicted in Figure 10.3. NFPA 70E determines the boundaries based on the voltage of the equipment. OSHA CFR1910.269 provides the same information. Table 10.5 summarizes the data in these standards.

The determination of all boundaries and the PPE must be communicated to the personnel working on or near the potentially hazardous equipment. NEC requires all equipment with arc flash potentials such as switchboards, panel boards, industrial control panels, and motor control centers to have visible warning labels. An example of such a label is shown in Figure 10.4. The NEC labeling requirements apply to any electrical equipment installed or modified after 2002.

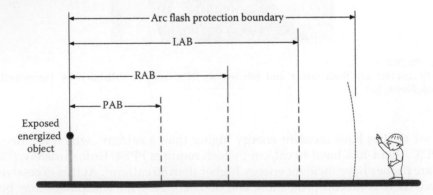

FIGURE 10.3
Protection boundaries.

TABLE 10.5

Approach Boundaries to Energized Electrical Object for Shock Protection

Nominal Voltage (Phase or Line)	LAB	RAB	PAB
51 V to 300 V	3 ft 6 in. (1.0668 m)	Avoid contact	Avoid contact
301 V to 750 V	3 ft 6 in. (1.0668 m)	1 ft 0 in. (0.3048 m)	0 ft 1 in. (0.0254 m)
751 V to 15 kV	5 ft 0 in. (1.524 m)	2 ft 2 in. (0.6604 m)	0 ft 7 in. (0.1778 m)
15.1 kV to 36 kV	6 ft 0 in. (1.8288 m)	2 ft 7 in. (0.7874 m)	0 ft 10 in. (0.254 m)
36.1 kV to 46 kV	8 ft 0 in. (2.4384 m)	2 ft 9 in. (0.8382 m)	1 ft 5 in. (0.4318 m)
46.1 kV to 72.5 kV	8 ft 0 in. (2.4384 m)	3 ft 3 in. (0.9906 m)	2 ft 1 in. (0.635 m)
72.6 kV to 121 kV	8 ft 0 in. (2.4384 m)	3 ft 2 in. (0.9652 m)	2 ft 8 in. (0.8128 m)
138 kV to 145 kV	10 ft 0 in. (3.048 m)	3 ft 7 in. (1.0922 m)	3 ft 1 in. (0.9398 m)
161 kV to 169 kV	11 ft 8 in. (3.556 m)	4 ft 0 in. (1.2192 m)	3 ft 6 in. (1.0668 m)
230 kV to 242 kV	13 ft 0 in. (3.9624 m)	5 ft 3 in. (1.6002 m)	4 ft 9 in. (1.4478 m)
345 kV to 362 kV	15 ft 4 in. (4.6736 m)	8 ft 6 in. (2.5908 m)	8 ft 0 in. (2.4384 m)
500 kV to 550 kV	19 ft 0 in. (5.7912 m)	11 ft 3 in. (3.429 m)	10 ft 9 in. (3.2766 m)
765 kV to 800 kV	23 ft 9 in. (7.239 m)	14 ft 11 in. (4.5466 m)	14 ft 5 in. (4.3942 m)

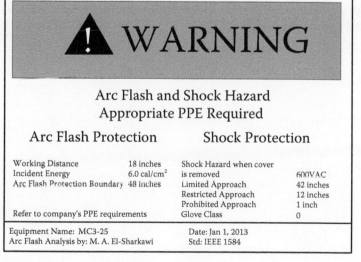

FIGURE 10.4
Example of arc flash and shock protection label.

Exercises

10.1 Consider 16 kV open air buses with bolted fault current of 10 kA, the gap between buses being 150 mm. Compute the arc flash current.

10.2 For the system in Exercise 10.1, calculate the incident energy of the arc at 100 mm distance. Assume the clearing time is 4 ms.

10.3 What is the PPE category that must be used for the case in Exercise 10.2?

10.4 Calculate the flash protection boundary for the system in Exercise 10.1.

10.5 For the system in Exercise 10.1, select the limited, restricted, and prohibited approach boundaries.

10.6 For a 480 V switch gear enclosure, compute the arc fault current if the bolted fault current is 5 kA.

10.7 For the system in Exercise 10.6, calculate the incident energy of the arc at 100 mm distance. Assume the clearing time is 100 ms and the system is grounded.

10.8 What is the PPE category that must be used for the case in Exercise 10.6?

10.9 Calculate the flash protection boundary for the system in Exercise 10.6.

10.10 For the system in Exercise 10.6, select the limited, restricted, and prohibited approach boundaries.

11

Atmospheric Discharge

Lightning is caused by the buildup of electrostatic charge in clouds. Scientists have found that the upper portion of the cloud builds up a positive charge and the lower portion a negative charge. The negative charges are concentrated in one or several clusters within the cloud as depicted in Figure 11.1. If the charges are dense enough, the negative charges may leap to the positive side of another cloud, or it may leap to the ground in the form of a lightning strike. The lightning strike is formed when the negative leader from the cloud travels toward earth while the highest grounded conductive object generates a positive leader that travels toward the negative leader leaping from the cloud. The two leaders travel at a speed of about 60 km/s and they meet at approximately one-third of the height of the cloud, creating a lightning channel which is a high-current flow. Each lightning strike is packed with 10–100 GJ of energy. Most of this energy is converted into thunder noise, flashing light, and heat. The channel's temperature can be as high as 30,000°C. Although the energy reaching the earth is a small fraction of the total lightning energy, it is sufficient to cause fires, make considerable damage to structures and equipment, disturb communications, and cause physical harm to living beings.

About 2 million lightning strikes occur every day worldwide. In the United States, the highest rate of lightning strikes is in the Tampa area in Florida, which receives about 12 lightning strikes per square kilometer annually. Central and South America, central Africa, and Malaysia may get over 200 lightning days per year.

11.1 Characteristics of Lightning Discharge

The waveform of any lightning strike is quite noisy. However, if you filter the noise, the current of the lightning strike looks like the one in Figure 11.2. The initial rise of the current is very rapid where it reaches the peak value in about 1 μs. The current then decays in about 50 μs. The peak current could reach 200 kA with a rate of rise of 200 kA/μs. Figure 11.3 shows the probability curve of lightning strikes, it shows the peak current versus the probability of its occurrence (P). It has been observed that 10% of lightning strikes peak at 250 kA and the average strike is about 20 kA.

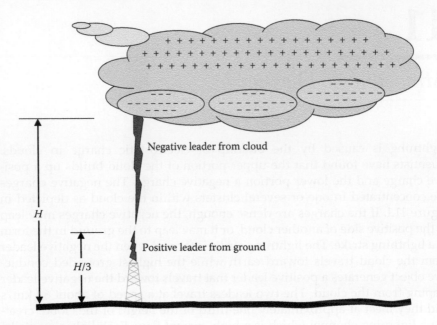

FIGURE 11.1
Physics of lightning.

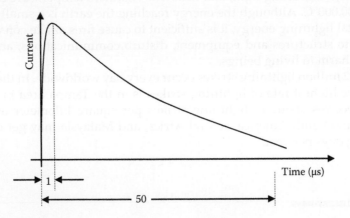

FIGURE 11.2
Lightning waveform.

From the human safety point of view, lightning strikes can be very hazardous. The following list summarizes some of these hazards:

- *Direct hit*: When lightning strikes a grounded object or living being, the extreme heat of the strike causes destruction, fire, and death.
- *Elevated step potential near the stricken area*: When lightning reaches the ground, its high current passes through the earth's crust, thus

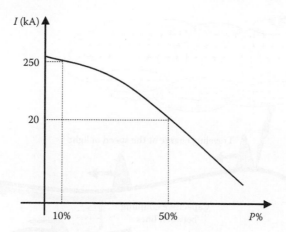

FIGURE 11.3
Probability of lightning strength.

elevating the step potential in nearby areas. The high step potentials could make it harder for people to maintain their balance, resulting in their falling to the ground. Once on the ground, the current could pass through their vital organs. See Example 11.1.

- *Hazardous ground potential rise*: The ground potential near the stricken area is elevated causing excessive touch potentials and ground-to-neutral potentials. Since the neutral could be grounded at remote distance, the potential difference between the neutral and the striken ground may reach a hazardous level for people who come in contact with the utility neutral while part of their body is at the striken ground.

- *Arcing from nearby striken objects*: Because of the elevated potential of the stricken object, a high potential difference could be created between the object and any nearby grounded metallic structure, causing arcing between them.

- *Hazardous voltage at remote areas due to traveling lightning wave along power lines*: When a lightning hits a power line, as shown in Figure 11.4, it travels along the line at the speed of light. The traveling wave is a high-voltage, high-frequency impulse. It elevates the voltage at any point reached by the impulse to hundreds of kilovolts above its nominal voltage. In addition, the lightning bolt induces hazardous voltage on adjacent lines or structures due to the electromagnetic field coupling between the lines. These phenomena can damage line insulators and equipment connected to the transmission lines.

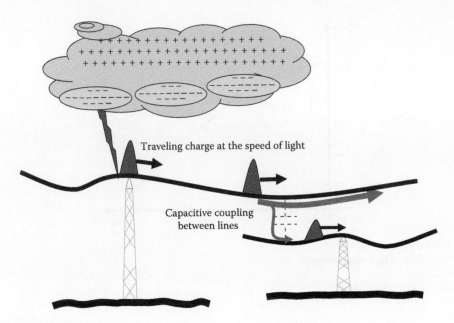

FIGURE 11.4
Traveling wave and electromagnetic coupling of lightning strike.

Example 11.1

During a weather storm, an atmospheric discharge hits a lightning pole. The pole is grounded through a hemisphere, and the maximum lightning current passing through the pole is 20 kA.

1. A person is playing golf 30 m away from the center of the hemisphere. The distance between his feet is 0.3 m, and his leg-to-leg resistance is 1.0 kΩ. Assume the soil surrounding the hemisphere is moist. Compute the current passing through the person and his step potential.
2. Another person is 3 m away from the center of the hemisphere. The distance between his feet is also 0.3 m, and his leg-to-leg resistance is 1.0 kΩ as well. Compute the current passing through the person and his step potential.

Solution:
For moist soil, $\rho = 100$ Ω-m.

1. Thevenin's voltage can be computed using Equation 4.10:

$$V_{th} = \frac{I\rho}{2\pi}\left[\frac{1}{r_a} - \frac{1}{r_b}\right] = \frac{20,000\times100}{2\pi}\left[\frac{1}{30} - \frac{1}{30.3}\right] = 105 \text{ V}$$

The ground resistance of the person is

$$R_f = 3\rho = 300 \ \Omega$$

The current passing through the man can be computed by Equation 5.9:

$$I_{man} = \frac{V_{th}}{2R_f + R_{man}} = \frac{105}{600+1000} = 65.63 \text{ mA}$$

The step voltage is the voltage between the person's feet:

$$V_{step} = I_{man}\times R_{man} = 65.63\times1000 = 65.63 \text{ V}$$

2. For the person 3 m away

$$V_{th} = \frac{I\rho}{2\pi}\left[\frac{1}{r_a} - \frac{1}{r_b}\right] = \frac{20,000\times100}{2\pi}\left[\frac{1}{3} - \frac{1}{3.3}\right] = 9.646 \text{ kV}$$

The current passing through the man can be computed by Equation 5.9:

$$I_{man} = \frac{V_{th}}{2R_f + R_{man}} = \frac{9646}{600+1000} = 6.03 \text{ A}$$

$$V_{step} = I_{man}\times R_{man} = 6.03\times1000 = 6.03 \text{ kV}$$

The step voltage for the second person, in particular, is extremely high, and it is unlikely the person can maintain his balance. If he falls on the ground, a lethal current could pass through his vital organs.

11.2 Protection from Lightning Strikes

There are several simple rules that could protect people from the hazards of lightning discharges. The best is to stay indoors during lightning storms. Those outside should avoid open fields, building roofs, swimming, or boating. Nonconductive umbrellas that do not attract lightning bolts should be used, and one must keep one's legs close together to limit the step potential. Any work on electric wires must be stopped.

Besides these simple safety rules, there are several protection devices that can be used to reroute the energy of the lightning bolt into ground, thus preventing it from reaching important structures or equipment. Among these devices are lightning poles, lightning discharge towers, overhead ground wires (OHGWs), spark gap, and surge arresters.

11.2.1 Lightning Pole and Lightning Discharge Tower

Because the resistance of air is dependent on the distance between the cloud and the stricken point, lightning tends to hit the highest grounded point in the area. To protect a building or a small area, metallic poles (masts) are placed higher than all nearby structures to channel the lightning bolt to the pole instead of the structure. The pole is well grounded to disperse the lightning energy into earth. Figure 11.5 shows lightning poles on two structures.

For large areas with vulnerable equipment such as substations, launching pads, or oil refineries, lightning discharge towers are used. These towers

FIGURE 11.5
Lightning pole on top of buildings.

FIGURE 11.6
Lightning discharge towers. (Courtesy of NASA, Washington, DC.)

are much bigger in size and height than the poles and can disperse a high amount of lightning energy into ground. Figure 11.6 shows lightning discharge towers protecting NASA's launching areas. Notice that the towers are much taller than the spacecraft.

The height of the lightning discharge tower depends on the volume of the desired protected zone. There is no explicit theoretical model that relates the height of the tower to the shape and size of the protective zone. However, engineers use field tests such as the one in Figure 11.7 to generate empirical

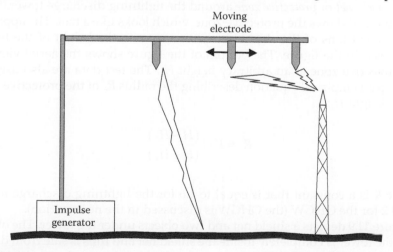

FIGURE 11.7
Test setup for lightning discharge tower.

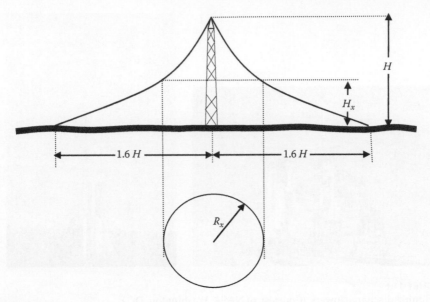

FIGURE 11.8
Shield net of lightning discharge tower.

models for the protection zone. The test system consists of an impulse generator and a movable electrode. At each position of the electrode, the impulse generator is activated to produce a discharge. Data are collected on the percentage of the discharges that strike the tower versus the ones that strike the ground. Based on this test, engineers develop empirically what is known as the *shield net* or *protective zone* around the lightning discharge tower.

Figure 11.8 shows the protective zone, which looks like a tent. The approximate dimensions of the shield net with respect to the height of the tower H is given in the figure. The bottom of the figure shows the aerial view of the protective zone at an arbitrary height H_x. The test data are also used to develop an empirical equation describing the radius R_x of the protective area at any height H_x:

$$R_x = KH \frac{(H - H_x)}{(H + H_x)} \tag{11.1}$$

where K is a constant that is equal to 1.6 for the lightning discharge tower and 1.2 for the OHGW (the OHGW is discussed in the next section).

Figure 11.9 depicts a shield net and two objects under the tower. The object near the tower is completely inside the shield net and thus is safe from direct lightning strikes. However, part of the other object is outside the shield net, which makes the object vulnerable to lightning strikes.

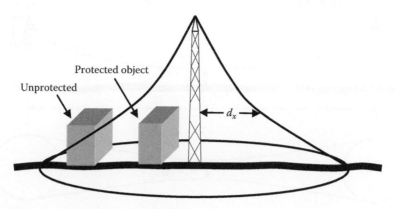

FIGURE 11.9
Protective zone of the lightning discharge tower.

Example 11.2

A 30 m tall lightning discharge tower is erected to protect substation equipment. The equipment is 10 m in height and 4 m wide. Compute the maximum lateral distance between the tower and this equipment.

Solution:

At 10 m height, use Equation 11.1 to compute the radius of the protective area R_x:

$$R_x = 1.6H \frac{(H-H_x)}{(H+H_x)} = 1.6 \times 30 \frac{30-10}{30+10} = 24 \text{ m}$$

Since the width of the equipment is 4 m, the inner side of the equipment should not be more than 20 m from the tower.

To protect a wide area, multiple lightning poles are used, as shown at the right-side photo in Figure 11.6. The shield net in this two-tower system is shown in Figure 11.10. The distance between the towers (D) is a function of the desired minimum height of the shield net at the middle points between the towers (H_0),

$$D = F(H - H_0) \tag{11.2}$$

where F is a constant that is equal to 7 for lightning discharge tower and 4 for OHGWs.

To compute the maximum distance between the two towers (D_{max}), we set $H_0 = 0$:

$$D_{max} = 7H \tag{11.3}$$

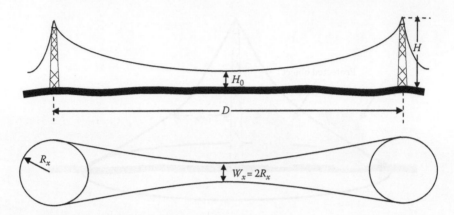

FIGURE 11.10
Shield net for two lightning discharge towers.

Equation 11.3 shows that when the distance between the two towers is equal or larger than $7H$, there will be unprotected areas between the towers.

The bottom part in Figure 11.10 shows the aerial view (horizontal section) of the shield net at an arbitrary height H_x. The width of this protective area at the midline between the two towers can be obtained by assuming that we have a tower at the middle point whose height is H_0. Hence, from Equation 11.1 we get

$$R_x = 1.6H_0 \frac{(H_0 - H_x)}{(H_0 + H_x)} \tag{11.4}$$

Example 11.3

Two lightning discharge towers of equal heights are separated by 210 m. Compute the height of the towers that protect 10 m cubical equipment at the midline between the towers.

Solution:

Direct substitution in Equation 11.2 yields the height of the tower:

$$D = 7(H - H_0)$$

$$210 = 7(H - 10)$$

$$H = 40 \text{ m.}$$

Now, we need to check if the width of the protective area at the ground level is at least 10 m. Use Equation 11.4 to compute R_x at zero height:

$$R_x = 1.6H_0 \frac{(H_0 - H_x)}{(H_0 + H_x)} = 1.6 \times 10 \frac{(10-0)}{(10+0)} = 16 \text{ m}$$

The width of the protective area is

$$W_x = 2R_x = 32 \text{ m}$$

The protective area at the ground level is wide enough to accommodate the 10 m cubical equipment. Hence, 40 m is an adequate height for the towers.

11.2.2 Overhead Ground Wire

Transmission lines pass through plain areas where the towers and their conductors are the highest conductive objects above ground. These towers are prime targets for lightning strikes; almost 20% of all outages worldwide are attributed to lightning damages. To protect transmission lines, utilities install OHGWs at the top of the towers and for the length of the transmission lines. These OHGWs are grounded at the substations and at all towers. Since the OHGWs are the highest grounded objects along the transmission lines, they collect the lightning bolts and disperse them into the ground, thus protecting the transmission line conductors below them from direct lightning strikes. Figure 11.11 shows two OHGWs on the top of a transmission line tower.

Figure 11.12 shows the protective zone of an OHGW. At any arbitrary height H_x, the horizontal section of the protective area along the line is shown at the bottom of the figure. The length of the protective area is the length of the OHGW itself, and the width at any arbitrary height can be computed by using Equation 11.1, where $K = 1.2$.

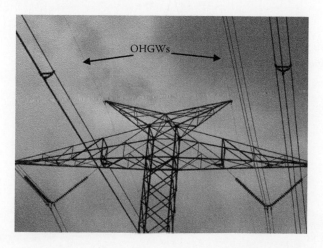

FIGURE 11.11
Two OHGWs.

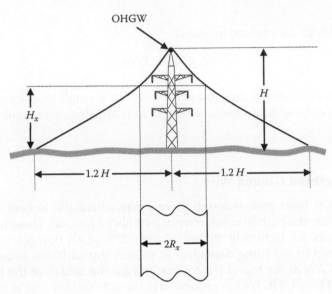

FIGURE 11.12
Protective zone of an OHGW.

Example 11.4

For the transmission line tower shown in Figure 11.13, compute the height of a single OHGW wire that protects all phases.

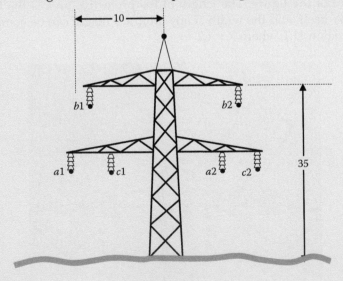

FIGURE 11.13
Tower structure.

Solution:

If we protect the top cross arms, all other conductors are also protected. Equation 11.1 relates the width of the protective zone to the height of the ground wire:

$$R_x = KH\frac{(H-H_x)}{(H+H_x)}$$

where $K = 1.2$ for OHGW. For conductor b, $H_x = 35$ m and $R_x = 10$ m. Hence,

$$10 = 1.2 \times H\frac{(H-35)}{(H+35)}$$

Solving the previous equation leads to an OHGW's height of 49.25 m, which would require the OHGW to be about 15 m above the top cross arm. This is an unrealistic tower size and more than one OHGW is needed.

As seen in the previous example, one OHGW at the center of the tower may not be enough to protect all transmission line conductors. In this case, we need to use two OHGWs, as shown in Figures 11.11 and 11.14.

The height of the sag of the protective zone between the two OHGWs H_0 in Figure 11.14 is a function of the distance between them and can be computed using empirical Equation 11.2 when $F = 4$.

$$D = 4(H - H_0) \tag{11.5}$$

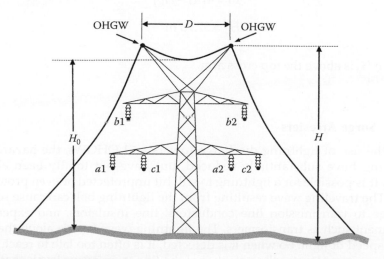

FIGURE 11.14
Protective zone of a double OHGW.

Example 11.5

For the transmission line tower shown in Figure 11.14, compute the height and separation between the two OHGWs. Use the conductor configuration in Figure 11.13.

Solution:

There are two conditions in this problem: the first is to protect all conductors and the second is to ensure that the sag of the protective zone H_0 is above the upper crossbar of the tower.

Let us assume that the height of the OHGW is 41 m. Use Equation 11.1 to check the width of the protective zone at the elevation of the top conductor:

$$R_x = KH\frac{(H-H_x)}{(H+H_x)} = 1.2 \times 41 \times \frac{(41-35)}{(41+35)} = 3.88 \text{ m}$$

At the selected height, the top cross arm is protected if we place the OHGW at a horizontal distance of 3.5 m away from the tip of the top cross arm. Hence, the horizontal distance of the OHGW with respect to the center axis of the tower is 6.5 m, and the distance between the two OHGWs is 13 m. Use Equation 11.5 to compute the sag of the protective zone:

$$D = 4(H-H_0)$$

$$13 = 4(41-H_0)$$

$$H_0 = 37.75 \text{ m}$$

since H_0 is above the top cross arm.

11.2.3 Surge Arresters

With the use of lightning discharge towers and OHGWs, the hazards of lightning have substantially reduced, but have not totally been eliminated. It is possible for a lightning to still hit unprotected or even protected lines. The traveling wave resulting from the lightning bolt can cause severe damage to transmission line conductors, line insulators, and expensive equipment such as transformers. The lightning wave travels along the line at the speed of light. So when it is detected, it is often too late to react. This traveling wave elevates the potential of the line to excessive levels that may cause the insulators of the transmission line to flashover (discharges the energy along the surface of the insulator) or breakdown (discharge the energy through the material of the insulator). If insulators experience flash-overs, they tend to recover once the lightning subsides. However, break-down causes permanent damage to insulators and must be replaced. Until they are replaced, the transmission line is out of service. If the wave reaches a substation, it could damage expensive equipment such as transformers because of the failure of their insulators. To protect the transmission line and all vulnerable equipment, surge arresters are placed at key locations within the substations. These devices operate very similarly to the zener diode whose resistance is inversely proportional to the applied voltage, as shown in Figure 11.15.

A photo of a surge arrester is shown at the far left of Figure 11.16. The top end of the arrester is connected to the high-voltage terminal of the equipment to

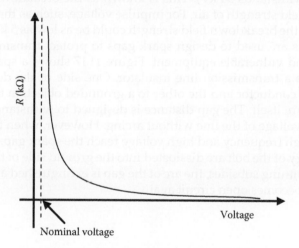

FIGURE 11.15
Resistance of surge arrester as a function of applied voltage.

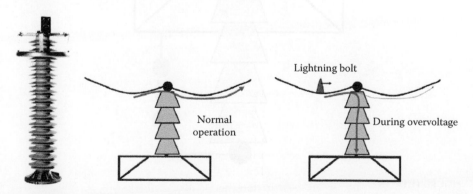

FIGURE 11.16
Surge arrester and its operation.

be protected. The lower end is connected to a good grounding system. During normal operation, the arrester is seen as an open circuit due to its extremely high resistance at nominal voltage. The line current in this case continues along the line as if the arrester does not exist. However, when lightning hits, the excessive voltage of the bolt causes the resistance of the arrester to drop substantially, making the surge arrester a short-circuit path for the lightning current, thus dispersing the energy of the bolt into ground and thus protecting all devices in its downstream.

11.2.4 Spark Gap

At 25°C, air can withstand electric field strength of approximately 30 kV/cm (1 cm of air withstands 30 kV). This is known as the breakdown voltage or breakdown field strength of air. For impulse voltage, such as the one caused by lightning, the breakdown field strength could be as low as 5 kV/cm. These characteristics are used to design spark gaps to protect transmission lines, insulators, and vulnerable equipment. Figure 11.17 shows a spark gap connected across a transmission line insulator. One side of the device is connected to the conductor and the other to a grounded object, in this case, the tower structure itself. The gap distance is designed to withstand more than the nominal voltage of the line without arcing. However, when the traveling wave with high frequency and high voltage reach the spark gap, the gap arcs and the energy of the bolt are dissipated into the ground side of the insulator. When the lightning subsides, the arc of the gap is extinguished and the spark gap arrester becomes open circuit again.

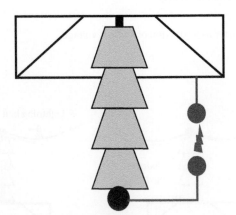

FIGURE 11.17
Spark gap across insulator.

11.3 Safe Distance from Lightning Protection Devices

When a lightning is discharged into earth, either directly or through a lightning protection device, several problems occur at the ground level due to the high current of the lightning bolt, including:

- Excessive ground potential rise.
- Excessive step potentials, especially near the device discharging the bolt into earth.
- Excessive potential for the device discharging the bolt into ground. The potential can be high enough to cause arcing between the device and any nearby objects.

The first two phenomena are analyzed earlier in this book. The last phenomenon needs more attention.

The lightning discharge tower protects all equipment placed inside the protective zone of the tower. Hence, one would wrongly think it should be placed very close to the tower. Consider the configuration in Figure 11.18. The object to be protected is at a horizontal distance d_x from the tower. The grounds of the tower and that of the object are separated by a distance d_g. As mentioned earlier, the breakdown voltage of air for impulse frequency is about 5 kV/cm. Hence, the separation between the tower and the object

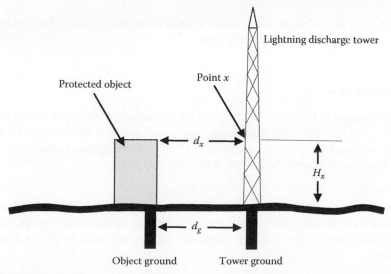

FIGURE 11.18
Safe distance from lightning discharge tower.

should be wide enough to prevent any arcing between them when a lightning bolt hits the tower.

The voltage at point x in Figure 11.18 is a function of the current i through the tower, the inductance and resistance of the section of the tower between point x and ground, and the resistance R_g of the tower's ground resistance. The resistance of the tower itself can be ignored as it is much smaller than R_g:

$$v_x = iR_g + L\frac{di}{dt} \tag{11.6}$$

where L is the inductance of the section of the tower between point x and ground plus the ground electrode's inductance. Hence, the minimum distance $d_{x\text{-}min}$ at elevation x is

$$d_{x\text{-}min} = \frac{v_x}{E_{BD\text{-}air}} \tag{11.7}$$

where $E_{BD\text{-}air}$ is the breakdown electric field strength of air, which is 5 kV/cm for the impulse lightning bolt.

Inside the soil, the potential difference between the two grounds v_g is the difference between their ground potential rises (ΔGPR). Hence, the minimum distance $d_{g\text{-}min}$ between the two grounds is

$$d_{g\text{-}min} = \frac{v_g}{E_{BD\text{-}soil}} \tag{11.8}$$

where $E_{BD\text{-}soil}$ is the breakdown electric field strength of soil, which is about 3 kV/cm for the impulse lightning bolt.

Example 11.6

For the object in Figure 11.18, compute the minimum distance between the object and the lightning discharge tower. Assume that the inductance of the tower and the ground electrode is 1.7 µH/m, the ground resistance of the tower is 20 Ω, and the height of the object is 10 m.

Solution:

The maximum value of the lightning current varies widely. However, 50% of the lightning discharges are about 20 kA, as shown in Figure 11.3.

As seen in Figure 11.2, the current reaches its peak in about 1 µs. Hence,

$$\frac{di}{dt} = 20\,\text{kA/s}$$

Use Equation 11.6 to compute the voltage of the tower at 10 m above the ground (which is the height of the object):

$$v_x = iR_g + L\frac{di}{dt} = 20 \times 30 + (1.7 \times 10) \times 20 = 940 \text{ kV}$$

The safe distance at 10 m above ground can be computed by Equation 11.7:

$$d_{x\text{-}min} = \frac{v_x}{E_{BD\text{-}air}} = \frac{940}{5} = 188 \text{ cm}$$

The distance between the ground electrodes can be computed by Equation 11.8:

$$d_{g\text{-}min} = \frac{v_g}{E_{BD\text{-}soil}} = \frac{\Delta GPR}{3}$$

The GPR of the object is zero since no current is flowing through the structure before the arcing:

$$d_{g\text{-}min} = \frac{\Delta GPR}{3} \approx \frac{iR_g}{3} = \frac{20 \times 20}{3} = 133.33 \text{ cm}$$

The distances must be higher than the calculated minimum values. Repeat the calculations assuming that the maximum lightning current is 250 kA, which occurs about 10% of the time.

Exercises

11.1 Compute the height of a tower that protects a substation's equipment that is 10 m in height and 4 m wide. The distance between the equipment and the center of the tower must not be less than 26 m.

11.2 Is it a good idea to have the equipment to be protected very close to the lightning discharge tower?

11.3 Compute the minimum height of two lightning discharge towers and the minimum lateral distance between them to protect 10 m cubical equipment at the midline between the towers. The separation between the center of the tower and the equipment must not be less than 30 m.

11.4 For the transmission line in the following figure, compute the height of a single OHGW wire that protects all phases. Is it realistic to use a single OHGW to protect all conductors?

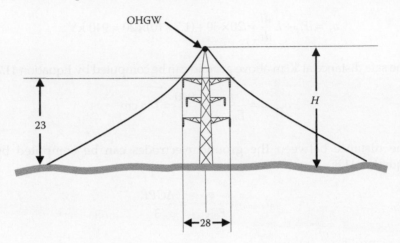

12

Stray and Contact Voltages

Stray voltage is a term used to describe the elevated voltage *from the normal delivery and/or use of electricity* on grounded neutral conductors. It is also known as *neutral to earth voltage* (NEV) and is associated with low-voltage magnitudes.

One of the main reasons for stray voltage is the bonding of the primary and secondary neutrals of service transformers. This neutral is grounded locally at the customers' premises, normally in the service panel. The National Electric Code (NEC) requires all metal casings of the electrical equipment on the premises to be bonded by *equipment grounding conductors* (EGC), which are also bonded to the utility neutral at the service panels. Thus, any metallic frame connected to this EGC will be at the neutral voltage of the premises.

Stray voltage can also be caused by cable televisions and phone lines. These services tend to create a bridge between the primary and the secondary neutrals of the service transformer. Metallic water pipes can also cause a leakage current from old or faulty equipment on one site to elevate the potential of the EGC on another site.

Besides the stray voltage, bonding the utility neutral to the EGC at the service panels could cause high voltage on metallic casings of electrical equipment if the neutral is deteriorated or severed, or when high impedance fault occurs between the hot wire and the neutral.

More recently, stray voltage has taken on a broader definition that includes leakage currents, electromagnetic coupling, and variations in ground potential rise. It is important, however, that we do not confuse stray voltage with *contact voltage* or *transient overvoltage*. The term "contact voltage" is used to describe a situation when the hot wire contacts the metal casing of equipment due to poor workmanship or insulation failure. The transient overvoltage is due to the presence of faults, lightning, or switching that causes the line to neutral voltage to temporarily increase.

Before we proceed, we need to understand two important terms: grounding and bonding. Grounding (or earthing) connects a noncurrent-carrying conductive object to a reference potential point on earth. This is done to limit the voltage on the conductive object with respect to a defined ground point. Bonding is a method to connect conductive objects together. The objects could be at a potential that is different from the local ground.

To understand the stray voltage problems and appreciate the solutions, we need to be aware of the National Electric Safety Code (NESC) that governs utility safety practices and the NEC that regulates the installations on customers' premises. This chapter addresses the issues in these codes that are related to stray voltage.

12.1 Neutral versus Ground

One of the most confusing issues in electric safety is the difference between the neutral and the ground wires. They are both grounded, so why do we use both of them? To answer this question, let us examine the generic representation of the electric equipment shown in Figure 12.1. The equipment consists of an internal electric circuit housed inside an enclosure. If the enclosure is conductive (chassis), the internal circuit is electrically isolated from the chassis. Because of the metal chassis, there are three concerns:

1. Any two adjacent conductive objects with different potentials have parasitic (leakage) capacitance between them. The parasitic capacitance is the unavoidable and unwanted capacitance that exists between closely spaced metallic elements with different potentials. In the generic equipment of Figure 12.1, the potential of the internal electric circuit is different than the potential of the conductive enclosure. Hence, a capacitive coupling (parasitic) between the internal circuit and the conductive enclosure exists, which can elevate the potential of the chassis.

2. The current of the internal circuit produces magnetic fields that link the conductive enclosure, thus inducing voltage on the conductive enclosure.

3. Damaged insulations could result in elevating the voltage of the conductive enclosure.

During normal operation, if a person standing on a grounded object touches the conductive enclosure as shown in Figure 12.2, the person completes an

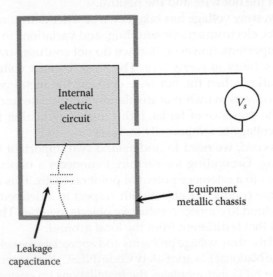

FIGURE 12.1
Generic representation of an appliance.

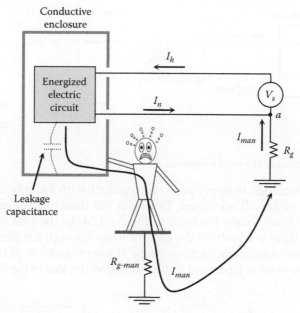

FIGURE 12.2
A person touching floated conductive enclosure.

electrical circuit. The current from the hot wire I_h energizes the circuit and then branches into two paths: one through the neutral current I_n and the other through the person I_{man}. The current passing through the person is the leakage capacitance current. At node a on the source side, the two currents are summed up and become equal to the hot wire's current:

$$I_h = I_n + I_{man} \tag{12.1}$$

The current passing through the person is often small, and the person may feel nothing or just skin sensations, except in some hospital environments when the skin of the patient is penetrated by medical equipment. The hospital case is described later in this chapter.

Before we proceed to hazardous scenarios, let us consider the generic representation of the household circuit shown in Figure 12.3. The equipment is powered from the secondary of the service transformer (xfm) through two wires: a hot wire, whose resistance is R_h, and a neutral wire, whose resistance is R_n. The transformer is grounded through $R_{g\text{-}xfm}$. At the entrance of the customer's dwelling, a service panel is installed to branch the power to the various loads. Each branch (circuit) is served through a dedicated circuit breaker (CB) designed to isolate (disconnect) the branch in case of faults at the equipment side. In the United States, panel circuit breakers serving 120 V loads are often rated at 10, 15, and 20 A. For heavier loads served by 240 V, the breakers are often rated at 30 and 40 A.

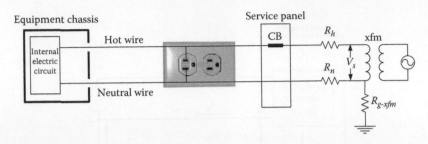

FIGURE 12.3
A typical connection of electrical equipment.

The NEC demands that any person in contact with metallic chassis must be protected under all scenarios. To assess the hazards under all possible scenarios, we must follow the logic in Figure 12.4. In the first step, we need to find out if there is a path for the current to go through the person. If one is found, we need to estimate its magnitude. If the magnitude of the current is hazardous, we need to figure out if the protection devices of the faulted circuit

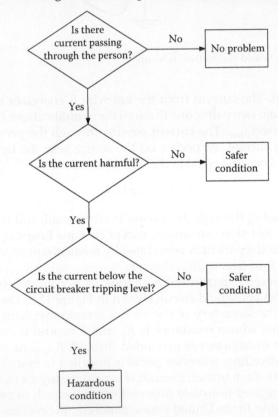

FIGURE 12.4
Assessment of hazards.

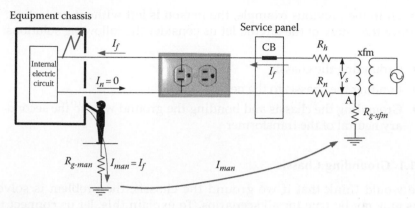

FIGURE 12.5
A person touching a floating chassis of faulty equipment.

(CB, fuses, etc.) are capable of interrupting the current. If the protection equipment cannot trip the circuit quickly, the situation is extremely hazardous. The fundamental rule for protection is "when a fault occurs, the fault current must be as high as possible to ensure quick tripping of the protection devices."

When the equipment is faulted and the hot wire comes in contact with the chassis, NEC demands that any person in contact with the metallic chassis must be protected. Consider the case in Figure 12.5 when the hot wire comes in contact with the chassis of equipment due to insulation failure. If a person standing on a grounded object touches the conductive enclosure, the person completes an electrical circuit and a current passes through his body. What makes this problem hazardous is that the current through the person could be deadly, but still below the clearing level of the circuit breaker of the affected branch, as shown in the next example.

Example 12.1

For the case in Figure 12.5, assume that the circuit breaker is rated 20 A, the source voltage is 120 V, $R_n = R_h = 0.5\ \Omega$, $R_{g\text{-}man} = 500\ \Omega$, $R_{man} = 1000$ Ω, and $R_{g\text{-}xfm} = 20\ \Omega$. Compute the current passing through the person touching the chassis.

Solution:

The current passing through the person is the fault current:

$$I_{man} = I_f = \frac{V_s}{R_h + R_{man} + R_{g\text{-}man} + R_{g\text{-}xfm}} = \frac{120}{0.5 + 1000 + 500 + 20} = 78.92\ \text{mA}$$

This level of current is high enough to cause ventricular fibrillation. But, unfortunately, the current is way below the rating of the circuit breaker, and it will not interrupt the fault.

As seen in the previous example, the person is left without protection. To improve the safety of the circuit, let us consider the following solutions:

- Grounding the chassis
- Bonding the chassis to the neutral wire
- Grounding the chassis and bonding the ground wire to the secondary neutral of the transformer

12.1.1 Grounding Chassis

One would think that if we ground the chassis, the problem is solved. This may not be true for all scenarios. To explain this, let us connect the metallic enclosure to the local ground at the service panel, as shown in Figure 12.6. The conductor that connects the chassis to the local ground is called *equipment grounding conductor* (EGC). The local ground can be established by a ground rod outside the dwelling, a wire connected to a well-grounded metal water pipe, or both. In this system, when fault occurs, the fault current is branched into two paths, as shown in Figure 12.6: one through the EGC wire (I_{EGC}) and the other through the person (I_{man}). The current passing through the person could still be high but below the interruption level of the CB.

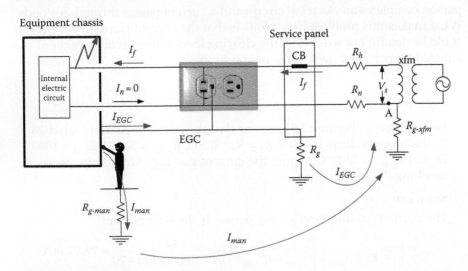

FIGURE 12.6
A person touching a chassis with EGC during fault.

Example 12.2

For the case in Figure 12.6, assume that the circuit breaker is rated 20 A and the source voltage is 120 V. The system parameters are $R_n = R_h = 0.5\ \Omega$, $R_{g\text{-}man} = 500\ \Omega$, $R_g = 20\ \Omega$, $R_{man} = 1000\ \Omega$, and $R_{g\text{-}xfm} = 20\ \Omega$. Compute the fault current and the current passing through the person touching the chassis.

Solution:

The equivalent circuit of the system in Figure 12.6 is shown in Figure 12.7.

Since R_g is in parallel with the person plus his ground resistance, the equivalent resistance of this combination is

$$R_{eq} = \frac{R_g \left(R_{man} + R_{g\text{-}man} \right)}{R_g + \left(R_{man} + R_{g\text{-}man} \right)} = \frac{20 \times (1000 + 500)}{20 + 1000 + 500} = 19.74\ \Omega$$

The fault current is then

$$I_f = \frac{V_s}{R_h + R_{eq} + R_{g\text{-}xfm}} = \frac{120}{0.5 + 19.74 + 20} = 3.98\ \text{A}$$

This fault current is still below the interruption level of household circuit breakers. Thus, the fault will not be cleared.

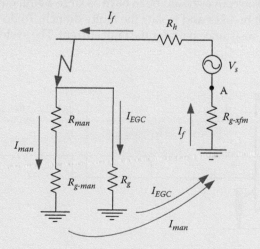

FIGURE 12.7
Equivalent circuit for the system in Figure 12.6.

The current passing through the person is

$$I_{man} = I_f \frac{R_g}{R_g + (R_{man} + R_{g\text{-}man})} = 3.98 \times \frac{20}{20 + (1000 + 500)} = 39.2 \text{ mA}$$

Because of grounding the chassis, the current passing through the person is reduced as compared with that in Example 12.1. However, it is still hazardous.

Based on the previous example, we can conclude that grounding the chassis alone is not protecting the person during faults. This is because of two major problems:

- Current passing through the person is still hazardous
- Fault current is below the activation level of the circuit breaker

The low value of the fault current is due to the high-resistance path through the local ground resistance R_g and the ground resistance at the transformer $R_{g\text{-}xfm}$. Intuitively, we need to have a low-resistance path for the fault current. This is our next system.

12.1.2 Bonding Chassis to Neutral

During faults, the objective should be to have as large a fault current as possible to open the circuit breaker and isolate the faulty branch. To do this, it is important to have a low-resistance path for the fault current. The system in Figure 12.8

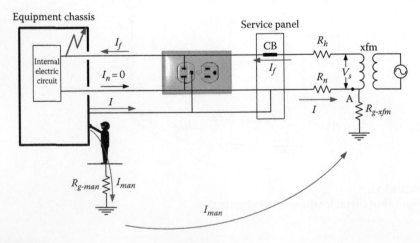

FIGURE 12.8
A person touching a chassis bonded to system neutral.

satisfies this objective. In this system, the chassis is bonded to the neutral wire at the service panel. This should be the best way to clear faults, as seen in the next example.

Example 12.3

For the case in Figure 12.8, assume that the circuit breaker is rated at 20 A and the source voltage is 120 V. The system parameters are $R_n = R_h = 0.5\ \Omega$, $R_{g\text{-}man} = 500\ \Omega$, $R_{man} = 1000\ \Omega$, and $R_{g\text{-}xfm} = 20\ \Omega$. Compute the fault current.

Solution:

The equivalent circuit for the case in Figure 12.8 is shown in Figure 12.9.
   Notice that R_n is in parallel with $R_{g\text{-}man} + R_{man} + R_{g\text{-}xfm}$. The equivalent resistance of this combination R_{eq} is

$$R_{eq} = \frac{R_n\left(R_{man} + R_{g\text{-}man} + R_{g\text{-}xfm}\right)}{R_n + \left(R_{man} + R_{g\text{-}man} + R_{g\text{-}xfm}\right)} = \frac{0.5\times\left(1000 + 500 + 20\right)}{0.5 + 1000 + 500 + 20} \approx 0.5\ \Omega$$

The fault current is then

$$I_f = \frac{V_s}{R_h + R_{eq}} = \frac{120}{0.5 + 0.5} = 120\ \text{A}$$

This current is very high and will certainly activate the circuit breaker. Thus, the person is quickly isolated from the hazard.

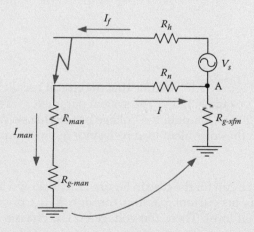

FIGURE 12.9
Equivalent circuit for Figure 12.8.

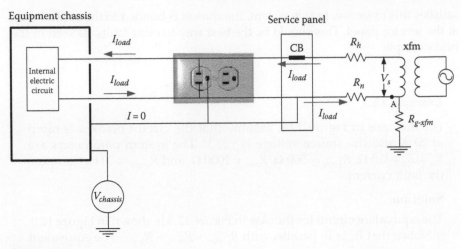

FIGURE 12.10
Chassis voltage of heavily loaded equipment.

Although the system in Figure 12.8 is very effective at clearing faults, it has a weakness. Bonding the chassis to the neutral may cause elevated voltage on the chassis even during normal operation, especially for heavily loaded branch and long feeders (high resistance). Consider the system in Figure 12.10 where the equipment is assumed to be drawing heavy current that is still below the circuit breaker clearing limit. If we measure the chassis voltage $V_{chassis}$ using a voltmeter between the chassis and ground, the voltage could be elevated, as seen in the next example.

Example 12.4

For the case in Figure 12.10, assume that the circuit breaker is rated 20 A and the source voltage is 120 V. The system parameters are $R_n = R_h = 0.5$ Ω, and $R_{g\text{-}xfm} = 20$ Ω. Compute the voltage of the chassis during normal operation when the equivalent load resistance of the equipment is 6 Ω.

Solution:

The equivalent circuit for the case in Figure 12.10 is shown in Figure 12.11.
Since there is no current passing through $R_{g\text{-}xfm}$, point A is at the ground potential level. Then, the voltage on the chassis is the voltage drop across R_n.

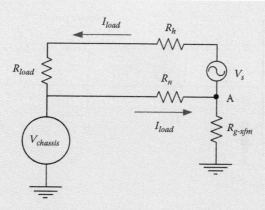

FIGURE 12.11
Equivalent circuit for Figure 12.10.

Let us compute the load current:

$$I_{load} = \frac{V_s}{R_n + R_h + R_{load}} = \frac{120}{0.5 + 0.5 + 6} = 17.14 \text{ A}$$

The current is within the operating range of the 20 A circuit breaker. The voltage across the chassis is

$$V_{chassis} = I_{load}R_n = 17.14 \times 0.5 = 8.57 \text{ V}$$

This level of voltage on a chassis is considered high for household appliances. Repeat the example for a 240 V system and 40 A breaker.

12.1.3 Grounding Chassis and Bonding Ground to Neutral

To address the problem associated with the elevated chassis voltage during normal operation and to ensure that the fault current is high enough to trip the circuit breaker, electrical engineers have combined the systems in Figures 12.6 and 12.8. This is shown in Figure 12.12, where the EGC is grounded locally and is also bonded to the neutral conductor. This is the system adapted in the United States and most of the world. The system ensures that the circuit breaker trips when internal fault occurs, as well as reduces the voltage on any equipment chassis during normal operation. These two advantages are examined in the following examples.

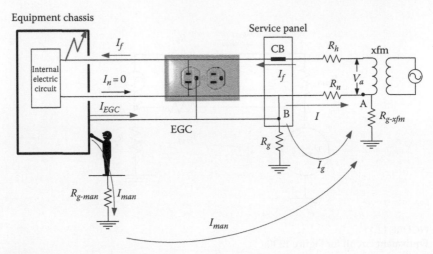

FIGURE 12.12
Wiring system in the United States and most of the world.

Example 12.5

For the case in Figure 12.12, assume that the circuit breaker is rated at 20 A and the source voltage is 120 V. The system parameters are $R_n = R_h = 0.5\ \Omega$, $R_{g\text{-}man} = 500\ \Omega$, $R_{man} = 1000\ \Omega$, $R_g = 10\ \Omega$, and $R_{g\text{-}xfm} = 20\ \Omega$. Compute the fault current.

Solution:

The equivalent circuit of the system in Figure 12.12 is shown in Figure 12.13.

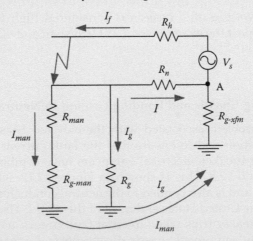

FIGURE 12.13
Equivalent circuit for the system in Figure 12.12.

R_g is in parallel with the person plus his ground resistance:

$$R_{eq1} = \frac{R_g\left(R_{man} + R_{g\text{-}man}\right)}{R_g + \left(R_{man} + R_{g\text{-}man}\right)} = \frac{10 \times (1000 + 500)}{10 + 1000 + 500} = 9.93\ \Omega$$

R_{eq1} is in series with $R_{g\text{-}xfm}$:

$$R_{eq2} = R_{g\text{-}xfm} + R_{eq1} = 20 + 9.93 = 29.93\ \Omega$$

The equivalent resistance representing R_n in parallel with R_{eq2} is

$$R_{eq} = \frac{R_n R_{eq2}}{R_n + R_{eq2}} = \frac{0.5 \times 29.93}{0.5 + 29.93} = 0.492\ \Omega$$

The fault current is then

$$I_f = \frac{V_s}{R_h + R_{eq}} = \frac{120}{0.5 + 0.492} = 120.97\ \text{A}$$

This fault current is very high and the circuit breaker will clear the fault.

Example 12.6

In the case in Figure 12.14, assume that the circuit breaker is rated at 20 A and the source voltage is 120 V. The system parameters are $R_n = R_h = 0.5\ \Omega$, $R_{g\text{-}man} = 500\ \Omega$, $R_{man} = 1000\ \Omega$, $R_g = 10\ \Omega$, and $R_{g\text{-}xfm} = 20\ \Omega$. Compute the voltage of the chassis during normal operation when the load resistance is 6 Ω.

Solution:

The equivalent circuit for the case in Figure 12.14 is shown in Figure 12.15.

The voltage of the chassis is the voltage drop across R_g. But first, let us compute the equivalent resistance of R_n, R_g, and $R_{g\text{-}xfm}$:

$$R_{eq} = \frac{R_n\left(R_g + R_{g\text{-}xfm}\right)}{R_n + R_g + R_{g\text{-}xfm}} = \frac{0.5 \times (10 + 20)}{0.5 + 10 + 20} = 0.492\ \Omega$$

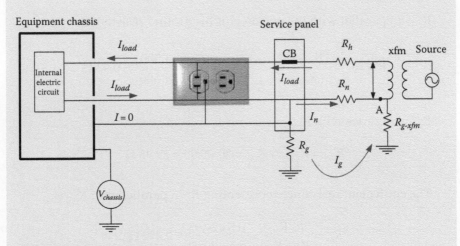

FIGURE 12.14
Chassis voltage of heavily loaded equipment with local ground.

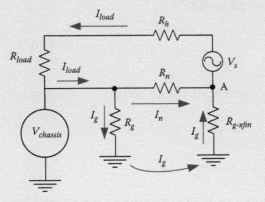

FIGURE 12.15
Equivalent circuit for Figure 12.14.

The load current is

$$I_{load} = \frac{V_s}{R_{eq} + R_h + R_{load}} = \frac{120}{0.492 + 0.5 + 6} = 17.16 \text{ A}$$

The current is within the normal operating value for the 20 A circuit breaker.

The current passing through R_g is

$$I_g = I_{load} \frac{R_n}{R_n + R_g + R_{g\text{-}xfm}} = 17.16 \frac{0.5}{0.5 + 10 + 20} = 0.281 \text{ A}$$

The voltage across the chassis is

$$V_{chassis} = I_g R_g = 0.281 \times 10 = 2.81 \text{ V}$$

This level of voltage is one-third of that computed in Example 12.4. Keep in mind that the voltage of the chassis will never be at zero potential. Our objective is to make it as small as possible.

12.1.4 Receptacles and Plugs

In the United States and most of the world, the wires coming from the service transformer enter the dwelling through a utility meter and then into a distribution panel, as shown in Figure 12.16. There are two functions to this panel: (1) distribute the power to the various loads inside the dwelling and (2) protect the individual circuits from internal fault by tripping the corresponding circuit breaker. In the United States, the dwelling is fed by three wires from the service transformer: two wires are at 120 V and one is neutral. The neutral wire, which is grounded at the transformer pole, is also grounded at the entrance of the dwelling to provide the local ground required by the NEC. The forms of local grounds are

- Local exposed ground rod
- Concrete encased electrode (rod, pipe, plate, braided wire, etc.)

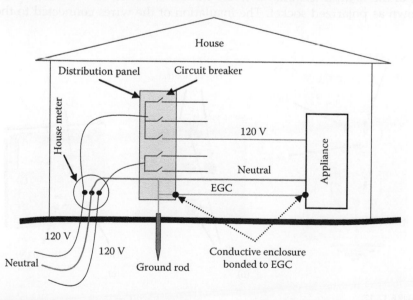

FIGURE 12.16
Electric circuit distribution inside a dwelling.

- Metallic water pipe with at least 10 ft (3.048 m) section that is in contact with earth
- Ground ring or grid
- Grounded steel reinforcement frame of building

Since metal pipes corrode over time or are replaced by plastic pipes, the NEC mandates that three other low-resistance paths to ground be utilized. The NEC also requires that local grounds be 5 ft (1.524 m) or less from the point of entry (distribution panel).

To bring all conductive enclosures inside the dwelling to the local ground level, an EGC wire (green wire) is used, as shown in Figure 12.16. The EGC is bonded to the local ground as well as the neutral wire.

The power receptacles (outlets or sockets) come in various types based on the local standards. There are about 13 different types of receptacles worldwide, but most of them are exclusively used by a few countries. Some of the common ones are shown in Figure 12.17. Each of the outlets in the figure has three terminals: hot, neutral, and ground. The hot terminal in these sockets is connected to the hot side of the service transformer (100–240 V depending on the country's standard). The neutral terminal is grounded at the service transformer. The EGC is grounded locally and bonded to the neutral at the service panel.

In North America, Type 1, the hot and neutral terminals accept plugs with flat prongs, and the EGC terminal accepts a rounded prong. Also, the opening of the neutral terminal is wider than that for the hot terminal. This is known as polarized socket. The insulation of the wires connected to these

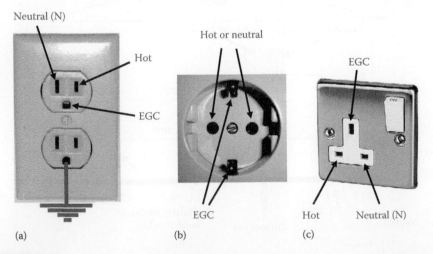

FIGURE 12.17
Household power outlet in three regions: (a) Type 1: most of the American continent, (b) Type 2: most of Europe, and (c) Type 3: England and parts of the Middle East and Asia.

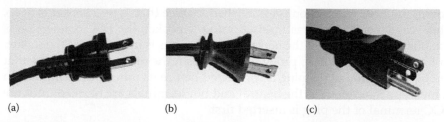

FIGURE 12.18
Common household plugs in North America: (a) Type 1: unpolarized, (b) Type 1: polarized, and (c) Type 1: three prongs.

terminals are color-coded; hot is black or red, neutral is white, and EGC is either green or uninsulated.

For Type 2, the socket has two EGC terminals, one at the top of the socket and the other at the bottom. The hot and neutral terminals are round-shaped. For the British system, Type 3, all terminals are flat.

Household plugs come in various shapes to match the types of receptacles. For Type 1 sockets, the common plugs are shown in Figure 12.18. Plugs for Type 2 and 3 sockets are shown in Figure 12.19. In Figure 12.18a, the plug is unpolarized and its prongs have equal width. In this case, either prong can be connected to the hot terminal of the outlet. This plug is suitable for nonconductive chassis. In Figure 12.18b, the plug has two prongs as well, but it is polarized where the neutral prong is wider than the hot prong. This way, the neutral of the plug can only fit in the neutral side of the outlet. This plug is also used with nonconductive chassis. Figure 12.18c shows the three-terminal plug, where two of them are flat and the third is rounded. The EGC prong is longer than the other two. In this configuration, when the plug in inserted into the outlet, the EGC terminal is connected first. Also, in this configuration, the neutral and hot terminals cannot be interchanged. This is the plug need for equipment with conductive chassis.

The international plug versions are shown in Figure 12.19. In Figure 12.19a, the prongs are rounded and unpolarized; suitable for nonconductive chassis. In Figure 12.19b, the EGC terminals are imbedded on the sides of the plug. This way, when the plug is inserted into the socket either straight or upside

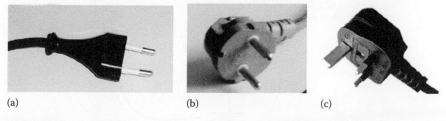

FIGURE 12.19
Common international household plugs: (a) Type 2: unpolarized, (b) Type 2: two prongs with EGC, and (c) Type 3: three prongs.

down, the EGC connection is established. Notice that Type 2 socket is deep. This way when the plug is inserted into the socket, the EGC connection is established before any of the neutral or hot prongs is inserted into the socket. In Figure 12.19c, the EGC prong is longer than the other two to ensure that the EGC terminal of the plug is connected before its other two terminals. In some Type 3 receptacles, the neutral and hot terminals are blocked until the EGC terminal of the plug is inserted first.

12.1.5 Ground Fault Circuit Interrupter

One of the most common electric shock scenarios is when electric equipment is mixed with water. Tap water is conductive, and its intrusion inside electric equipment allows the circuit components to be electrically connected to the enclosure. This is a hazardous situation as the chassis would be at or near the potential of the internal circuit. Even when the enclosure is nonconductive, the presence of water on the enclosure makes its surface conductive.

Assume that a hair dryer is plugged into an electric outlet. If the dryer falls into a sink filled with water, the high-voltage wires and the circuit components come in contact with the dryer's enclosure. If a person touches the wet dryer while part of his or her body is in contact with a grounded object, such as faucets, the person could be electrocuted, as depicted in Figure 12.20. Since the current through the person passes through two ground resistances, its magnitude could be below the interruption rating of the circuit breaker. For this reason, *ground fault circuit interrupter* (GFCI) is used in wet areas such as kitchens, bathrooms, gardens, etc.

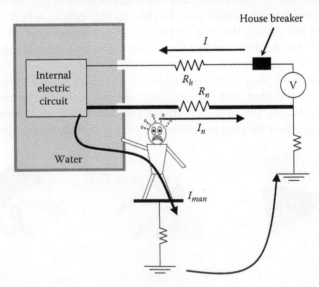

FIGURE 12.20
Water inside a device can create hazardous condition.

FIGURE 12.21
GFCI outlet.

The GFCI receptacle is similar in shape to the regular outlet as shown in Figure 12.21, but has two buttons and a built-in circuit interrupter. The circuit interrupter of the GFCI is shown in Figure 12.22. It consists of a core, an activation circuit, and a relay switch. The hot and neutral wires go through the ring of the core. The core has winding wrapped around it. If the current

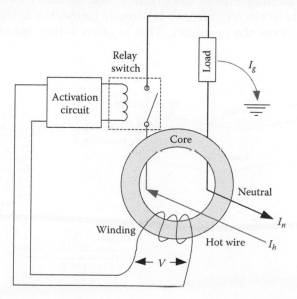

FIGURE 12.22
Current differential detector of the GFCI circuit.

in the hot wire I_h is exactly equal to the neutral current I_n, the total magnetic field produced by both wires is zero. Hence, no voltage is induced across the winding. When a ground current (I_g) is present, I_h is higher than I_n and the total magnetic field is nonzero. Hence, a voltage V is induced across the winding of the core. This voltage through the activation circuit energizes a relay that disconnects the hot wire. The GFCI has two buttons: one is used to test the circuit for functionality and the other is used to reset the circuit after interruption.

The normal trip current of the GFCI is about 10–20 mA and its speed is about 20–40 ms. Keep in mind that the GFCI is designed to protect a person from electric shock and is not designed to remove stray voltages.

12.2 Service Transformer

A typical house in the United States is served by one phase of a three-phase distribution feeder, as shown in Figure 12.23. The voltage of the feeder is often 15 kV class (7.9–8.6 kV phase to ground). At the service site, a distribution transformer is installed to step down the line voltage to customers' level (120/240 V).

Figure 12.24 shows a typical connection of the distribution transformer. It consists of a primary winding (N_1) and two secondary windings (N_2 and N_3). The neutral wire of the primary side (primary neutral), which carries the return electric current, is directly connected to the secondary neutral that serves the customer. This is often called auto-transformer connection.

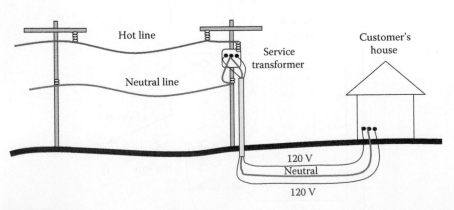

FIGURE 12.23
House service.

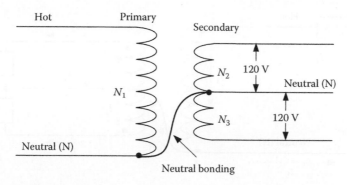

FIGURE 12.24
Connection of a typical distribution transformer.

The two main reasons for using auto-transformer are:

1. Utility would be able to clear faults that impact the secondary of the transformer at the customer's side. In case of internal transformer failure that leads to a direct current path between the primary and secondary windings as shown in Figure 12.25, the fault current in the secondary winding will go back to the primary through the neutral connection. This is a low impedance path that causes the fault current to be high enough for detection and clearance by the protection equipment of the distribution system (CB, fuse, etc.). Similarly, if the primary hot wire should drop into the secondary hot wire, the neutral connection provides a path for the current that can be detected and cleared by the protection equipment.
2. If all neutrals are grounded, they provide access to more grounding systems that help dissipate any lightning energy that may hit the system.

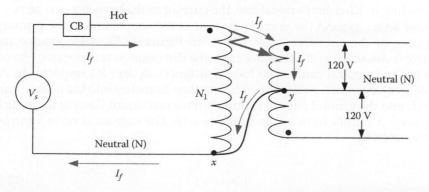

FIGURE 12.25
Faulted transformer.

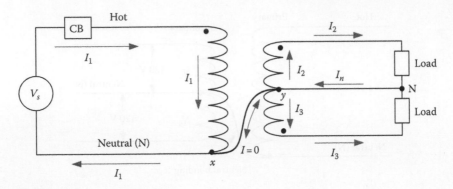

FIGURE 12.26
Isolated neutral system.

This connection plus the grounding of the neutrals is the main cause of stray voltage on the consumer's side. To understand the cause of the stray voltage, let us first assume that the neutral of the system is not connected to any ground, as shown in Figure 12.26. The two nodes of the neutral wire are labeled x and y in the figure. The current of the source I_1 flows inside the primary windings N_1. The secondary windings carry the two load currents I_2 and I_3. The neutral current at the secondary side of the transformer is the sum of these two load currents:

$$I_n = I_1 + I_2 \tag{12.2}$$

According to Kirchhoff's nodal low, the current in the branch x–y is zero.

Now let us ground the primary neutral at the source and at the primary winding of the transformer, as shown in Figure 12.27. Also, assume the source (substation) is at a distance from the distribution transformer. Hence, the hot and neutral conductors have resistances R_h and R_n, respectively. At node x, the current I_1 enters the node and then branches into the neutral current I_N and the ground current I_g. Still we have no current flows in the branch x–y since it cannot form a loop to the source. The current at node x can be represented by

$$I_1 = I_N + I_g \tag{12.3}$$

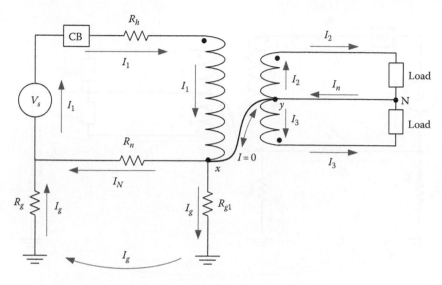

FIGURE 12.27
Grounded neutral system at the primary side.

At point x, the current I_g passes through the resistance of the grounded neutral R_{g1} and returns back to the source through the source ground resistance R_g. Keep in mind that if the neutral wire resistance R_n is zero, no ground current will exist. However, unfortunately, the neutral wire will always have some resistance; thus, the ground current is unavoidable and the secondary neutral at the customer's side will retain some voltage.

As discussed earlier, the neutral at the customer site must be bonded to the local ground through grounding rod, metal pipes, or metal grounded structures. The NEC, article 250.24 states

> Service equipment supplied from a grounded system must have the neutral conductor grounded.

The rule is represented in Figure 12.28 by adding R_{g2}. This added ground will cause additional current loop for part of I_1. The current I_1 of the source goes to the primary winding and reaches node x. At x, part of the primary current flows through the neutral connection from point x to point y, and then to the ground resistance R_{g2} and back to the source through R_g. This current, which is labeled I_{g2} in the figure, is one of the main stray voltage problems, as will be seen later in this chapter.

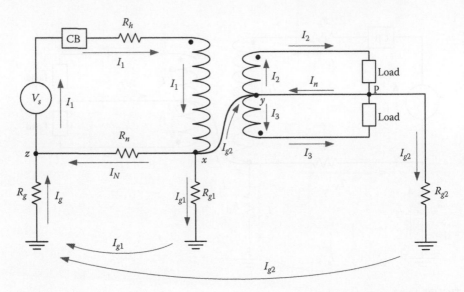

FIGURE 12.28
Grounded neutral system at the primary and secondary of the service transformer.

The current of all nodes in Figure 12.28 can be written as

At node x $I_1 = I_N + I_{g1} + I_{g2}$

At nodes y and p $I_{g2} + I_n = I_2 + I_3$ (12.4)

At node z $I_1 = I_g + I_N = I_{g1} + I_{g2} + I_N$

Example 12.7

For the system in Figure 12.28, AAC-type conductor is used for the hot and neutral wires of the distribution line. The resistance of the wire is 0.3 Ω/km and its rated current is 300 A. Assume that $R_{g1} = R_{g2} = 20\ \Omega$ and the substation grounding $R_g = 5\ \Omega$. The service transformer is 5 km away from the substation. The transformer is rated at 25 kVA, 7.8 kV/240 V. At full load, compute the neutral voltage at the customer's side.

Solution:

The full load current of the transformer primary is

$$I_1 = \frac{S}{V} = \frac{25}{7.8} = 3.2 \text{ A}$$

The wire resistance for the 5 km is

$$R_h = R_n = 5 \times 0.3 = 1.5 \ \Omega$$

The equivalent circuit of the system in Figure 12.28 is shown in Figure 12.29. The circuit shows the primary side of the transformer and the neutral groundings.

The problem can be solved using the current divider method. But first, compute the parallel resistance of the grounds R_{g1} and R_{g2}:

$$R = \frac{R_{g1} R_{g2}}{R_{g1} + R_{g2}} = 10 \ \Omega$$

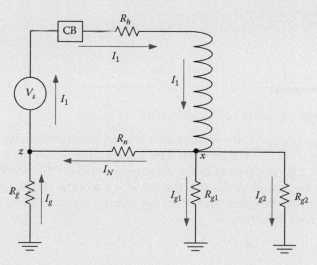

FIGURE 12.29
Equivalent circuit of the system in Figure 12.28.

The current in the neutral wire is then

$$I_N = I_1 \frac{R+R_g}{R+R_g+R_n} = 3.2 \frac{10+5}{10+5+1.5} = 2.91 \text{ A}$$

The current I_g is then

$$I_g = 3.2 - 2.91 = 290 \text{ mA}$$

The neutral voltage (stray voltage) of the neutral wire at the customer's side is

$$V_{stray} = I_g R = 290 \times 10 = 2.9 \text{ V}$$

This stray voltage can be problematic or even harmful as will be seen later in this chapter. Moreover, during faults, the neutral voltage can reach hazardous levels (repeat the example with higher current). Also, keep in mind that when the distance between the substation and the transformer is longer, you should expect higher stray voltage.

12.3 Voltage on Neutral Conductor

In distribution networks, the neutral wire is used to return the current back to the source, as shown in Figure 12.30. The voltage drop across the neutral wire between two adjacent towers is the difference between the neutral voltages with respect to ground of the two towers. If we assume that the source side is at the left of the diagram, the towers are equally separated, and the neutral wire is not grounded, then

$$I R = V_2 - V_1 = V_3 - V_2 = V_4 - V_3 = ... \tag{12.5}$$

where
  I is the return current
  R is the resistance of the neutral wire between two adjacent towers
  V_1, V_2, V_3, V_4 are the voltages of the neutral wire with respect to ground at
    towers 1, 2, 3, and 4, respectively

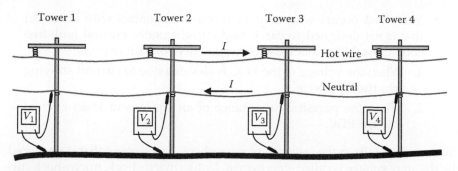

FIGURE 12.30
Voltage drop on neutral wire.

Hence

$$V_4 > V_3 > V_2 > V_1 \tag{12.6}$$

If you assume that the voltage V_1 at tower 1 is 1 V and the voltage drop between any two towers at a given current condition is 0.5 V, then

$$V_2 = 1.5 \text{ V}$$

$$V_3 = 2 \text{ V} \tag{12.7}$$

$$V_2 = 2.5 \text{ V}$$

... ...

As seen, the voltage of the neutral wire is continuously elevated as we move away from the source. This voltage is known as the neutral to earth voltage (NEV). At some of these towers, a service transformer is used to energize customers' loads. Because it is a common practice to have the neutrals of the primary and secondary of the transformer bonded, this NEV is also seen at customers' sites. The NEV at customers' neutral is often referred to as stray voltage.

To reduce the NEV, utility ground the neutral wires at poles to bring down the voltage of the neutral. This does not eliminate the NEV at customers' sites because the impedance of the neutral wire plus all grounding resistances of the poles are still nonzero. Thus, grounding will not eliminate stray voltage.

12.4 Stray Voltage

Electric shocks are often categorized into *macroshock* or *microshock*.

- *Macroshock* occurs when a person comes in contact with an object that is designed to carry load current.

- *Microshock* occurs when a person comes in contact with an object that *is not* designed to carry load current under normal fault-free conditions. Microshock can occur due to two main reasons:
  1. When the voltage of the EGC is elevated due to current straying into the EGC wire
  2. When the parasitic capacitance of an equipment leaks current into the EGC.

Stray voltage falls in the category of microshock. The current that is produced by the stray voltage is called *stray current*. Unlike macroshock, microshock current is often small in magnitude. However, microshock could be hazardous under certain conditions such as in swimming pools and hospitals. It could also impact negatively farm productions.

- Young and inexperienced swimmers were reported to have drowned when stray currents passed through their bodies.
- In hospitals, if a stray current is injected directly into sensitive organs, such as the heart, through pacemaker lead or a saline-filled catheter, the patient may die if the current exceeds just 10 μA. Additionally, the stray voltage can affect sensitive equipment such as cardiograms that rely on small voltages to measure the health condition of patients.
- In farms, stray voltage can cause permanent change in animal behavior that affect farm production. It also makes it more difficult to control animals.

The IEEE working group on *Voltages at Publicly and Privately Accessible Locations* defined stray voltage in 2008 as

> Stray voltage is a voltage resulting from the normal delivery and/or use of electricity (usually smaller than 10 V) that may be present between two conductive surfaces that can be simultaneously contacted by members of the general public and/or their animals. Stray voltage is caused by primary and/or secondary return current, and power system induced currents, as these currents flow through the impedance of the intended return pathway, its parallel conductive pathways, and conductive loops in close proximity to the power system. Stray voltage is not related to power system faults, and is generally not considered hazardous.

> Note: "Conductive surfaces" as used in this definition are intended to include the earth and/or extensions of the earth such as concrete sidewalks and metal floor drains.

12.4.1 Power Distribution of a Dwelling

A schematic of typical distribution circuits of a dwelling in the United States is shown in Figure 12.8. The dwelling is fed by three wires from the service transformer: two wires are at 120 V and one wire is neutral. With this arrangement, the dwelling receives 120 V for all outlets except for heavy

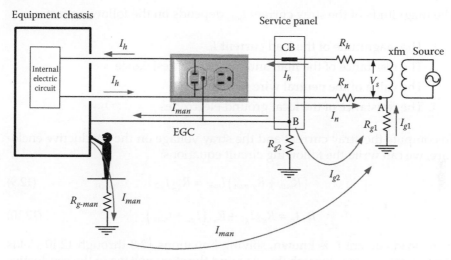

FIGURE 12.31
Stray current.

loads such as water heaters and dryers where they are powered by 240 V (120 + 120 V). As shown in Section 12.1.4, the neutral wire is grounded at the transformer pole and may also be grounded at other poles. At the entrance of the dwelling, the neutral wire is grounded as required by the NEC.

The system in Figure 12.8 provides good protection against internal faults. However, it is not designed to protect against microshocks due to stray voltages. Unfortunately, contrary to popular belief, more grounding and lower grounding resistances will not remove the stray voltage. These two points will become apparent in the rest of this chapter.

Figure 12.31 shows a case where stray voltage could exist. The ground rod of the house (represented by R_{g2}) is connected to the neutral wire as well as to the EGC. Assume the conductive enclosure is physically isolated from the ground by robber wheels, dry wood floor, etc. The service transformer (source) is assumed at a distance from the house and the neutral wire resistance is R_n. The neutral wire at the source side is grounded, and its ground resistance is R_{g1}. At the load side, the conductive enclosure is connected to the EGC, which is bonded to the neutral at the service panel. The EGC is electrically grounded through the grounding resistance R_{g2}. Assume that the load current of the hot wire is I_h. When this current leaves the load, it branches out into three paths: one through the neutral wire I_n, the second (I_{g2}) through the local ground (ground rod at the service box), and the third though the EGC to the man (I_{man}). This is a stray current that goes through the person touching the conductive enclosure. These three currents add up to I_h at node A at the source side:

$$I_h = I_n + I_{g2} + I_{man} \tag{12.8}$$

The magnitude of the stray current I_{man} depends on the following factors:

1. The magnitude of the load current I_h
2. The resistance of the man and his ground resistance
3. The value of the neutral wire resistance R_n
4. The resistance of the local ground electrodes

To compute the stray current and the stray voltage on the conductive enclosure, we can write the following circuit equations:

$$(R_{man} + R_{g\text{-}man})I_{man} = R_{g2}I_{g2} \tag{12.9}$$

$$R_n I_n = R_{g2}I_{g2} + R_{g1}(I_{g2} + I_{man}) \tag{12.10}$$

If the load current I_h is known, solving Equations 12.8 through 12.10 yields the current passing through the man and the stray voltage of the conductive enclosure. You can rearrange the equation in the following matrix form:

$$\begin{bmatrix} I_h \\ 0 \\ 0 \end{bmatrix} = \begin{bmatrix} 1 & 1 & 1 \\ (R_{man} + R_{g\text{-}man}) & -R_{g2} & 0 \\ R_{g1} & (R_{g1} + R_{g2}) & -R_n \end{bmatrix} \begin{bmatrix} I_{man} \\ I_{g2} \\ I_n \end{bmatrix} \tag{12.11}$$

Hence

$$\begin{bmatrix} I_{man} \\ I_{g2} \\ I_n \end{bmatrix} = \begin{bmatrix} 1 & 1 & 1 \\ (R_{man} + R_{g\text{-}man}) & -R_{g2} & 0 \\ R_{g1} & (R_{g1} + R_{g2}) & -R_n \end{bmatrix}^{-1} \begin{bmatrix} I_h \\ 0 \\ 0 \end{bmatrix} \tag{12.12}$$

Example 12.8

For the system in Figure 12.31, the neutral wire resistance is 1 Ω, the ground resistance at the source side is 10 Ω, and the ground resistance at the load side is 15 Ω. Assuming the load current is 20 A, compute the stray voltage of the conductive enclosure.

Solution:

Without the presence of the man, we can compute the stray voltage across the conductive enclosure by modifying Equations 12.8 and 12.10.

$$I_h = I_n + I_{g2} \quad \text{or} \quad 20 = I_n + I_{g2}$$

$$R_n I_n = (R_{g2} + R_{g1})I_{g2} \quad \text{or} \quad I_n = 25 I_{g2}$$

Hence

$$I_{g2} = 770 \text{ mA}$$

The stray voltage is

$$V_{stray} = R_{g2} I_{g2} = 15 \times 0.77 = 11.55 \text{ V}$$

The stray voltage is high enough to cause harm to humans and animals.

Example 12.9

For the system Example 12.8, compute the stray current passing through a person touching the conductive enclosure and standing on a grounded object. Assume the ground resistance of the man is 30 Ω.

Solution:

Assume the resistance of the man's body is 1000 Ω. A direct substitution in Equation 12.12 yields

$$\begin{bmatrix} I_{man} \\ I_{g2} \\ I_n \end{bmatrix} = \begin{bmatrix} 1 & 1 & 1 \\ (R_{man} + R_{g\text{-}man}) & -R_{g2} & 0 \\ R_{g1} & (R_{g1} + R_{g2}) & -R_n \end{bmatrix}^{-1} \begin{bmatrix} I_h \\ 0 \\ 0 \end{bmatrix}$$

$$\begin{bmatrix} I_{man} \\ I_{g2} \\ I_n \end{bmatrix} = \begin{bmatrix} 1 & 1 & 1 \\ (1000 + 30) & -15 & 0 \\ 10 & (10 + 15) & -1 \end{bmatrix}^{-1} \begin{bmatrix} 20 \\ 0 \\ 0 \end{bmatrix}$$

Hence

$$I_{man} = 11.13 \text{ mA}$$

This level of stray current passing through the man is high enough to cause harm, especially if the load current is high.

The stray current does not disappear if the enclosure is grounded; it only reduces its magnitude. Take the case in Figure 12.32. The current in the hot wire I_h goes from the source to the load. When it leaves the load, it branches into four paths at the distribution panel: the neutral current I_n (which is the most of I_h) the current in the panel grounding system I_{g2}; the current passing through

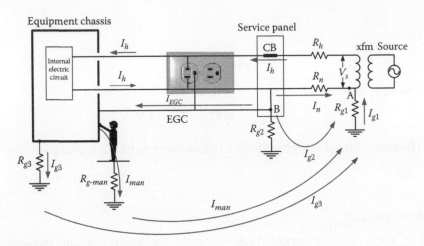

FIGURE 12.32
Stray current in grounded enclosures.

the enclosure to its ground resistance I_{g3}; and the current passing through the person touching the enclosure I_{man}. All ground currents I_{g2}, I_{g3}, and I_{man} return to the source through the ground resistance R_{g1}:

$$I_{g1} = I_{g2} + I_{g3} + I_{man} \tag{12.13}$$

As can be seen from the figure, the current passing through the person will always exit unless the ground resistance R_{g2} or R_{g3} is zero, which is impossible to achieve as the soil resistivity is always nonzero. This point is explained later in this chapter.

12.4.2 Stray Voltage in Farms

Livestock avoid drinking or eating from metal containers having stray voltages and are stressed when forced to enter areas if the stray current passing through them is as little as 3 mA. These behaviors make livestock management difficult. In addition, animal weight and milk production are reduced. Several studies made by the US Department of Agriculture (USDA) and other organizations have found that as little as 5 mA may cause livestock to refuse eating and drinking or to produce less milk. A summary of these studies is given in Table 12.1.

TABLE 12.1

Effect of Stray Current on Livestock

Stray Current (mA)	Effect on Livestock
1–3	Signs of awareness by livestock, but no milk production is lost
3–4	Animal may become more difficult to manage
5–6	Short-term changes in feed/water consumption or milk production
>6	Long-term changes in feed/water consumption or milk production

Example 12.10

For the system in Figure 12.33, the neutral wire resistance is 1 Ω, the ground resistance at the source side is 10 Ω, and the ground resistance at the load side is 15 Ω. Assume that a cow is touching the conductive enclosure while standing on a conductive surface whose ground resistance is 20 Ω. Assume that the body resistance of the cow is 500 Ω. Compute the stray current passing through the animal.

Solution:

The stray current passing through the farm animal can be computed by using the circuit in Figure 12.33. There are two parallel paths for the stray current at point B: one through ground resistance R_{g2} and the other through the cow.

Hence

$$I_{g2} + I_{cow} = I_h \frac{R_n}{R_n + R_{g1} + R_{g2} \, // \, (R_{cow} + R_{g3})}$$

$$= 10 \frac{1}{1 + 20 + 30 \, // \, (500 + 20)} = 0.203 \text{ A}$$

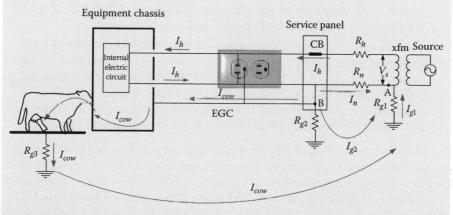

FIGURE 12.33
Stray current in farm animal.

The stray current passing through the cow is

$$I_{cow} = 0.203 \frac{R_{g2}}{R_{g2} + R_{g3} + R_{cow}} = 0.203 \frac{30}{30 + 20 + 500} = 11\,\text{mA}$$

According to Table 12.1, this level of stray current will have long-term changes in the animal's feed and water consumption as well as milk production.

Keep in mind that reducing the ground resistance at the service panel will not eliminate the stray voltage, as seen in Example 12.20.

12.4.3 Stray Voltage in Swimming Pools

Swimming pools have one of the most annoying stray voltage problems. Although the stray voltage in most cases is small causing just uncomfortable shock, it can cause panic to marginal swimmers and children, thus reducing their ability to stay afloat. More importantly, under fault condition, the level of voltage could be lethal.

Most swimming pools are made of steel walls with polymer lining such as vinyl. Others are made of steel reinforcement concrete with fiberglass lining. These two lining materials are high in resistivity and can electrically isolate the water of the pool from the surrounding soil. These swimming pools can have two electric safety problems:

1. The conductive chassis of any equipment immersed in the pool's water such as pumps, immersed light, or sound fixtures are bonded to the EGC. This will cause the water of the pool to be at the voltage of the EGC. Any person standing on local ground at the pool and touching the water will be exposed to this EGC voltage.

2. For swimming pools without immersed fixtures bonded to the EGC, if an electric device with two-prong plug is accidentally immersed in the water, the GFCI will not interrupt the circuit because the current in the neutral wire will always be equal to the current in the hot wire; there is no alternative path to ground. If the internal short caused by the water is not cleared by the circuit breaker, the water will be at elevated potential.

The electrical connection of a typical pool is shown in Figure 12.34. The substation is assumed to be at some distance from the pool so the resistances of the line conductors are considered. The service transformer is auto connected where the neutrals of the primary and secondary windings are bonded. The neutral is then connected to the ground electrode (and water pipes as well) at the service box of the site.

Let us assume that the swimming pool has a fixture immersed in water. The conductive enclosure of the fixture is connected to the EGC.

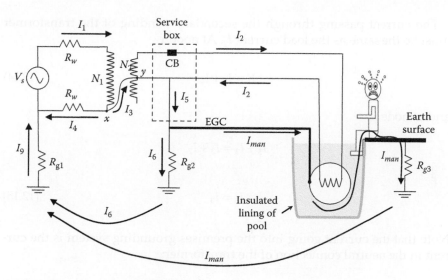

FIGURE 12.34
A scenario for swimming pool shock.

Also assume that the pool lining is made of high-resistance material such as fiberglass and that the adjacent surface to the pool (deck) is in contact with the surrounding soil. If a person touches the water of the pool and at the same time is in contact with the deck, the person could receive an electric shock. To analyze this situation, we can use the equivalent circuit of the swimming pool shown in Figure 12.35.

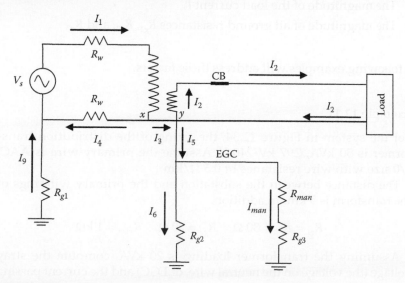

FIGURE 12.35
Equivalent circuit of swimming pool system in Figure 12.34.

The current passing through the secondary winding of the transformer must be the same as the load current I_2. At node x

$$I_1 = I_3 + I_4 \tag{12.14}$$

and at node y

$$I_5 + I_2 = I_3 + I_2$$

$$I_5 = I_3 \tag{12.15}$$

Note that the current going into the premises grounding system is the current in the neutral connection of the transformer:

$$I_3 = I_5 = I_6 + I_{man} = I_9 \tag{12.16}$$

Since the current passing through the person I_{man} is part of I_5, the existence and severity of the electric shock depend on the following factors:

- The bonding of the transformer's neutrals; without bonding $I_{man} = 0$
- The resistance of the conductors between the source and the transformer R_w, or between the transformer and the pool
- The magnitude of the load current I_2
- The magnitude of all ground resistances R_{g1}, R_{g2}, and R_{g3}

The following examples will address these factors.

Example 12.11

For the system in Figure 12.34, the rating of the distribution transformer is 50 kVA, 7.97 kV/240 V. Assume the primary wire is AAC, 3/0 size with wire resistance of 0.5 Ω/km.

The distance between the substation and the primary windings of the transform is 5 km. In addition,

$$R_{g1} = R_{g2} = 30\ \Omega, \quad R_{g3} = 20\ \Omega, \quad R_{man} = 1\ k\Omega$$

Assuming the transformer loading is 20 kVA, compute the stray voltage (the voltage on the neutral wire, or EGC) and the current passing through the man.

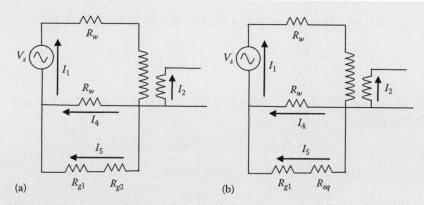

FIGURE 12.36
Distribution of currents (a) without the man touching the EGC and (b) with the man.

Solution:

The wire resistance is

$$R_{wire} = 0.5 \times 5 = 2.5 \ \Omega$$

The primary current of the transformer is

$$I_1 = \frac{20}{7.97} = 2.5 \text{ A}$$

The stray voltage is the voltage of the EGC (neutral) before the person touches the EGC. This is the voltage across R_{g2}. To compute the stray voltage, we can use the equivalent circuit shown in Figure 12.36a.

I_5 can then be computed by using the current divider method:

$$I_5 = I_1 \frac{R_w}{R_w + R_{g1} + R_{g2}} = 2.5 \times \frac{2.5}{2.5 + 30 + 30} = 100 \text{ mA}$$

The stray voltage of the neutral wire (EGC) at the customer's side is

$$V_{stray} = I_5 R_{g2} = 100 \times 30 = 3 \text{ V}$$

To compute the current passing through the person, we need to use the circuit at the right side in Figure 12.36, where R_{eq} is the equivalent resistance of all ground paths branching from the service box. Hence

$$\frac{1}{R_{eq}} = \frac{1}{R_{g2}} + \frac{1}{R_{g3} + R_{man}} = \frac{1}{30} + \frac{1}{20 + 1000}$$

$$R_{eq} = 29.14 \ \Omega$$

I_5 can then be computed by using the current divider method:

$$I_5 = I_1 \frac{R_w}{R_w + R_{g1} + R_{eq}} = 2.5 \times \frac{2.5}{2.5 + 30 + 29.14} = 101.8 \text{ mA}$$

Now, we can calculate the voltage of the neutral wire (EGC) at the customer's side:

$$V_n = I_5 R_{eq} = 101.8 \times 29.14 = 2.966 \text{ V}$$

The current passing through the man is then

$$I_{man} = \frac{V_n}{R_{g3} + R_{man}} = \frac{2.966}{20 + 1000} \approx 2.96 \text{ mA}$$

Repeat the example assuming the person is wet and his body resistance is 200 Ω.

The next example will address the effect of reduced system grounding on the stray voltage.

Example 12.12

For the system in Example 12.11, assume that it is possible to substantially reduce the ground resistances to $R_{g1} = R_{g2} = 10 \text{ Ω}$. Compute the stray current passing through the man.

Solution:

$$\frac{1}{R_{eq}} = \frac{1}{R_{g2}} + \frac{1}{R_{g3} + R_{man}} = \frac{1}{10} + \frac{1}{20 + 1000}$$

$$R_{eq} = 9.9 \text{ Ω}$$

I_5 can then be computed by using the current divider method:

$$I_5 = I_1 \frac{R_w}{R_w + R_{g1} + R_{eq}} = 2.5 \times \frac{2.5}{2.5 + 10 + 9.9} = 280 \text{ mA}$$

Now, we can calculate the voltage of the neutral wire at the customer's side:

$$V_n = I_5 R_{eq} = 280 \times 9.9 = 2.77 \text{ V}$$

Note that the neutral to ground voltage is slightly reduced, but not eliminated. The current passing through the man is then

$$I_{man} = \frac{V_n}{R_{g3} + R_{man}} = \frac{2.77}{20 + 1000} \approx 2.77 \text{ mA}$$

The stray current is slightly reduced when we *substantially* reduce the ground resistances of the system, but not eliminated. Repeat the example assuming the person is wet and his body resistance is 200 Ω.

Example 12.13

Repeat Example 12.12, but assume that the transformer operates near its rating.

Solution:

The primary current of the transformer is

$$I_1 = \frac{50}{7.97} = 6.27 \text{ A}$$

The current I_5 is then

$$I_5 = I_1 \frac{R_w}{R_w + R_{g1} + R_{eq}} = 6.27 \frac{2.5}{2.5 + 10 + 9.9} = 700 \text{ mA}$$

The voltage of the neutral wire at the customer's side is

$$V_n = I_5 R_{eq} = 700 \times 9.9 = 6.933 \text{ V}$$

The current passing through the man is

$$I_{man} = \frac{V_n}{R_{g3} + R_{man}} = \frac{6.933}{20 + 1000} \approx 6.933 \text{ mA}$$

This level of voltage and current are within the uncomfortable level for most people. Repeat the example assuming the person is wet and his body resistance is 200 Ω. Can you draw any conclusion?

In the previous example, we conclude that the heavier the loading of the transformer, the higher is the stray voltage and the stray current.

Example 12.14

Repeat Example 12.12, but assume that a fault inside the primary winding of the transformer resulted in a fault current on the high-voltage wire of 100 A.

Solution:

The current I_5 is

$$I_5 = I_1 \frac{R_w}{R_w + R_{g1} + R_{eq}} = 100 \times \frac{2.5}{2.5 + 10 + 9.9} = 11.16 \text{ A}$$

The voltage of the neutral wire at the customer's side is

$$V_n = I_5 R_{eq} = 11.16 \times 9.9 = 110.5 \text{ V}$$

The current passing through the man is

$$I_{man} = \frac{V_n}{R_{g3} + R_{man}} = \frac{110.5}{20 + 1000} \approx 110 \text{ mA}$$

This level of voltage and current are definitely hazardous. Repeat the example assuming the person is wet and his body resistance is 200 Ω.

12.4.4 Stray Voltage in Outdoor Showers

Outdoor showers in places such as beaches, swimming pools, and campgrounds can have stray voltage problems. Take, for example, the case in Figure 12.37. The water pipe feeding the shower is often made of copper or copper alloy and is bonded to the ground rod and the EGC at the service box of the building. The water pipe is then routed inside the house's walls until it reaches the shower area. Hence, the pipe is only in contact with the ground near the service box and the potential of the pipe is the potential of the EGC. If a person is standing away from the service box and is touching the pipe, his feet will be at his local ground potential and his hand at the potential of the EGC. In this scenario, part of the primary current I_3 goes through the water pipe to the man, and then to his local ground and back to the source through R_{g1}.

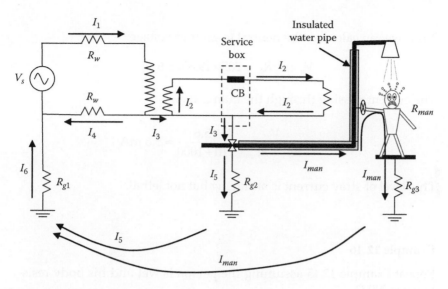

FIGURE 12.37
Stray voltage in outdoor showers.

Example 12.15

For the system in Figure 12.37, the transformer is a single-phase auto-transformer rated at 7.97 kV/240 V. The primary current $I_1 = 4.5$ A, $R_{g1} = R_{g3} = 10$ Ω, $R_{g2} = 20$ Ω and $R_w = 2$ Ω. Compute the current passing through the man, assuming his body resistance is 1 kΩ.

Solution:

The equivalent resistance R_{eq} of all ground resistances from the service box to the man is

$$\frac{1}{R_{eq}} = \frac{1}{R_{g2}} + \frac{1}{R_{g3} + R_{man}} = \frac{1}{20} + \frac{1}{10 + 1000}$$

$$R_{eq} = 19.6 \ \Omega$$

I_6 can be computed by using the current divider method:

$$I_6 = I_1 \frac{R_w}{R_w + R_{g1} + R_{eq}} = 4.5 \frac{2}{2 + 10 + 19.6} = 285 \text{ mA}$$

Now, we can calculate the neutral to ground voltage:

$$V_n = I_3 R_{eq} = 285 \times 19.6 = 5.6 \text{ V}$$

The current passing through the man is then

$$I_{man} = \frac{V_n}{R_{g3} + R_{man}} = \frac{5.6}{10 + 1000} \approx 5.6 \text{ mA}$$

This level of stray current is annoying but not lethal.

Example 12.16

Repeat Example 12.15 assuming the person is wet and his body resistance is 200 Ω.

Solution:

The equivalent resistance R_{eq} of all ground resistances from the service box to the man is

$$\frac{1}{R_{eq}} = \frac{1}{R_{g2}} + \frac{1}{R_{g3} + R_{man}} = \frac{1}{20} + \frac{1}{10 + 200}$$

$$R_{eq} = 18.26 \ \Omega$$

I_6 can be computed by using the current divider method:

$$I_6 = I_1 \frac{R_w}{R_w + R_{g1} + R_{eq}} = 4.5 \times \frac{2}{2 + 10 + 18.26} = 297.4 \text{ mA}$$

Now, we can calculate the neutral voltage:

$$V_n = I_3 R_{eq} = 297.4 \times 18.26 = 5.43 \text{ V}$$

The current passing through the man is then

$$I_{man} = \frac{V_n}{R_{g3} + R_{man}} = \frac{5.43}{10 + 500} \approx 10.65 \text{ mA}$$

This level of stray current is within the let go level.

12.4.5 Stray Voltage in Hospitals

Hospitals have a specific environment that needs special safety measures.

- Patients are often placed on conductive metal tables.
- Patients are often surrounded by multiple sensitive electrical devices.
- Patients could be immobile and unable to free himself from annoying or hazardous situations.
- The areas are often wet.
- There may be volatile agents such as flammable anesthetics.
- Some medical procedures require the intrusion of medical instruments, such as intravenous catheters or biopotential electrode, into the skin of the patient. Since the outer layer of the skin is the first line of defense against electric shock as it provides most of the body resistance, these procedures reduce or even eliminate skin resistance, which increases the risk of electrical shock. If catheters are used, fibrillation can occur for as little as 50 µA. Thus, the safety limit for hospitals is often set to 10 µA.

The electrical circuits of hospitals are specially designed to eliminate the hazards of micro- and macroshocks within the premises. The circuits of the operating rooms are carefully designed to eliminate hazards of stray currents during surgeries and to ensure that arcs that excite the flammable anesthetics are not generated. Several systems designs are addressed in the next sections.

12.4.5.1 *Microshock due to Grounded System*

Figure 12.38 shows the parasitic capacitance of medical equipment for nonisolated circuits (when the neutral wire at the secondary of the service transformer is grounded and bonded to the chassis through the EGC). Figure 12.39 shows

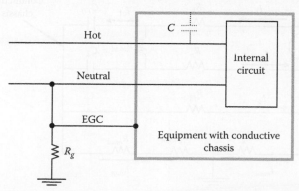

FIGURE 12.38
Parasitic capacitance.

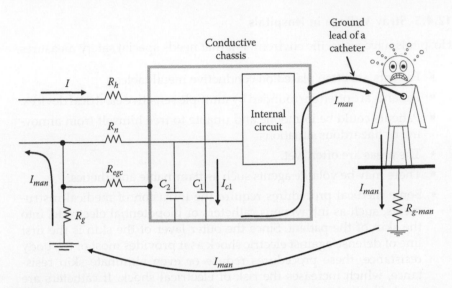

FIGURE 12.39
Microshock scenario.

a patient with catheter lead bonded to the metallic chassis of a medical equipment. The bed is assumed to be bonded to the ground through $R_{g\text{-}man}$. In this case, we are including the resistances of the conductors feeding the equipment.

Even if the medical equipment is not tuned on, there is a path for current from the source to the parasitic capacitance, to the chassis, to the patient, to the grounds, and finally back to the source.

The equivalent circuit for the case in Figure 12.39 is shown in Figure 12.40. R_{load} in the circuit represents the resistance of the internal circuit of the equipment.

Keep in mind that the patient resistance is very low as the catheter is connected to his internal organ.

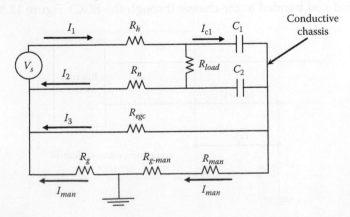

FIGURE 12.40
Equivalent circuit of Figure 12.39.

Example 12.17

For the system in Figure 12.39, assume $C_1 = C_2 = 50$ nF, $R_h = R_n = R_{egc} = 1.0 \, \Omega$, $R_g = R_{g\text{-}man} = 20 \, \Omega$. Assume the instrument circuit is disconnected ($R_{load} = \infty$). Compute the current passing through the person.

Solution:

Since the catheter is connected to the internal organ, assume the resistance of the man's body is 100 Ω:

$$\overline{Z}_1 = R_h + \frac{1}{j\omega C_1} = 1 - j5.3 \times 10^4 \, \Omega$$

$$\overline{Z}_2 = R_n + \frac{1}{j\omega C_2} = 1 - j5.3 \times 10^4 \, \Omega$$

$$\overline{Z}_3 = R_{egc} = 1.0 \, \Omega$$

$$\overline{Z}_4 = R_g + R_{g\text{-}man} + R_{man} = 140 \, \Omega$$

$$\overline{Z}_{eq} = \overline{Z}_1 + \left(\overline{Z}_2 \, // \, \overline{Z}_3 \, // \, \overline{Z}_4 \right) = 2 - j5.3 \times 10^4 \, \Omega$$

$$\overline{I}_1 = \frac{\overline{V}_s}{\overline{Z}_{eq}} = \frac{120}{\overline{Z}_{eq}} = 8.5 \times 10^{-2} + j2262 \, \mu A$$

The current passing through the man can be computed by writing the equation of the outer loop:

$$\overline{I}_1 \overline{Z}_1 + \overline{I}_{man} \overline{Z}_4 = \overline{V}_s$$

Hence, the current passing through the patient is

$$I_{man} = \left| \frac{\overline{V}_s - \overline{I}_1 \overline{Z}_1}{\overline{Z}_4} \right| = 16 \, \mu A$$

The current is above 10 μA limit.

12.4.5.2 Microshock in Isolated System

To address the microshock problem in the previous example, it would seem that using isolation transformer should correct the problem. This is not always true. Consider, for example, the case in Figure 12.41 where medical equipment is powered through isolation transformer. At the load side, there is no neutral wire (both are hot wires), and the EGC is not bonded to any of the load conductors. R_{load} in the figure represents the load of the medical equipment.

The chassis of the medical equipment must still be bonded to the EGC. Because of the isolation transformer, part of the secondary current I_2 passes through the capacitors (I_c in Figure 12.42) and returns through the other hot wire. The EGC carries none of this current.

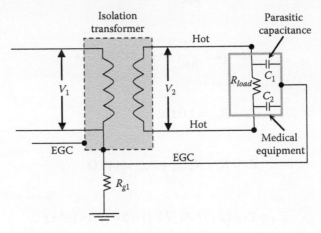

FIGURE 12.41
System with isolation transformer.

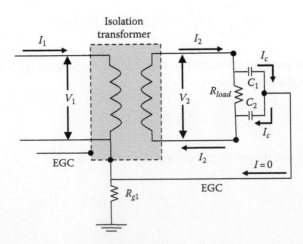

FIGURE 12.42
Current in parasitic capacitance of medical equipment with isolation transformer.

The current passing through the parasitic capacitors is

$$I_c = V_2 \omega C_{eq} \qquad (12.17)$$

where
V_2 is the voltage across the load
$\omega = 2\pi f; f$ is the supply frequency
C_{eq} is the equivalent capacitance of the series combination of C_1 and C_2

$$C_{eq} = \frac{C_1 C_2}{C_1 + C_2} \qquad (12.18)$$

When the hot wires are closely spaced, we can assume $C_1 \cong C_2$:

$$C_{eq} = 0.5 C_1 \qquad (12.19)$$

Because of the isolation transformer, the macroshock can be avoided. Let us assume the shock scenario in Figure 12.43 where a patient is in contact with one of the hot wires while part of his body is at the EGC potential (or connected to the ground electrode of the equipment). For simplicity, assume that the equipment is turned off; hence, $R_{load} = \infty$. The current of the secondary of the transformer will split into two parts, one through C_1 and the other through the patient. The current passing through the patient reaches the chassis of the equipment and is combined with the current in C_1 to return to the secondary winding of the transformer through C_2.

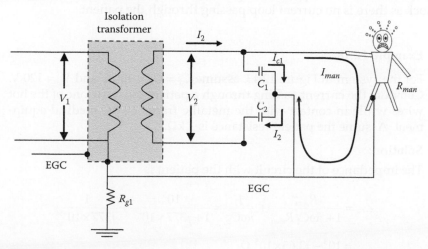

FIGURE 12.43
Electric shock with isolation transformer.

To estimate the current passing through the patient, we compute the equivalent impedance of the patient and the capacitive reactance of C_1 as a parallel combination:

$$\bar{Z}_1 = \frac{R_{man}}{1 + j\omega C_1 R_{man}} \tag{12.20}$$

The total impedance of the circuit is Z_1 plus the capacitive reactance of C_2 as a series combination:

$$\bar{Z}_{eq} = \bar{Z}_1 + \frac{1}{j\omega C_2} \tag{12.21}$$

The total current is then

$$\bar{I}_2 = \frac{\bar{V}_2}{\bar{Z}_{eq}} \tag{12.22}$$

The current passing through the patient can be computed by the current divider equation:

$$\bar{I}_{man} = \frac{\bar{I}_2}{1 + j\omega C_1 R_{man}} \tag{12.23}$$

Note that without the parasitic capacitance, the patient does not receive any shock as there is no current loop passing through the patient.

Example 12.18

For the system in Figure 12.43, assume $C_1 = C_2 = 10$ nF and $V_2 = 120$ V. Compute the current passing through a patient touching one of the hot wires while in contact with the metallic frame of the medical equipment. Assume the patient resistance is 1 kΩ.

Solution:

The impedance of the circuit with the patient is

$$\bar{Z}_{eq} = \frac{R_{man}}{1 + j\omega C_1 R_{man}} + \frac{1}{j\omega C_2} = \frac{10^3}{1 + j377 \times 10^{-5}} + \frac{1}{j377 \times 10^{-5}}$$

$$= 10^3 - j2.65 \times 10^5 \ \Omega$$

The current in the secondary's hot wire is

$$\bar{I}_2 = \frac{\bar{V}_2}{\bar{Z}_{eq}} = \frac{120}{\bar{Z}_{eq}} = 1.706 \times 10^{-6} + j4.52 \times 10^{-4} \approx 452\ \mu A$$

The current passing through the patient is

$$I_{man} = \frac{\bar{I}_2}{1 + j\omega C_1 R_{man}} = \frac{1.706 \times 10^{-6} + j4.52 \times 10^{-4}}{1 + j377 \times 10^{-6}} \approx 452\ \mu A$$

In this example, because of the isolation transformer, the current is within tolerable limit.

Let us assume another shock scenario. In Figure 12.44, the patient is in contact with the chassis of the equipment while part of his body is at the local ground potential. Assume the secondary of the transformer is heavily loaded. Note that the current in the secondary windings of the transformer I_2 will not go through the patient as the patient does not close any loop with the secondary windings. Hence

$$I_2 = I_{load} + I_c \tag{12.24}$$

However, the current of the primary winding is split into three components: the first is through the neutral wire of the primary I_n; the second is

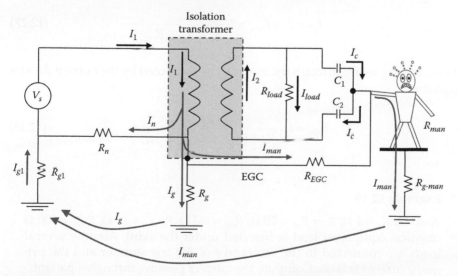

FIGURE 12.44
Microshock due to parasitic capacitance of the medical equipment with isolation transformer.

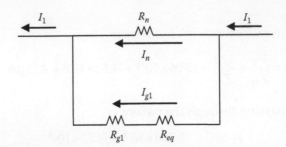

FIGURE 12.45
Electric circuit of system in Figure 12.44.

through the local ground of the transformer I_g and then back to the source; and the third is through the EGC to the patient then back to the source:

$$I_1 = I_n + I_g + I_{man} \tag{12.25}$$

We can use the circuit in Figure 12.44 to estimate the current passing through the patient. First, we compute the equivalent resistance of the man, the ground, and the EGC (Figure 12.45):

$$\frac{1}{R_{eq}} = \frac{1}{R_g} + \frac{1}{R_{man} + R_{g\text{-}man} + R_{EGC}} \tag{12.26}$$

The total ground current is

$$I_{g1} = I_g + I_{man} = I_1 \frac{R_n}{R_n + R_{g1} + R_{eq}} \tag{12.27}$$

The current passing through the man can be computed by the current divider equation:

$$I_{man} = I_{g1} \frac{R_g}{R_g + R_{EGC} + R_{man} + R_{g\text{-}man}} \tag{12.28}$$

Example 12.19

Assume $R_n = 1\ \Omega$, $R_g = R_{g1} = 20\ \Omega$, $R_{egc} = 0.2\ \Omega$, $R_{g\text{-}man} = 20\ \Omega$, $R_{man} = 500\ \Omega$ (medical equipment lead is inserted under the skin). Assume several loads are connected to the secondary of the transformer and the primary current is 30 A. Compute the current passing through a patient.

Solution:

Compute the equivalent resistance:

$$\frac{1}{R_{eq}} = \frac{1}{R_g} + \frac{1}{R_{man} + R_{g\text{-}man} + R_{EGC}} = \frac{1}{20} + \frac{1}{500 + 20 + 0.2}$$

$$R_{eq} = 19.26 \, \Omega$$

The total ground current is

$$I_{g1} = I_g + I_{man} = I_1 \frac{R_n}{R_n + R_{g1} + R_{eq}} = 30 \times \frac{1}{1 + 20 + 19.26} = 745 \text{ mA}$$

The current passing through the man is

$$I_{man} = I_{g1} \frac{R_g}{R_g + R_{EGC} + R_{man} + R_{g\text{-}man}} = 745 \times \frac{20}{20 + 0.2 + 500 + 20} = 27.6 \text{ mA}$$

This is extremely high current. The severity of this shock will depend on several factors:

- The loading condition of the transformer secondary—the higher the load current, the higher the shock current.
- The resistance of the neutral wire—the higher the neutral resistance, the higher the shock current.
- The transformer ground resistance—the higher the transformer ground resistance, the higher the shock current.

12.5 Detection of Stray Voltage

Utilities and government organizations have recently become more concerned about the impact of stray and conduction voltages on public safety. The New York State Public Service Commission (PSC), for example, requires the serving utility, Con Edison, to manually test all of its assets annually. All objects detected with a stray voltage of one or more volt are reported.

The detection can be done by one of two methods: direct and remote. The direct method is just manual measurement of the voltage on an object with respect to a reference point, which is often the local ground. This method requires a person to do the measurement, which is very burdensome on human resources, very slow, and costly. In addition, manual testing on painted or rusted object is unreliable.

The remote method relies on measuring the electric field. The measured electric field strength indicates the magnitude of the stray voltage on the object. This is done by sensors scanning wide areas fairly quickly and efficiently. The sensors can be mounted on mobile vehicles that scan the electric fields from both sides of the vehicle. The system requires elaborate algorithms to filter the electric fields from power lines, cellular fields, etc. Some of these systems are able to detect objects energized with as little as 1 V, and they have a range as long as 10 m on either side of the sensor.

The sensors come in various forms, but the most common ones are the *free-body* and the *electro-optic* sensors shown in Figure 12.46. The free-body sensor consists of a probe with two electrodes and insulation between them. When placed near charged object, the electric field at the sensor is

$$E_x = \frac{q}{2\pi\varepsilon_0 \, x} \tag{12.29}$$

where
  E_x is the electric field at the sensor
  q is the charge on the object that is producing the electric field
  x is the distance from the object

This probe produces the highest reading when it is aligned perpendicular to the electric field lines. This can be used to identify the location of the charged object.

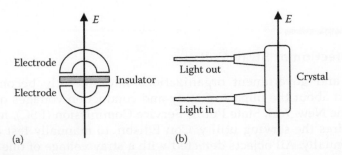

FIGURE 12.46
(a) Free-body and (b) electro-optic sensors.

Since the charge on the object oscillates, the electric field oscillates as well; hence

$$\frac{dE_x}{dt} = \frac{1}{K}\frac{dq}{dt} \tag{12.30}$$

K is equal to the denominator in Equation 12.29.

The probe is connected to electronic circuit that allows the current to flow between the two electrodes. Since the electric field oscillates, the current flowing from one electrode to the other is

$$i = \frac{dq}{dt} \tag{12.31}$$

Hence,

$$i = K\frac{dE_x}{dt} \tag{12.32}$$

If the electric field is sinusoidal, the instantaneous current is

$$i = K\frac{d}{dt}\left(E_{max}\sin\omega t\right) = K\omega E_{max}\cos\omega t \tag{12.33}$$

In RMS

$$I = K\omega\frac{E_{max}}{\sqrt{2}} \tag{12.34}$$

Once the current of the probe is measured, the electric field at distance x is known.

The electro-optic sensor utilizes the *Pockels effect* (electro-optic effect). When light passes through certain types of crystals, it is split into two rays: *ordinary* rays (perpendicular polarizations) and *extraordinary* rays (parallel polarizations). Several crystal materials, including potassium titanium oxide phosphate, lithium niobate, and gallium arsenide, can produce the Pockels effect.

The difference in speed between these two rays, known as refractive index Δn, is directly related to the electric field strength exerted on the crystal:

$$\Delta n = n_e - n_o \tag{12.35}$$

where
n_e is the speed of the extraordinary rays
n_o is the speed of the ordinary rays
Δn is the refractive index of the rays

12.6 Mitigation of Stray Voltage

Depending on the source of the stray voltage, there are several techniques that can reduce or eliminate the problem:

- Double-bushing transformers
- Isolation transformer
- Neutral isolator
- Four-wire system
- Isolation of cable TV, phone lines, etc.
- Equipotential area
- Lower ground resistance at the service panel (the least effective method)

12.6.1 Double-Bushing Transformers

Service transformers are typically single-phase with either single-bushing or double-bushing type. Figure 12.47 shows the single-bushing type, which is also known as auto-transformer. A typical wiring connection is shown in the middle of the figure and the circuit diagram is shown at the right. The primary of the transformer is connected between one phase of the feeding circuit and the neutral. The neutral terminals of the primary and secondary windings are bonded. The frame and neutral are grounded locally. In the United States, the primary voltage is often 15 kV-class (13.8 kV) for most residential loads. If the primary is connected in grounded Y, the voltage V_1 applied to the primary winding is about 7.97 kV. If the primary is grounded delta, the voltage applied to the primary winding is about 13.8 kV.

The double-bushing transformer is shown in Figure 12.48. The primary winding is connected between two different phases of a three-phase circuit

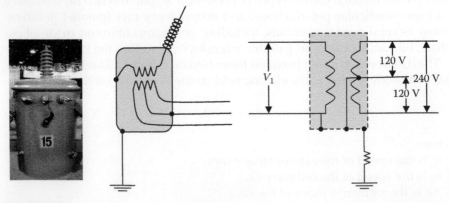

FIGURE 12.47
Single-bushing transformer.

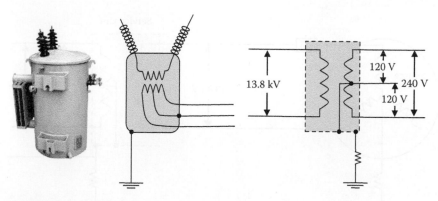

FIGURE 12.48
Double-bushing transformer.

(line-to-line). A typical wiring connection is shown in the middle of the figure and the diagram is shown on the right. The primary in the figure does
not have neutral terminal. The secondary winding is bonded to the metal
enclosure of the transformer that is grounded locally.

As we have discussed in this chapter, one of the major stray voltage problems is caused by bonding the primary and secondary neutrals of single-
bushing transformers (auto-transformers). This connection is important for
system protection as the utility would be able to clear transformer faults, as
shown in Figure 12.25. The source voltage in this case is a single phase of a
three-phase circuit. To satisfy the NEC, the neutrals are grounded, as shown
in Figure 12.28. This system, although very good from the circuit protection
point of view, causes stray voltage on customers' premises. One of the obvious
methods to address this problem is to eliminate the neutral from the primary
of the transformer by using double-bushing transformers. This transformer
is connected between two phases of a grounded Y or grounded delta system,
as shown in Figure 12.49. With this transformer, the neutral of the secondary
winding is isolated from the primary winding. The secondary neutral is connected to the local ground and the EGC. With this connection, there is no direct
path for the primary current to reach the neutral or the EGC of the secondary.
Thus, the stray voltage problem caused by the transformer is eliminated.

The system in Figure 12.49 can still provide the protection needed when the
primary and secondary windings are faulted. The path for the fault current in
the double-bushing system is shown in Figure 12.50. This fault is a single-line-
to-ground fault that can be detected and cleared by the circuit breaker.

Although double-bushing transformers can effectively eliminate the stray
voltage associated with the auto-transformer, there are two major challenges:

- Most of the distribution transformers in the United States are of the
  single-bushing type. It is costly to replace them with double-bushings
  and rewire the system.
- Most rural loads are energized by a single-phase distribution feeder.

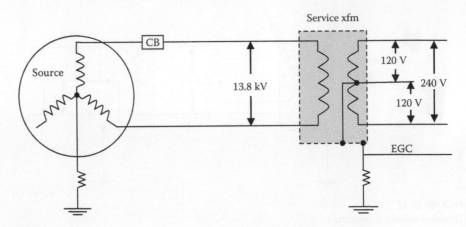

FIGURE 12.49
Double-bushing transformer connected to a grounded Y source.

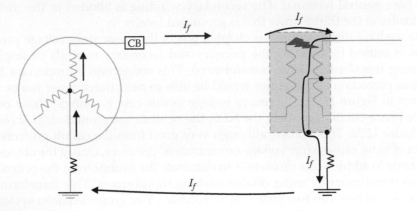

FIGURE 12.50
Protection of a double-bushing transformer.

12.6.2 Isolation Transformer

The isolation transformer can be used between a single-bushing transformer and the customer's circuit, as shown in Figure 12.51. The turn ratio of the isolation transformer is often 1:1. The center of the secondary winding of the isolation transformer is bonded to the local ground and EGC. By this connection, there is no direct path for the neutral current of the high-voltage circuit to reach the neutral (or EGC) of the load.

Keep in mind that the isolation transformer does not protect the person from the stray current caused by the various parasitic capacitances. This point has been explained earlier in this chapter.

In swimming pools, the use of isolation transformer, or double-bushing transformer, can eliminate the stray voltage, as shown in Figure 12.52.

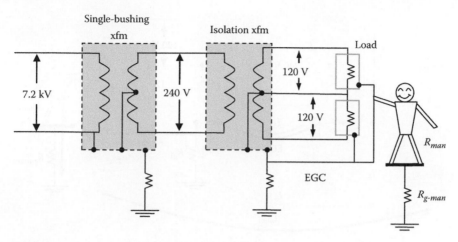

FIGURE 12.51
Single-bushing transformer with isolation transformer.

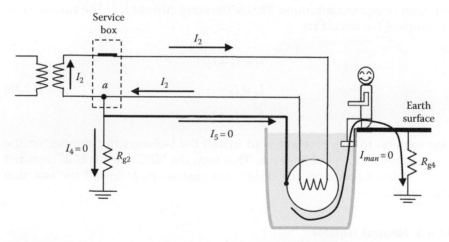

FIGURE 12.52
A solution to stray voltage in swimming pools.

Because the secondary winding of the transformer carries the entire load current I_2, the ground currents I_4 and I_5 are virtually eliminated according to Kirchhoff's nodal law at node a.

One major hurdle in using the isolation transformer is that it cannot be installed on customers' premises. This is because the NEC requires that all non-current-carrying metal parts be bonded (through EGC). If the frames of the two transformers are bonded, the isolation transformer is useless, as shown in Figure 12.53. Part of the primary current of the single-bushing transformer will find a path through the EGC connections to any person

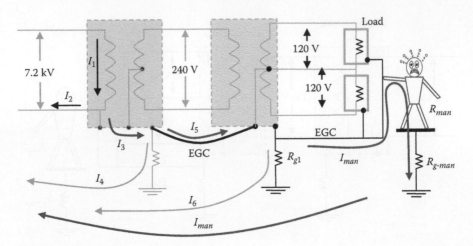

FIGURE 12.53
Single-bushing transformer with isolation transformer and NEC requirement.

touching an equipment frame. This is the stray current I_{man}. The current relationships of the circuit are

$$I_1 = I_2 + I_3$$

$$I_3 = I_4 + I_5 \tag{12.36}$$

$$I_5 = I_6 + I_{man}$$

One solution to this problem is to install the isolation transformer on the pole and away from public reach. This way, the NEC rule for EGC does not apply. Doing so requires the utility, not customers, to install the isolation transformer.

12.6.3 Neutral Isolator

The neutral isolator is a device that is installed between the primary and the secondary neutrals, as shown in Figure 12.54. It is a nonlinear resistance similar in function to the metal oxide varistor (MOV) used for overvoltage protection in substations and distribution networks. It has a very high resistance when the voltage across it is <9 V. The resistance drops rapidly to a few milliohms when the voltage reaches 11 V or higher. This way, the two neutrals are isolated at all times, thus mitigating the stray voltage problem. When a fault causes a path between the hot primary terminal and the secondary of the transformer, the high voltage of the secondary neutral will cause the neutral isolator to close, thus activating the circuit breaker (CB).

Under normal operating conditions (the neutral isolation is open), there is no path for any current to reach the EGC, as shown in Figure 12.55. The hot

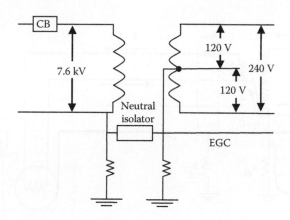

FIGURE 12.54
Neutral isolator.

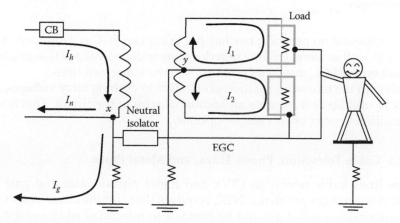

FIGURE 12.55
Current loops in transformer with neutral isolator.

wire current I_h will split into two components at point x: the majority is the neutral current I_n and the ground current I_g. There will be no current passing to the EGC. On the secondary side, the currents I_1 and I_2 will circle their own loops; no current leaves point y to the EGC.

12.6.4 Four-Wire System

Figure 12.56 shows the four-wire system. The EGC in this system is provided by the utility as a fourth wire from the substation. In addition, the neutrals at the transformer and at the service box are separate from the EGC. By this connection, there is no current from the transformer passing through the EGC. At node x, the hot wire current I_h branches into I_n and I_g. The current I_2 has no path back to the source; hence, $I_2 = 0$. The load current I_1 loops in its

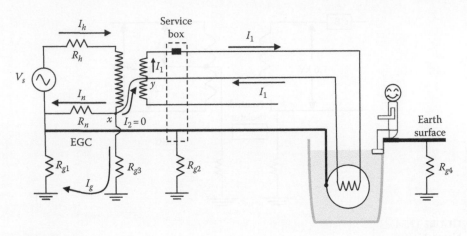

FIGURE 12.56
Four-wire system.

own circuit and no part of I_1 reaches the EGC. The EGC is grounded at the source as well as along the line to reduce any induced voltage that might be present on the EGC due to its proximity to the energized lines.

Although the four-wire solution is effective in curbing stray voltages, it is costly to utilities as it requires an additional wire in the circuit. This is why some utilities prefer other cheaper options.

12.6.5 Cable Television, Phone Lines, and Metal Pipes

Phone lines, cable television (TV), and metal pipes (water and gas) can cause stray voltage problems. NEC requires that communication services and distribution panel ground be bonded to minimize equipment damage due to potential differences between the various systems, especially during lightning storms. A typical circuit distribution in dwellings is shown in Figure 12.57. Because of this bonding, any of these services can elevate the potential of the EGC even when the utility install isolators or isolation transformers. Take, for example, the system in Figure 12.58. In this system, the utility installed a neutral isolator to separate the two neutrals of the transformer. This will eliminate the stray voltage coming from the power grid. However, communication cables have shield meshes that are grounded at the transformer pole and at the customer's site. Because these services are grounded at several places along their service area, they bypass the neutral isolator rendering it ineffective. Metal pipes (water or gas) can cause the same problem. Any of these services creates a path between the two neutrals of the transformer causing a stray voltage problem.

Metal pipes can cause stray voltage in another form. Since the ground conductors at the customer's sites are often bonded to metal pipes, they

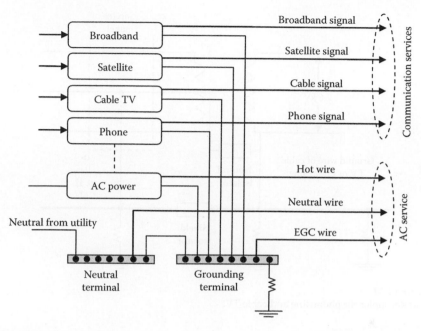

FIGURE 12.57
Typical circuit distribution in the United States.

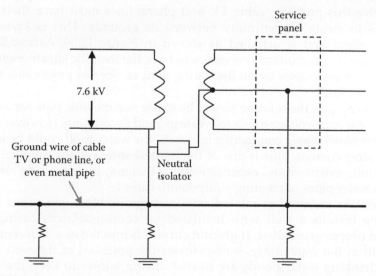

FIGURE 12.58
Phone line, cable TV, and water pipe systems.

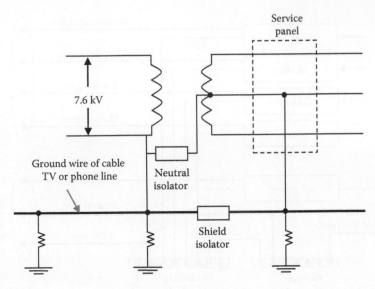

FIGURE 12.59
Shield decoupler for phone line and cable TV.

link all EGCs bonded to the pipe. So if faulty equipment at one customer's site injects current into a water pipe buried in high-resistivity soil, stray currents can be expected through every EGC connected to the pipe.

To solve this problem, cable TV and phone lines must have their own isolators to create discontinuity between its grounds. This is known as shield isolator and is installed as shown in Figure 12.59. Article 800 of NEC requires communications cables to have the metallic sheath members grounded *or interrupted* by an insulating joint as close as practicable to the point of entrance.

For water pipes, the solution would be to use non metallic pipe sections in areas where stray voltage exists to create ground decoupling. However, since the water may include conductive minerals, the water itself could be a path for the stray current. This is one of the hardest stray voltage problems and may require extensive and expensive efforts that may include using nonconductive water pipes or creating equipotential areas.

In rural areas, another related stray voltage problem can occur when someone installs a well with insulated (or semi-insulated) casing and without proper grounding. If ground current is injected at a different location within the water table, it can elevate the potential of the well. Also, when drinking water ponds are heated during winter to keep the water from freezing, the metal casing of the heater that is bonded to the EGC can elevate the voltage of the water. In these cases, when a person or animal stands on the local ground and touches the water, stray current could pass through its body.

12.6.6 Equipotential Area

When it is difficult to remove stray voltage in a premise, an *equipotential area* (plane) can be constructed to equalize the voltage between any two points within the area. Inside the equipotential area, all conductive objects within direct or indirect reach are bonded. This method is suitable for farms and swimming pools and is similar to the equipotential zone used by utilities during power line works.

In farms, the National Electric Code (NEC) requires in Article 547.10 the establishment of an *equipotential plane* in areas where stray voltage could be present.

> An equipotential plane must be installed at indoor and outdoor concrete confinement areas where metallic equipment is located that may become energized and is accessible to animals.

The code defines this equipotential plane as follows:

> An area where wire mesh or other conductive elements are embedded in or placed under concrete, bonded to all metal structures, and fixed nonelectrical equipment that may become energized, and connected to the electrical grounding system to prevent a difference in voltage from developing within the plane.

The code identifies the applications that require equipotential planes as

> Outdoor and concrete surface confinement areas such as feedlots, and indoor or outdoor dirt surface areas such as horse stalls and feedlots shall have equipotential planes installed around metallic equipment that is accessible to animals and likely to become energized. The equipotential plane shall encompass the area around the equipment where the animal will stand while accessing the equipment.

The code requires the bonding of the equipotential area to ground:

> The equipotential planes shall be bonded to the building or structure electrical grounding system. The bonding conductor shall be copper, insulated, covered, or bare, and not smaller than eight AWG. The means of bonding to wire mesh or conductive elements shall be by pressure connectors or clamps of brass, copper, copper alloy, or an equally substantial approved means.

Figure 12.60 shows the equipotential area in and around barns. An *equipotential grid* (EG) is imbedded in the concrete slab of the barn. The EG must cover the barn floor and extended outward from all direction to ensure that any animal contacting the barn from any side is fully positioned over the EG. All conductive elements of the barn such as utility service boxes, EGCs of all outlets, feeding and drinking containers must be bonded to the EG. This way, the entire area is free from unequal potentials between any two points within the EG area. The EG can be further grounded by ground rods.

The obvious question is how to get the animal in and out of the area without experiencing electric shock due to step potential? While stepping in or stepping out, part of their body is at the EG potential and the other part

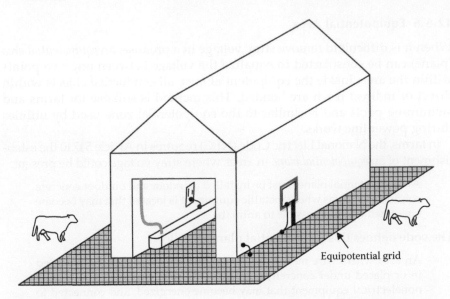

FIGURE 12.60
Equipotential grid in farm.

is at the local ground potential. To solve this problem, we need to establish a *voltage ramp* where the voltage is gradually changing to the EG level. Indeed, the NEC requires that a *voltage gradient* (also known as voltage ramp) be established at the animal entrance or exit to a building to minimize the step potential between the ground and the concrete. This way, animals do not experience an abrupt voltage/current sensation when moving from one potential to another. Figure 12.61 shows a way to establish the voltage ramp. Near the edge of the slab, the EG is placed deep in the concrete and gradually elevated toward the surface. Near the deep EG, the potential of the concrete is near the ground level and the animal can step in and out without feeling

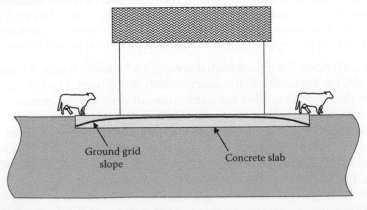

Ground grid
slope   Concrete slab

FIGURE 12.61
Voltage ramp.

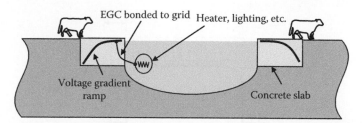

FIGURE 12.62
Equipotential plane for water ponds.

a noticeable difference in potential. Within the ramp, the change in potential must not be abrupt. A 45° slope for the EG is a good rule of thumb.

In farms located in cold environment, drinking water ponds are often heated in winter. The casing of the water heater must be connected to the EGC according to the NEC. Since the EGC could be a few volts above the ground potential, the water will be energized at that potential. If an animal is standing on ground potential and touching the water, it will be exposed to this stray voltage. One solution is to construct EG around the pond with voltage ramp, as shown in Figure 12.62. The EG must also be bonded to the casing of the water heater (EGC).

NEC article 680 addresses outdoor water installations used by the public such as pools, outdoor spas, and outdoor hot tubs. If these water installations have conductive shells, the structural reinforcing steel can be used as part of the EG. The area of the EG must extend 3 ft (~0.914 m) horizontally beyond the inside walls of a pool, outdoor spa, or outdoor hot tub, including unpaved, paved, and poured concrete surfaces, as shown in Figure 12.63. The EG must be bonded to the following parts:

- All fixed metal parts within a horizontal distance of 5 ft (~1.52 m) from the pool wall. The exception is when they are behind barriers, or if located more than 12 ft (~3.66 m) vertically above the maximum water level.
- Underwater metal-forming shells, mounting brackets for luminaires and speakers, metal-sheathed cables and raceways, metal piping, etc.
- Metal fittings that penetrate into the water, such as ladders and handrails.
- Electrical equipment associated with the pool, outdoor spa, or outdoor hot tub water circulating system such as water heaters and pump motors. The exception is when they are behind barriers, located more than 5 ft (~1.52 m) horizontally from the inside walls, or if located more than 12 ft (~3.66 m) vertically above the maximum water level.

A view of NEC 680.26 is shown in Figures 12.64 and 12.65. In both figures, all metal parts are bonded to the EG. In Figure 12.64, the equipotential area includes the water, pool ladder, diving board, water circulation pumps, heaters, fences, and lighting fixtures. In Figure 12.65, the water circulation pumps are behind a barrier so the EG does not have to be extended beyond the barrier.

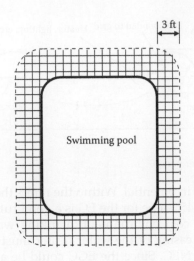

FIGURE 12.63
Equipotential area for pools.

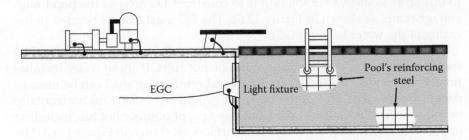

FIGURE 12.64
Horizontal view of equipotential area for pools.

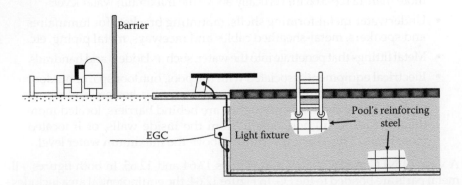

FIGURE 12.65
Horizontal view of equipotential area for pools with separated electrical equipment.

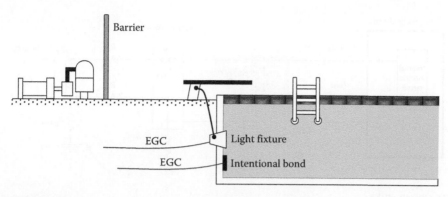

FIGURE 12.66
Intentional grounding of water in insulated pool shell.

For fiberglass pools, the water must be bonded to the local ground, as shown in Figure 12.66. An intentional bond of a minimum conductive surface of 9 in.² (~58.1 cm²) must be in contact with the water in the pool. This is done to equalize the voltage of the water, ladder, diving board, etc. If a lighting fixture has conductive surface of adequate area, it can be used as the intentional bond.

The obvious question is why not keep the water isolated since it is in fiberglass shell? It would seem that if the ladder and diving board are made of fiberglass, the system will be totally isolated from the electrical system. The reason is to protect people from electric shock when a two-wire (hot and neutral) electrical equipment accidentally comes in contact with the water of the pool. Unless the water is grounded, all of the load current of the immersed equipment will return to the source through the neutral wire. None of this current returns through ground. The GFCIs in this case cannot interrupt the circuit. Therefore, the water will be at an elevated voltage. If someone is standing on the local ground and touches the water, he or she will receive an electric shock. However, if the water is bonded to the EGC, some of the load current of the immersed equipment will go through the ground, thus triggering the GFCI to trip the circuit.

12.6.7 Reducing Grounding Resistance

This is the least effective method to solve the stray voltage problem. Although reducing the ground resistance of the EGC will allow less current to pass through a person touching the metal frame of electrical equipment, it will not eliminate stray voltage. This is because the ground resistance cannot be lowered to near zero even for the lowest soil resistivity. Consider the system in Figure 12.67. The electric equipment is connected to the source through a distribution line.

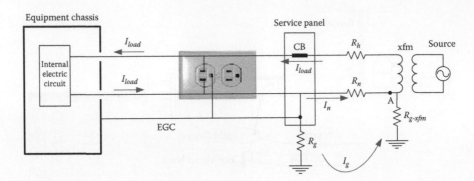

FIGURE 12.67
Stray voltage on EGC.

The neutral and EGC are bonded at the service panel. The load current I_{load} leaving the electric circuit branches into two paths: one through the neutral wire I_n and the other I_g through R_g, then $R_{g\text{-}xfm}$. The voltage of the EGC is

$$V_{EGC} = I_g R_g \tag{12.37}$$

where I_g is the current in the local ground R_g, which can be computed by the current divider method:

$$I_g = I_{load} \frac{R_n}{R_n + R_g + R_{g\text{-}xfm}} \tag{12.38}$$

Hence, the voltage of the conductive enclosure (the voltage of the EGC) is

$$V_{EGC} = I_{load} \frac{R_n R_g}{R_n + R_g + R_{g\text{-}xfm}} \tag{12.39}$$

Equation 12.39 shows that lowering the ground resistance will reduce the stray voltage but will not eliminate it. This is because R_g cannot be lowered to zero. Electricians often make the wrong assumption that by reducing the ground resistance, the stray voltage disappears (see the following example).

Example 12.20

Compute the ground resistance of the ground rod of a house panel that reduces the stray voltage on the EGC to 1 V. Assume the load current I_{load} = 50 A and R_n = 1 Ω. Assume the ground resistance at the transformer $R_{g\text{-}xfm}$ is just 10 Ω.

Solution:

The voltage of the EGC is

$$V_{EGC} = I_{load} \frac{R_n R_g}{R_n + R_g + R_{g\text{-}xfm}}$$

$$1 = 50 \frac{R_g}{11 + R_g}$$

Hence,

$$R_g = 0.22 \ \Omega$$

This resistance is very low and cannot be obtained. If you assume that a ground rod of 4 cm diameter is inserted in a good soil with low soil resistivity, the length of the ground rod would be unrealistic. As given in Chapter 5, the ground rod resistance is

$$R_{rod} = \frac{\rho}{2\pi L_{rod}} \left[\ln\left(\frac{4L_{rod}}{r} \right) - 1 \right]$$

Assuming soil resistivity of 100 Ω-m, we get

$$0.22 = \frac{100}{2\pi L_{rod}} \left[\ln\left(\frac{4L_{rod}}{r} \right) - 1 \right]$$

Solving the equation yields

$$L_{rod} \approx 780 \ \text{m}$$

This is about a 0.5 mile long ground rod, which is totally unrealistic.

12.7 Mitigation of Stray Voltage in Hospitals

Besides direct shocks, hospitals need to protect their patients against stray voltages. The basic protection techniques against all shocks include the following:

- Equipotential grounding
- Neutral isolation
- A combination of both

Unfortunately, even when a hospital implements extensive isolation systems to eliminate stray voltages, anyone who connects some type of household devices, even if Underwriters Laboratories (UL) approved, to hospital circuits can defeat the isolation system. Thus, additional equipment are needed to monitor the presence of non-hospital-standard equipment on their system and interrupt the corresponding circuits. These issues are discussed in the next sections.

12.7.1 Equipotential Grounding

An effective method to eliminate most electric shocks in hospitals is to establish an equipotential area where all metallic objects are bonded to a single-point ground system, as shown in Figure 12.68. All EGCs and conductive surfaces in the vicinity of the patient are connected to a common grounding strip. This grounding strip is connected to the reference grounding point (building ground). This single-point ground system brings the potential between any two metallic objects to near zero.

Although it is very effective, the equipotential area can have elevated voltage during faults. With line-to-ground faults, the equipotential area will carry some of the fault current depending on the circuit configuration and the various impedances in the branches. The current flows within the area may create elevated potential differences between points

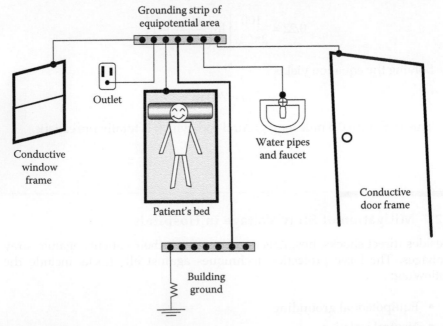

FIGURE 12.68
Equipotential area in patient's room.

in the area, which could be harmful. To reduce the potential differences, the grounding conductors are sized large enough to reduce their wire resistance.

12.7.2 Neutral Isolation

Isolating the neutral of the grid from the local ground is an effective technique to protect people against direct shocks. The isolation can be achieved by using *isolated-neutral transformer* (INT) similar to the one shown in Figure 12.69. The wires of the secondary of the INT are isolated from the EGC or any grounded object.

Consider the case in Figure 12.70 where a person is touching a metal chassis while contacting a grounded object. In the secondary loop of the INT, the current is the same everywhere. Hence, the current passing through the

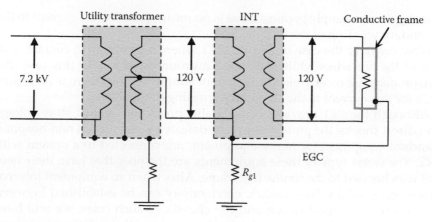

FIGURE 12.69
Isolated neutral transformer for sensitive installations.

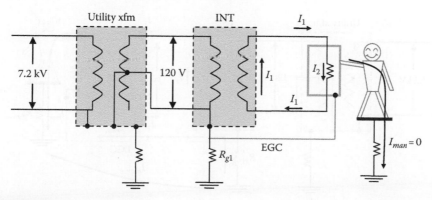

FIGURE 12.70
INT eliminates stray voltage.

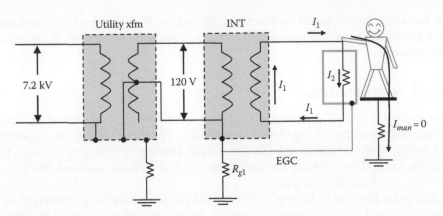

FIGURE 12.71
INT to protect against direct electric shock.

man I_{man} is zero simply because there is no return path for this current to the secondary winding of the INT.

Now consider the case in Figure 12.71 where a person is in contact with one of the hot wires while contacting a grounded object. In this case, the person does not receive electrical current simply because there is no return path for this current to the secondary winding of the INT.

Although the INT provides considerable protection against stray voltage and direct shocks, the protection can be defeated when certain non-hospital-standard equipment (or faulty equipment) are connected to a system with INT. The worst type of these equipments are the ones that have their neutral wire bonded to the conductive frame. Also, when an equipment internal insulation partially fails, electric conductivity can be established between the energized internal circuit and the chassis. In such cases, we will have two problems: (1) there will be stray current passing through the person and (2) the magnitude of the stray current may not trip the circuit breaker.

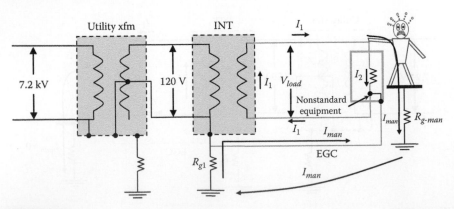

FIGURE 12.72
Impact of nonstandard equipment.

Consider the case in Figure 12.72. When a person touches the hot wire while touching the grounded object or the chassis of the equipment, the current passing through the person, I_{man}, is

$$I_{man} = \frac{V_{load}}{R_{man} + R_{g\text{-}man} + R_{g1}}$$

(12.40)

Example 12.21

For the system in Figure 12.72, compute the current passing through the person if V_{load} = 120 V, the patient resistance is 1 kΩ, the ground resistance of the patient $R_{g\text{-}man}$ = 100 Ω, and the ground resistance R_{g1} = 20 Ω

Solution:

The current passing through the man is

$$I_{man} = \frac{V_{load}}{R_{man} + R_{g\text{-}man} + R_{g1}} = \frac{120}{1000 + 100 + 20} = 107 \text{ mA}$$

This current is lethal but is too low to activate the panel's circuit breaker.

Because non-hospital-standard equipment can create lethal conditions, hospitals monitor their circuits and take actions when faulty or nonstandard equipment is plugged into their circuits. The monitoring device is called *line isolation monitor* (LIM). There are several designs for the LIM. Some are simply impedance matching bridges and others are more sophisticated time domain reflectometers. Some of the LIM can monitor multiple circuits and come with warning signals that can be used to trip the affected outlet.

The simplest LIM is based on impedance matching technique. Note that the presence of the person or the nonstandard equipment altered the impedance between any hot wire and the EGC of the equipment. For the case in Figure 12.72, the load impedance is seen between the upper hot wire and the EGC, while there is zero impedance between the lower hot wire and the EGC. These unequal impedances can be detected by the LIM.

An impedance matching LIM is shown in Figure 12.73. It is a four-terminal circuit consisting of four impedances (Z_1, Z_2, Z_3, and Z_4) in a bridge configuration. A voltmeter measures the voltage at the midpoints of the bridge V. The voltage V is zero only when

$$\frac{Z_1}{Z_2} = \frac{Z_3}{Z_4}$$

(12.41)

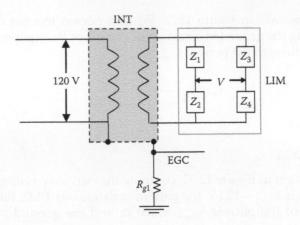

FIGURE 12.73
Line isolation monitor.

The impedances of the LIM are large enough to limit its current. If the condition in Equation 12.41 is not met, $V \neq 0$.

Now let us insert the LIM in the circuit where the nonstandard equipment is installed, as shown in Figure 12.74. Because one of the leads of the load of the equipment is connected to the EGC, the impedance Z_4 is short-circuited and the condition in Equation 12.41 is not met. Hence, the value of V is nonzero, which activates audio and visual signals to alert the staff of the potential hazard. In some cases, the voltage triggers automatic tripping actions.

Since this LIM is connected to the EGC, its impedances must be extremely high. This is important to ensure that any direct shock, as the one shown in Figure 12.75, produces extremely low current.

Another detection method is based on time-domain reflectometry. The main blocks of the device are shown in Figure 12.76. Two identical signals

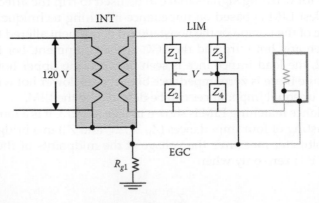

FIGURE 12.74
LIM with nonstandard equipment.

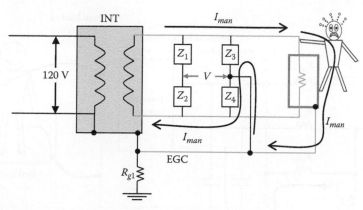

FIGURE 12.75
Hazardous current loop of LIM.

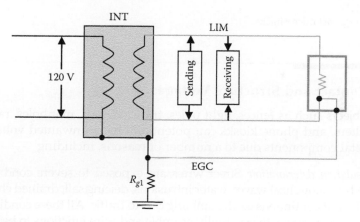

FIGURE 12.76
LIM time domain reflectometry.

are injected into the two hot wires at exactly the same time by the sending circuit. These signals are of high frequency to cause the signals to travel in all directions and reflect back. The reflected signal is received by the receiving block. If the two received signals are shifted, it indicates the presence of unsymmetrical path caused by the nonstandard equipment.

12.7.3 Protection against Microshock

The system in Figure 12.69 cannot protect patients against microshocks associated with the parasitic capacitors. For the isolated or the nonisolated systems, equipotential grounding must be established by bonding the patient to the chassis of the equipment as well as the bed frame, as shown in Figure 12.77.

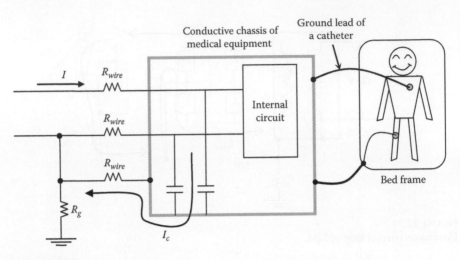

FIGURE 12.77
Protection against microshocks.

12.8 Contact and Structure Voltages

Street objects such as fences, light poles, traffic signals, manholes, railways, bus shelters, and phone kiosks can potentially have unwanted voltages on their metal components due to a number of reasons, including

- *Insulation deterioration*: Street wires are exposed to severe conditions such as snow, heat waves, water intrusions, deicing salt, drained chemicals, and continuous earth rumbling due to traffic. All these conditions cause insulations to eventually crumble and wire junctions to become loose. The older a city is, the more prone to this problem.

- *Shoddy workmanship*: It is not uncommon for unskilled workers to swap wires or place exposed parts of wires near conductive enclosures.

- *Injection of large currents into ground*: Injection of large currents into the ground, for example, during faults, can elevate the potential of nearby objects.

- *Isolated low-potential current-carrying conductor*: When the return current conductor is isolated from the ground, its potential could be elevated.

- *Induction voltage*: If conductive objects are near a power line, the line induces voltage on the objects due to electromagnetic coupling.

Although these problems are not technically stray voltages, some engineering, and certainly public, circles are adding them to the stray voltage list of problems. In this section, some of the most common problems are discussed.

12.8.1 Light Rail Systems

Light rail traction (LRT), including street cars and light rails, are very popular in major cities. These systems are generally powered by a dc *traction electrification system* (TES) similar to the one shown in Figure 12.78. The system is monopole *overhead contact power line* (OCPL). At the *traction power substation* (TPS), the ac high voltage from the serving utility is converted into lower dc voltage through a 12-pulse, 3-phase ac/dc converter. The converter makes the OCPL positive polarity and the tracks negative polarity. To prevent corroding utility pipes, the return current is not injected into the ground but is returned to the source through the tracks. Hence, the tracks are often insulated from earth. This is known as the *negative floating return system* (NFRS). The insulation of the tracks is done by elevating them above ground or by encasing the track slab with insulating membrane. From electric safety point of view, this insulation creates three problems:

1. Detection of downed OCPL conductor
2. Leakage current into earth
3. Rail-to-earth potential

For the first problem, when the OCPL conductor is downed, but not in contact with the track, the fault current does not have a path to the source. Hence, it cannot be detected and cleared by the protection devices at the substation. To address this problem, a grounding diode is installed at the substation, as shown in Figure 12.79. The fault current loop in this case starts from the secondary of the transformer to the converter, to the OCPL conductor, to ground, to the grounding diode, then back to the secondary winding to close the current loop. This way, we are providing a low impedance path for the fault current to the substation. Thus, the magnitude of the fault current is

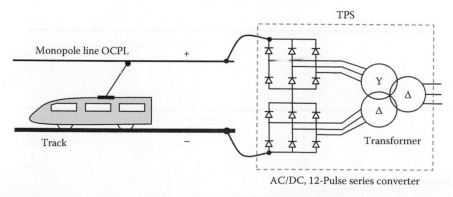

FIGURE 12.78
Traction electrification system of light rails.

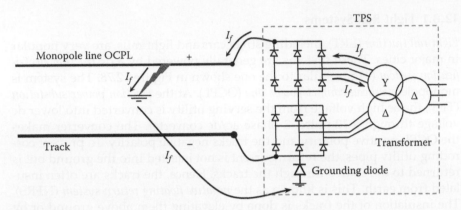

FIGURE 12.79
Fault detection of light rail system.

high, making it easier to detect and clear. This scenario is equivalent to the three-phase short circuit on a three-phase transformer.

Solving the first problem creates the second problem, leakage current. Consider the system in Figure 12.80. Each of the street cars can have a typical load of 500 kW; thus a load current of several hundreds of amps is expected from each car. This large current must return to the substation through one or both tracks. Now let us assume that the equivalent resistance of the tracks is R_{track} and the ground resistance of the tracks at the point of contact is R_g. The ground resistance is very high as the track is enclosed by isolating material. The current of the car is then branched into two components, most goes through the track, I_{track}, and small amount through ground, $I_{leakage}$.

Assuming the grounding diode resistance in the forward direction is very small compared with R_g, we can compute the leakage current as

$$I_{leakage} = I_{load} \frac{R_{track}}{R_{track} + R_g} \tag{12.42}$$

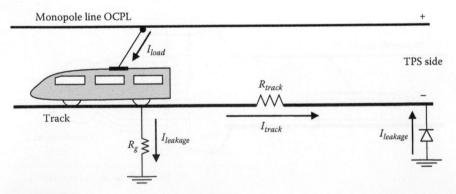

FIGURE 12.80
Leakage current of light rail system.

The voltage of the track with respect to remote ground at the point of contact is

$$V_{track} = I_{leakage} R_g \qquad (12.43)$$

Example 12.22

An electric light rail train draws 500 A. Assume the equivalent track resistance from the train to the substation (TPS) is 0.05 Ω and the equivalent ground resistance of the track is 10 kΩ. Compute the leakage current and the voltage of the track at the train's location.

Solution:

Using Equation 12.42,

$$I_{leakage} = I_{load} \frac{R_{track}}{R_{track} + R_g} = 500 \frac{0.05}{0.1 + 10000} = 2.5 \text{ mA}$$

The voltage of the track with respect to ground at the train's location is

$$V_{track} = I_{leakage} R_g = 2.5 \times 10 = 25 \text{ V}$$

This voltage is very high, even though the leakage current is very low.

Keep in mind that the leakage current can increase when

- The train is farther away from the substation; this increases the track resistance
- Ground resistance of the track is reduced due to failure in the insulation material
- Load current increases when several of these trains are on the same track

As seen in the previous example, the voltage on the track is elevated at the location of the street car. This is the third problem. When a person touches the track near the train while standing on a grounded object, the person can receive an electric shock. A case is depicted in Figure 12.81, where a person with conductive walking cane touches the track while standing on ground potential.

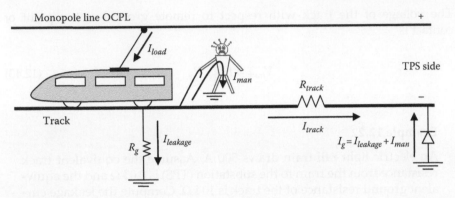

FIGURE 12.81
Electric shock in light rail system.

Example 12.23

For the system in Example 12.22, compute the current passing through a person touching the track near the train. Assume the ground resistance of the person is 500 Ω and his body resistance is 1 kΩ.

Solution:

The equivalent resistance of the person and the track ground resistance are

$$R_{eq} = \frac{R_g R_{man}}{R_g + R_{man}} = \frac{10 \times 1.5}{10 + 1.5} = 1.3 \text{ k}\Omega$$

Substituting in Equation 12.42, we can compute the total ground current:

$$I_g = I_{leakage} + I_{man} = I_{load}\frac{R_{track}}{R_{track} + R_{eq}} = 500 \times \frac{0.1}{0.1 + 1300} = 38.5 \text{ mA}$$

The current passing through the man is

$$I_{man} = I_g \frac{R_g}{R_g + R_{man}} = 38.5 \times \frac{10,000}{10,000 + 1,500} = 33.5 \text{ mA}$$

This current is hazardous.

To reduce the hazards of electric shock, the following solutions can be implemented:

- Have the LRT run on a dedicated right of way where public access is not possible.
- Reduce ground current by reducing the track resistance between the substation and the train. This can be done by using track material with very low resistivity.
- Reduce the distance between the TPS and the train. This is accomplished by using multiple TPSs with reduced spacing between them, as shown in Figure 12.82. In this case, the track resistance between the train and any substation is reduced and the track current is split and returned to the TPSs in both directions.
- Install low-voltage metal-oxide varistor (MOV) along the track, as shown in Figure 12.83. These devices have nonlinear resistance that is dependent on the voltage across them. When the voltage is high, the resistance is very small; when the voltage is low, the resistance is very high. The MOV will increase the stray current but will reduce the voltage of the track to safe levels.

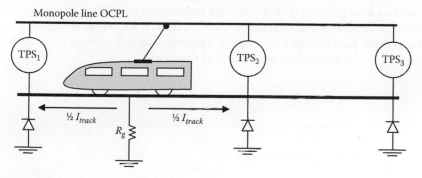

FIGURE 12.82
Multiple TPS systems.

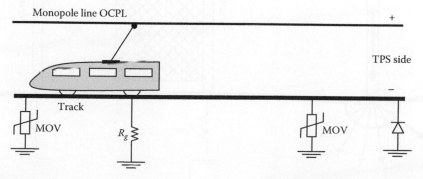

FIGURE 12.83
MOV for light rail system.

12.8.2 Fences and Gates

Voltages on metal fences and gates can occur due to various reasons such as the presence of ground current, electromagnetic coupling from nearby transmission lines, or unintentional contact with energized object. If these fences are not probably bonded and grounded, they can attain hazardous levels of voltage.

12.8.2.1 Effect of Ground Currents

Ground currents near metal gates and fences can cause potential difference between two separate conductive objects in contact with earth. This problem is particularly severe near substations and during faults where ground currents could be substantial. The ground current elevates the potential at the surface of the earth to a level that depends on its proximity to the object injecting current into earth. This is known as the *ground potential rise* (GPR). Any metal object in contact with earth will have its potential elevated to the GPR of its immediate soil.

An example of such a scenario is shown in Figure 12.84. Assume a current is injected into ground due to fault or imbalance in a three-phase system. Assume two posts are at distances r_a and r_b from the current injection point. Although the posts could be far apart, a person can still contact both of them through a metal gate mounted on one of the posts.

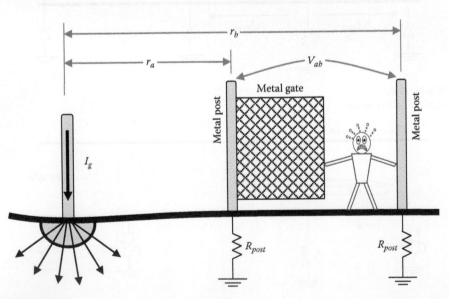

FIGURE 12.84
Stray voltage in metal gates.

As discussed in Equation 4.5, the potential difference between any two points on ground can be computed by

$$V_{ab} = \frac{\rho I}{2\pi}\left[\frac{1}{r_a} - \frac{1}{r_b}\right] \tag{12.44}$$

The level of V_{ab} depends on the magnitude of the injected current I, soil resistivity ρ, distance between the near post and injected current r_a, and the separation between the posts. The equation assumes the object injecting current into the ground is a hemisphere. However, for other shapes, the process is similar.

Example 12.24

The system in Figure 12.84 shows an object injecting a current into the ground. A nearby metal gate is mounted between two metal posts separated by 3 m. The nearest post to the ground hemisphere is 10 m away from the center of the hemisphere. The ground resistance of each post is 50 Ω. The soil resistivity is 500 Ω-m. If a current of $I_g = 10$ A is injected into the ground, compute the potential difference between the two posts when the gate is open. Also compute the hand-to-hand current passing through a person that touches the open gate and the post.

Solution:

The potential between the two posts without the person touching the gate is

$$V_{ab} = \frac{\rho I_g}{2\pi}\left[\frac{1}{r_a} - \frac{1}{r_b}\right] = \frac{500 \times 10}{2\pi}\left[\frac{1}{10} - \frac{1}{10+3}\right] = 18.4 \text{ V}$$

V_{ab} is Thevenin's voltage of the circuit. Thevenin's impedance R_{th} between the two posts is

$$R_{th} = 2R_{post} = 2 \times 50 = 100 \ \Omega$$

The current passing through the person (hand-to-hand) is

$$I_{man} = \frac{V_{th}}{R_{th} + R_{man}} = \frac{18.4}{100 + 1000} = 12.14 \text{ mA}$$

The current passing through the person is at the let-go level. This situation can be lethal if the current passing through ground is high such as in the case of faults. If the fault current is just 1000 A, the current passing through the person will be around 1.2 A.

To equalize the potential of the two posts, they should be bonded electrically by a conductor, as shown in Figure 12.85. However, the bonding alone is not a solution for the hand-to-feet shock, as shown in the next example.

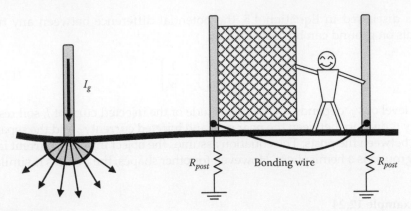

FIGURE 12.85
Equalizing gate voltage.

Example 12.25

Use the data in Example 12.24 and compute the hand-to-feet current of the man. Assume the person is 2 m away from the post.

Solution:

Figure 12.86 is the diagram of the problem at hand. The potential between the post and the feet of the person is

$$V_{th} = \frac{\rho I_g}{2\pi}\left[\frac{1}{r_a} - \frac{1}{r_b}\right] = \frac{500 \times 10}{2\pi}\left[\frac{1}{10} - \frac{1}{10+2}\right] = 13.27 \text{ V}$$

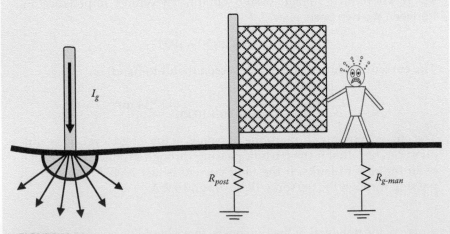

FIGURE 12.86
Gate hand-to-feet shock.

The ground resistance of the standing person as discussed in Equation 4.13 is

$$R_{g\text{-}man} = 1.5\rho = 1.5 \times 500 = 750 \ \Omega$$

Thevenin's impedance R_{th} between the post and the feet of the person is

$$R_{th} = R_{post} + R_{g\text{-}man} = 50 + 750 = 800 \ \Omega$$

The current passing through the person (hand-to-feet) is

$$I_{man} = \frac{V_{th}}{R_{th} + R_{man}} = \frac{13.27}{800 + 1000} = 7.4 \text{ mA}$$

Although the current might be tolerable, it could be lethal during faults. For 1000 A fault current, $I_{man} = 740$ mA.

Because hand-to-feet shock could be severe, ground grid under the gate sweep area is used. The grid should be bonded to the posts, as shown in Figure 12.87. Figure 12.88 shows posts that are bonded to the ground grid and a standing step inside a substation that is also bonded to the substation ground grid.

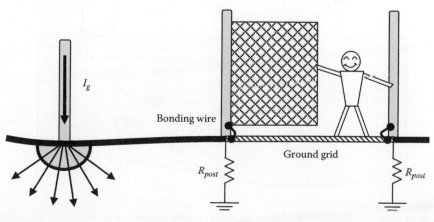

FIGURE 12.87
Equalizing gate and ground voltages.

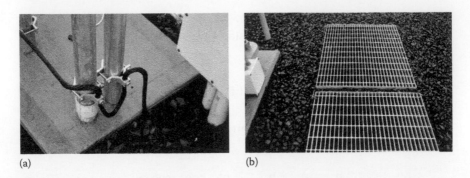

FIGURE 12.88
(a) Bonding gate to ground grid and (b) grounded standing platform.

12.8.2.2 Effect of Electric Coupling

When a poorly grounded metal fence runs in parallel with a transmission line, it acquires electric charge and its voltage is elevated. In Chapter 6, the induced voltage on de-energized metal conductor in parallel with transmission line is analyzed. Here, we can use the same procedure to compute the induced voltage on the fence assuming its rail is cylindrical, as shown in Figure 12.89.

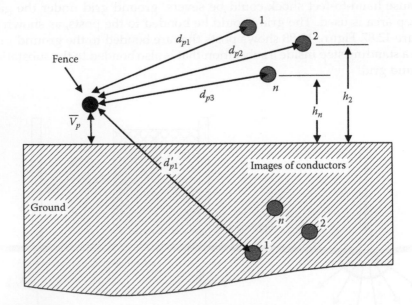

FIGURE 12.89
Induced voltage on fence due to electric field from nearby transmission line.

The capacitance matrix of this system is given in Equation 6.52 and is repeated here.

$$
\mathbf{C} = 2\pi\varepsilon
\begin{bmatrix}
\ln\left(\dfrac{2h_1}{r_1}\right) & \ln\left(\dfrac{d'_{12}}{d_{12}}\right) & \cdots & \ln\left(\dfrac{d'_{1p}}{d_{1p}}\right) \\[2ex]
\ln\left(\dfrac{d'_{21}}{d_{21}}\right) & \ln\left(\dfrac{2h_2}{r_2}\right) & \cdots & \ln\left(\dfrac{d'_{2p}}{d_{2p}}\right) \\[2ex]
\vdots & \vdots & \ddots & \vdots \\[2ex]
\ln\left(\dfrac{d'_{n1}}{d_{n1}}\right) & \ln\left(\dfrac{d'_{n2}}{d_{n2}}\right) & & \ln\left(\dfrac{2h_p}{r_p}\right)
\end{bmatrix}^{-1}
\tag{12.45}
$$

Equation 12.45 can be written as

$$
\mathbf{C} =
\begin{bmatrix}
C_{11} & C_{12} & \cdots & C_{1p} \\
C_{21} & C_{22} & \cdots & C_{2p} \\
\vdots & \vdots & \ddots & \vdots \\
C_{n1} & C_{n2} & & C_{pp}
\end{bmatrix}; \quad \text{where } C_{ij} = C_{ji}
\tag{12.46}
$$

The diagram that describes Equation 12.46 is given in Figure 12.90. The capacitance C_{pp} between the fence and the ground stores electric energy E, which that is equal to

$$
E = \frac{1}{2} C_{pp} V_p^2
\tag{12.47}
$$

where
 V_p is the induced voltage on the fence computed by the method described
  in Chapter 7
 C_{pp} is the capacitance between the fence and the ground

If someone touches this fence as shown in Figure 12.91, the energy stored in all capacitances coupling the fence will dissipate to ground through his body. Since the distance between the rail of the fence and the earth is much smaller than the distance between the rail and the conductors, C_{pp} is much larger than all other coupling capacitances. Hence, the initial shock will be mainly due to C_{pp}, which is in parallel with the person. To assess

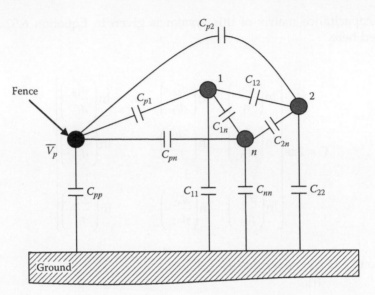

FIGURE 12.90
Capacitance of *n*-phase system with nearby fence.

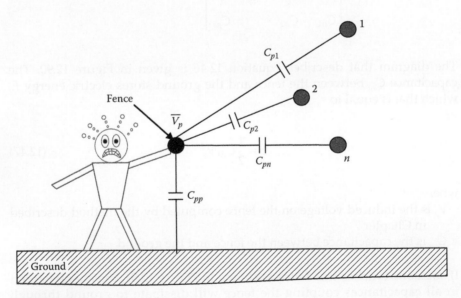

FIGURE 12.91
A person touching a fence.

the hazard of this electric shock, we need to use the Dalziel formula given in Chapter 3:

$$I_{vf} = \frac{K}{\sqrt{t}}$$ (12.48)

where
I_{vf} is the ventricular fibrillation threshold current in mA
t is the duration of exposure in seconds
K is a constant that depends on the weight of the person; K^2 is known as the energy constant (the unit of K^2 is A² s)

Equation 12.48 can be written for the current in amperes:

$$I_{vf}^2 t = 10^{-6} K^2$$ (12.49)

Multiply each side of the equation by the resistance of the person R_{man}:

$$I_{vf}^2 R_{man} t = R_{man} K^2 10^{-6}$$ (12.50)

The term on the left of Equation 12.50 represents the energy that can cause ventricular fibrillation E_{vf}:

$$E_{vf} = R_{man} K^2 10^{-6}$$ (12.51)

If the energy in the capacitor C_{pp} is equal or larger than E_{vf}, the shock is hazardous. The safe condition is when

$$\frac{1}{2} C_{pp} V_p^2 < E_{vf}$$ (12.52)

Example 12.26

Assume a 3 km fence parallel to the power line has an induced voltage of 2 kV (peak value). The capacitance of the fence with respect to ground is 1.0 nf/m. If a child touches the fence, compute the energy through his body.

Solution:

From Equation 12.47, we can compute the stored energy in C_{pp}:

$$E = \frac{1}{2} C_{pp} V_p^2 = \frac{1}{2} \left(10^{-9} \times 3000 \right) \times 4 \times 10^6 = 6.0 \text{ Ws}$$

Equation 12.50 gives the maximum limit of the energy through the child. As discussed in Chapter 3, K for a child is 61. Assume the body resistance is 1.0 kΩ:

$$E_{vf} = R_{body} \, K^2 \, 10^{-6} = 10^3 (61)^2 \, 10^{-6} = 3.72 \text{ W s}$$

Hence, $E > E_{vf}$.

This is as unsafe condition; the child will be exposed to ventricular fibrillation.

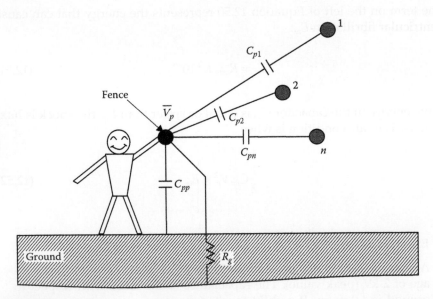

FIGURE 12.92
A person touching a grounded fence.

If the condition in Equation 12.52 cannot be met, ground rods or other grounding techniques should be implemented to bring the voltage of the fence to lower values. The fence should be grounded at several locations by several ground rods, where the total ground resistance is R_g, as shown in Figure 12.92. Because the voltage of the fence is reduced, the stored energy in C_{pp} is reduced as well. A 50% reduction in the induced voltage will result in 75% reduction in the stored energy.

12.8.3 Street Structures

Street structures such as manholes, light poles, traffic signals, bus shelters, phone kiosks, and access hatches may have elevated voltages on their metal parts. There are several reasons for this unintended voltage, including equipment aging, insulation failure, loose fittings, poor workmanship, or ground current.

Consider the case of a street light in Figure 12.93. The light pole is assumed conductive and the insulator of the internal circuit has partially failed causing stray current I_{stray} to go through the pole to the ground. The pole is grounded through $R_{g\,pole}$. If a person touches the conductive pole while standing on ground,

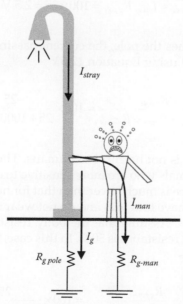

FIGURE 12.93
Electric shock in light poles.

the stray current will split into two parts—one goes to the tower ground I_g and the other through the person I_{man}, where

$$I_{man} = I_{stray} \frac{R_{g\,pole}}{R_{g\,pole} + R_{man} + R_{g\text{-}man}} \qquad (12.53)$$

where
  R_{man} is the body resistance of the man
  $R_{g\text{-}man}$ is the ground resistance of the man's feet

Example 12.27

Assume the stray current passing through the light pole in Figure 12.93 is 100 mA, the pole ground resistance is 25 Ω, the body resistance of the man is 1 kΩ, and his ground resistance is 100 Ω. Compute the GPR of the pole and the current passing through the man.

Solution:

The ground potential rise of the pole is the potential across the pole's ground resistance:

$$V_{pole} = I_{stray}\,R_{g\,pole} = 100 \times 25 = 2.5\ \text{V}$$

When a person touches the pole, the current passing through the person can be computed using Equation 12.53:

$$I_{man} = I_{stray} \frac{R_{g\,pole}}{R_{g\,pole} + R_{man} + R_{g\text{-}man}} = 100 \times \frac{25}{25 + 1000 + 100} = 2.2\ \text{mA}$$

This level of current is not harmful to humans. The scenario could be deadly for small animals who are more sensitive to electric current and whose body resistance is much lower than that for humans. In addition, since small animals have four legs and do not wear shoes, their ground resistance is also low. Assume that the body resistance of a puppy is 200 Ω and its ground resistance is 50 Ω. In this case, the current passing through the puppy is

$$I_{puppy} = I_{stray} \frac{R_{g\,pole}}{R_{g\,pole} + R_{puppy} + R_{g\,puppy}} = 100 \times \frac{25}{25 + 200 + 50} = 9.1\ \text{mA}$$

This level of current could be deadly for small animals.

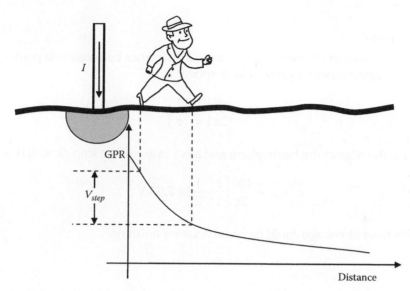

FIGURE 12.94
Step potential.

The simple solution to this problem is to install GFCI in every light pole. As seen in Section 12.1.5, the GFCI continuously check for current to ground and interrupt the circuit once a ground current is detected. Although simple, most utilities unfortunately do not use GFCI in their light poles. Another solution is to bond the pole to an EGC with good ground resistance.

If a structure is injecting current into the ground, it could elevate the step potential. The ground potential rise GPR at the injection point is at its maximum value. Moving away, the GPR is substantially reduced. If a person is walking near the point of injection as shown in Figure 12.94, he is placed at the high gradient region of the GPR curve. The step potential (potential between front and rear legs) would be high. Moving away from the point of injection, the step potential is substantially reduced.

Example 12.28

An object injects 1 A into the ground. The object ground has a hemisphere of 2 m in diameter inserted into soil of 100 Ω-m resistivity. Compute the step potential near the point of injection.

Solution:

As discussed in Chapter 5, the potential difference between two points on the ground near a ground hemisphere is

$$V_{ab} = \frac{\rho I}{2\pi} \left[\frac{1}{a} - \frac{1}{b} \right]$$

If a is the edge of the hemisphere and b is 1 m away, the step potential is

$$V_{step} = \frac{100}{2\pi} \left[\frac{1}{1} - \frac{1}{2} \right] = 7.97 \text{ V}$$

This level of voltage could be hazardous for animals.

12.9 Neutral Deterioration

The bonding of the EGC, neutral wire, and local ground at the service panel can make the system unsafe if the neutral wire between the secondary of the transformer and the load is damaged or deteriorated. Consider the case in Figure 12.95 where the neutral wire between the service transformer and the service panel is broken. In this case, the current loop is from the source to the load, to the service panel through the neutral wire of the facility, to the ground resistance of the panel, to the ground resistance of the transformer, and back to the source. This current can be high in magnitude if the ground resistances are low.

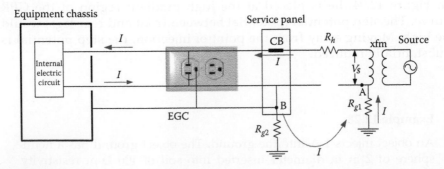

FIGURE 12.95
Ground current when neutral wire is broken.

Example 12.29

For the circuit in Figure 12.95, the load is 10 Ω resistive, the ground resistances $R_{g1} = 14$ Ω, $R_{g2} = 30$ Ω, $R_h = 1.0$ Ω, and the source voltage is 120 V. Calculate the voltage of the EGC.

Solution:

The current in the circuit with broken neutral is

$$I = \frac{V}{R_{load} + R_h + R_{g1} + R_{g2}} = \frac{120}{10 + 1 + 14 + 30} = 2.18 \text{ A}$$

The voltage of the EGC is

$$V = I R_{g2} = 2.18 \times 30 = 65.4 \text{ V}$$

This is a lethal voltage.

If a person touches the chassis of the equipment when the neutral wire is broken as shown in Figure 12.96, the current at the service panel (point B) is divided into two components, I_g and I_{man}, where

$$I_{man} = I \frac{R_{g2}}{R_{g2} + R_{man} + R_{g-man}} \tag{12.54}$$

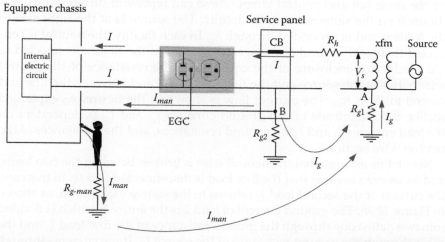

FIGURE 12.96
Hazard due to broken neutral conductor.

Example 12.30

Assume a person touches the chassis of the equipment in the previous example, compute the current passing through the person assuming his body resistance is 1 kΩ and his ground resistance is 100 Ω.

Solution:

The EGC voltage computed in the previous step is also Thevenin's voltage of the circuit between the chassis and the reference ground. Thevenin's impedance R_{th} between the chassis and remote ground is R_{g2} in parallel with the series combination $R_{g1} + R_h + R_{load}$:

$$R_{th} = \frac{R_{g2}\left(R_{g1}+R_h+R_{load}\right)}{R_{load}+R_{g1}+R_h+R_{g2}} = \frac{30(14+1+10)}{10+14+1+30} = 13.63\ \Omega$$

The current passing through the person is

$$I_{man} = \frac{V}{R_{man}+R_{g\text{-}man}+R_{th}} = \frac{65.4}{1000+100+13.63} = 58.7\ \text{mA}$$

This is a lethal level of current.

Broken or deteriorated neutral can be hazardous in local as well as remote areas connected to the same feeder. Take, for example, the system in Figure 12.97. The figure shows a system with two loads being powered by the same hot and neutral wires. These can represent different facilities (houses) on the same secondary circuits. The source is at the right side of the feeders and is grounded through R_g. In each facility, the neutral is connected to a local ground at the service panel and the EGC is connected to the conductive enclosure of the equipment. The resistance of the neutral wire between the source and the first load is R_{n1}, and between the first and second loads is R_{n2}. The current flow is shown in the figure. As explained earlier, the magnitude of the ground currents I_{g1} and I_{g2} is dependent on the load currents I_1 and I_2, all ground resistances, and the resistances of the neutral wire sections.

Now let us assume that the neutral wire is broken between the two loads, and let us even assume that the first load is disconnected ($I_1 = 0$). In this case, the current of the second load I_2 returns to the source via ground, as shown in Figure 12.98. The ground current of load 2 is the entire I_2 which is divided into two paths: one through the ground resistance of the first load I_{s1} and the other through the ground resistance of the source I_{s2}. If the currents through the ground resistances are high, the EGC voltage could be elevated to lethal levels. This scenario is analyzed in the next example.

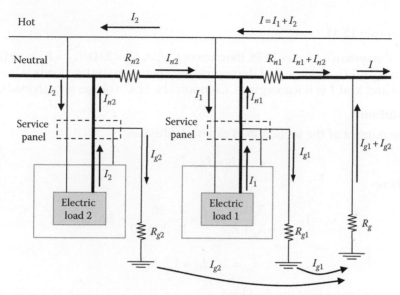

FIGURE 12.97
Ground current in system with bonded neutral and ground wires.

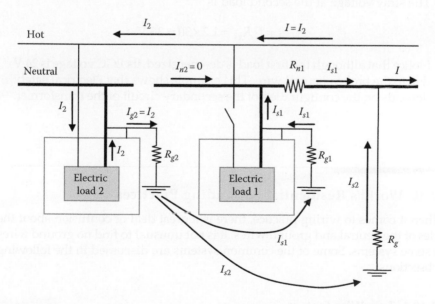

FIGURE 12.98
Ground current in system with bonded ground and neutral wires and with broken neutral wire.

Example 12.31

For the system in Figure 12.98, the current $I_2 = 2A$, $R_g = 20\,\Omega$, $R_{g1} = R_{g2} = 30\,\Omega$, and $R_{n1} = 1\,\Omega$. Assume that the neutral wire between the two loads is broken and load 1 is not energized. Compute the EGC voltage at both loads.

Solution:

The current of the second load can be written as

$$I_2 = I_{s1} + I_{s2}$$

where

$$I_{s1} = I_2 \frac{R_g}{R_{n1} + R_g + R_{g1}} = 2 \times \frac{20}{1 + 20 + 30} \approx 800\ \text{mA}$$

$$I_{s2} = I_2 - I_{s1} = 1.2\ \text{A}$$

The EGC voltage at the first load is

$$V_{EGC1} = I_{s1} R_{g1} = 0.8 \times 30 = 24\ \text{V}$$

The stray voltage at the second load is

$$V_{EGC2} = I_2 R_{g2} = 1.2 \times 30 = 36\ \text{V}$$

Notice that although the first load is de-energized, its EGC voltage is 24 V due to the broken neutral wire. This example shows that electrical safety depends on the configuration of the secondary circuit of the transformer.

12.10 World's Residential Grounding Practices

When it comes to wiring practice, there is a great deal of confusion about the roles of the neutral and ground wires. It is not unusual to find no ground wires in some systems. Some of the common systems are discussed in the following subsections.

12.10.1 Two-Wire System

The system with the worst safety configuration is the two-wire shown in Figure 12.99. In this system, no ground wire is available. This system is discussed earlier in Figure 12.5.

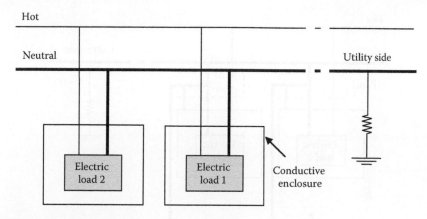

FIGURE 12.99
System without ground wires.

12.10.2 Two-Wire Bonded System

The two-wire bonded system is shown in Figure 12.100, the conductive enclosure is bonded to the neutral wire. As we have seen in Figure 12.8, this system is not safe.

12.10.3 Two-Wire EGC System

In the two-wire EGC system in Figure 12.101, the neutral is grounded at the service panel. This system is the one used in the United States. Sometimes, utilities elect to use water pipes as local grounds, as seen in Figure 12.102. This is quite common in areas where water pipes are made of copper or iron alloy and are buried in soils with low resistivity. This system is safe as long as the neutral integrity is maintained.

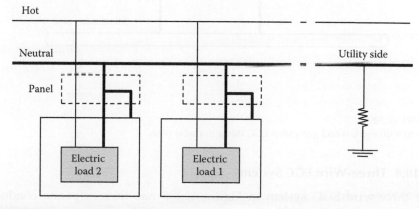

FIGURE 12.100
System with the neutral bonded to chassis.

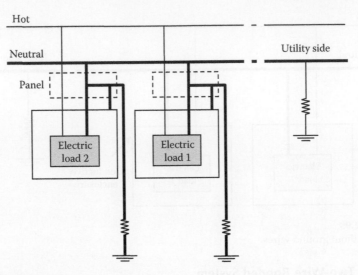

FIGURE 12.101
System with neutral and grounded EGC.

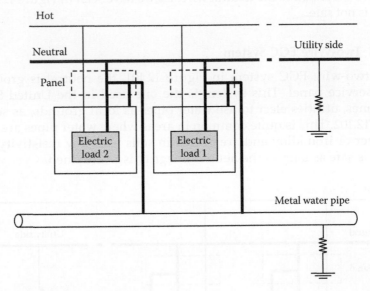

FIGURE 12.102
System with neutral and grounded EGC through metal pipe.

12.10.4 Three-Wire EGC System

The three-wire EGC system in Figure 12.103 has three separate conductors coming from the utility: hot, neutral, and EGC. The neutral and EGC wires are separated at the service panels but bonded at the secondary of

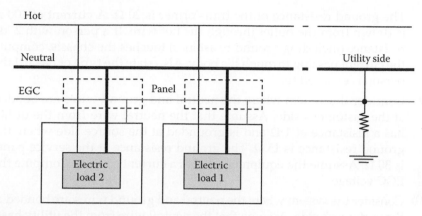

FIGURE 12.103
System with separated neutral and ground wires.

the transformer. This system is the safest but requires additional conductor. Most utilities eschew this system as it is more expensive.

Exercises

12.1 What are the disadvantages of the circuit in Figure 12.6?

12.2 Why is the neutral grounded at the service panel?

12.3 Why is the neutral of the primary and secondary of some distribution transformers bonded?

12.4 For the system in Figure 12.28, the resistance of the high-voltage conductor is 0.3 Ω/km. Assume that $R_{g1} = R_{g2} = 20\ \Omega$ and the substation grounding $R_g = 5\ \Omega$. The service transformer is rated at 7.96 kV/120–240 V and is 5 km away from the substation. When one of the 120 V loads draws 10 A, compute the voltage of the EGC.

12.5 Define macroshock.

12.6 Define microshock.

12.7 An electric circuit is powered by unpolarized 120 V, 60 Hz outlet. The circuit is inside an ungrounded metallic chassis. A 100 nf capacitance exists between the circuit and the chassis. If a man touches the chassis, compute the current passing through his body. Assume that the body resistance of the man plus his ground resistance is 1000 Ω.

12.8 An electric circuit is powered by a two-pronged polarized outlet through a feeder. The resistance of the neutral wire is 0.2 Ω. The chassis of the circuit is metallic and is connected to the neutral terminal of the outlet. The neutral is grounded only at the service transformer.

The ground resistance of the transformer is 20 Ω. A current of 200 A is drawn from the outlet through the hot wire. If a person with 2 kΩ resistance (including ground resistance) touches the chassis, compute the current passing through his body. Also state the type of hazard the person is exposed to.

12.9 Consider the system where the neutral and ground wires are bonded at the customer's side. Assume that the neutral wire from the utility has a resistance of 1 Ω and is grounded at the source side where the ground resistance is 15 Ω. The ground resistance at the service panel is 30 Ω. Assume the equipment draws a current of 10 A. Compute the EGC voltage.

12.10 Consider the system where the neutral and ground wires are bonded at the customer's side. Assume that the neutral wire from the utility has a resistance of 1 Ω and is grounded at the source side where the ground resistance is 15 Ω. The ground resistance at the service panel is 30 Ω. Assume the equipment draws a current of 15 A. Assume a cow is touching the equipment in the previous example and that the body resistance of the cow plus the ground resistance of the surface it is standing on is 300 Ω. Compute the stray current passing through the cow and its effect.

12.11 For the 120 V system in Figure 12.43, assume no isolation transformer is used. Compute the current passing through a patient touching one of the hot wires while in contact with the metallic frame of the medical equipment. Assume the patient resistance is 1 kΩ.

12.12 Name some techniques that can reduce or eliminate the stray voltage problem.

12.13 Name some techniques that can reduce or eliminate the stray voltage problem in hospitals.

12.14 For the system in Figure 12.41, assume $V_2 = 120$ V, $C_1 = C_2 = 1$ nf, $R_{g1} = 20$ Ω. Compute the microshock voltage.

12.15 Why do double-bushing transformers reduce stray voltage?

12.16 What is a voltage ramp? How can it be constructed?

12.17 Can reducing the ground resistance of the neutral wire eliminate stray voltage?

12.18 Stray voltage can be substantially reduced by installing isolation transformers between the service area and the distribution transformer. Can we install the isolation transformer at the customer premises? Why?

12.19 For the rail system in Figure 12.80, compute the leakage current and the voltage behind the train. Assume the train draws 500 A, the rail ground resistance is 100 kΩ, and the rail resistance from the train to the TPS is 0.5 Ω.

12.20 What are the options to reduce the shock hazards of railroad track?

12.21 For the system in Figure 12.84, a lightning hits a lightning pole with hemisphere ground. The lightning injects 10 kA into the ground. A metal gate post is 100 m away from the center of the hemisphere. Compute the GPR at the post. Assume the soil resistivity is 500 Ω-m.

12.22 A power line insulator partially fails and 10 A passes through the structure to the tower's ground. The tower's ground is a hemisphere with a radius of 0.5 m. The soil resistivity is 100 Ω-m. Assume that a man touches the tower while standing on the ground. Compute the current passing through the man, assuming his body resistance is 1 kΩ.

12.23 During a weather storm, an atmospheric discharge hits a lightning pole. The pole is grounded through a hemisphere. The maximum value of the lightning current passing through the pole is 10 kA. The soil of the area is moist. A man who is walking 20 m away from the center of the hemisphere experiences an excessive step potential. The man's body resistance is 1500 Ω. Assume that the step of the person is about 0.6 m. Compute the current passing through his legs and his step potential.

12.24 During a weather storm, an atmospheric discharge hits a lightning pole that is grounded through a hemisphere. The maximum value of the lightning current passing through the rod is 20 kA. The soil of the area is moist. A man is playing golf 50 m away from the center of the hemisphere. At the moment of the lightning strike, the distance between his two feet is 0.4 m. Compute the current passing through the person assuming that the resistance between his legs is 1500 Ω.

12.25 Repeat the previous exercise and assume the person is 5 m away from the center of the hemisphere. What is the effect of the proximity of the man to the grounding hemisphere?

12.26 For the system in Figure 12.100, assume the neutral wire is broken at the utility side. Compute the voltage at both loads assuming the load currents are $I_1 = 2$ A and $I_2 = 4$ A. The system has $R_g = 20\ \Omega$, $R_{g1} = R_{g2} = 30\ \Omega$, and $R_{n1} = R_{n2} = 1\ \Omega$. Also compute the ground potential rise at the utility side due to the two load currents.

12.27 What is the advantage of bonding the neutral and ground wires at the service panel?

12.28 What are the advantages and disadvantages of using water pipes as ground path?

13

Electric Safety under Power Lines

Transmission lines often pass near residential and commercial areas as well as over road crossings and parking lots, as shown in Figure 13.1. These energized transmission lines induce voltages on any conductive surface under or near them. If these surfaces are solidly grounded, their acquired charges from the power lines are continuously dissipated into ground and their potentials are low. However, if they are isolated or poorly grounded, they can accumulate enough charges to elevate their potentials to uncomfortable or even hazardous levels. In severe cases, it could reach levels that produce secondary shocks to any person touching the surface while standing on a grounded object. The magnitude of this induced voltage depends on several factors such as the voltage level of the energized lines, the proximity of the surface to the line, and the humidity in the air.

The induced voltage is due to the presence of electric and magnetic fields. The electric field is mainly due to the voltage of the energized line, and the magnetic field is due to the current carried by the conductor. Because the voltage of transmission lines is very high and the current is low, the induced voltage due to electric field is much higher than the one due to magnetic field. Therefore, we shall consider the electric field only in this chapter.

13.1 Electric Field Calculation

The electric fields surrounding transmission lines are produced by the electrical charges on the energized conductors. The electric field strength is directly proportional to the voltages of the lines and is inversely proportional to the distance from the energized conductors. The unit of the electric field strength is volts per meter (V/m). Since the voltage of the transmission line is fairly constant, the electric field around a transmission line is practically unchanged.

FIGURE 13.1
Power lines passing above a residential area.

The calculations of the electric field strength are made assuming balanced voltage and under two conditions:

- The highest operating voltage of the transmission line.
- The minimum clearance of the conductor to earth. This is the maximum sag at the highest temperature in summer or the worst icing condition in winter.

As seen in Chapter 6, any energized conductor induces voltage on any nearby conductive (metallic) object. If the conductive object is directly connected to ground, the acquired charges from the energized lines are dissipated into the ground. However, if the metallic objects are isolated from the ground, such as automobiles with rubber tires, the acquired charges from the energized line may not be totally dissipated into the ground. The remaining charges elevate the voltage of their bodies, which in most cases causes no physical harm to anyone touching the structure while standing on the ground. However, for large structures with extensive metallic frames or even long trucks, the induced voltages could be hazardous.

The induced voltage for objects with irregular shapes can be easily measured but is very difficult to compute. Alternatively, utilities use the electric field strength E as an indication of the severity of the induced voltage on conductive structure near power lines. The simulation of the

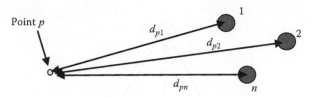

FIGURE 13.2
Electric field strength at a point in space.

electric field strength is a powerful method for issues such as right-of-way assessment and height of towers.

For the arrangement in Figure 13.2, the electric field strength E_p at point p in space is the phasor sum of the electric field strengths due to the charges in the three energized conductors. Hence

$$\bar{E}_p = \frac{\bar{q}_1}{2\pi\varepsilon_0 \, d_{p1}} + \frac{\bar{q}_2}{2\pi\varepsilon_0 \, d_{p2}} + \cdots + \frac{\bar{q}_n}{2\pi\varepsilon_0 \, d_{pn}} \tag{13.1}$$

We can generalize Equation 13.1 as

$$\bar{E}_p = \frac{1}{2\pi\varepsilon_0} \sum_{i=1}^{n} \frac{\bar{q}_i}{d_{pi}} \tag{13.2}$$

where d_{pi} is the distance between point p and the center of conductor i.

Using Equation 13.2, the electric field strength can be computed at any point under or in the vicinity of power lines. This electric field strength at the arbitrary point p is a function of several factors; the most dominant ones are

- Voltage of the transmission line. The electric field strength increases with the increase in voltage (more charge is more voltage).
- Distance from the energized conductor. Shorter distance results in stronger electric field.
- Humidity that could provide a low-resistance path for the charge to reach the object. It could also provide a low-resistance path for the accumulated charges on the object to dissipate into ground.
- Three-phase arrangement where the electric field strength from the three phases is less than the electric field strength from just one or two phases. This is because the total electric field strength is the phasor sum of the electric fields due to the charges of each phase.
- The proximity of the three conductors to each other can reduce the electric field. In three-phase cables, the electric field strength outside the cable is very small when the voltages of the phases are nearly balanced.

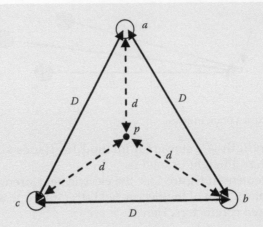

FIGURE 13.3
Three-phase conductors arranged as equilateral triangle.

Example 13.1

Assume a balanced three-phase system whose conductors are arranged as equilateral triangle, as shown in Figure 13.3. Compute the electric field at point p located at the orthocenter of the triangle. Assume that the charges on the conductors are balanced.

Solution:

The orthocenter of a triangle is a point inside the triangle where the distances to all corners are equal. Rewrite Equation 13.2 for three phases:

$$\bar{E}_p = \sum_{i=1}^{n} \frac{\bar{q}_i}{2\pi\varepsilon_0\, d_{pi}} = \frac{\bar{q}_a}{2\pi\varepsilon_0\, d} + \frac{\bar{q}_b}{2\pi\varepsilon_0\, d} + \frac{\bar{q}_c}{2\pi\varepsilon_0\, d} = \frac{\bar{q}_a + \bar{q}_b + \bar{q}_c}{2\pi\varepsilon_0\, d}$$

Since the system is balanced, the charges are balanced. Thus, the phasor sum of the charges is zero:

$$\bar{q}_a + \bar{q}_b + \bar{q}_c = 0$$

Hence, the electric field strength at the orthocenter is zero.

Another scenario where the electric field strength is diminished near high-voltage lines is given in Example 13.2. In this case, the three phases are close to each other.

Example 13.2

Compute the electric field strength at 5 m from a three-phase cable whose voltage is balanced.

Solution:

Equation 13.2 can be written as

$$\bar{E}_p = \sum_{i=1}^{n} \frac{\bar{q}_i}{2\pi\varepsilon_0 d_{pi}} = \frac{\bar{q}_a}{2\pi\varepsilon_0 d_{pa}} + \frac{\bar{q}_b}{2\pi\varepsilon_0 d_{pb}} + \frac{\bar{q}_c}{2\pi\varepsilon_0 d_{pc}}$$

For cable, we can assume that the distances between any conductor and point p 5 m away are equal:

$$d_{pa} \approx d_{pb} \approx d_{pc} = d = 5$$

Since the voltages of the conductors are balanced, the charges on the conductors are also balanced. Hence

$$\bar{q}_a + \bar{q}_b + \bar{q}_c = 0$$

Hence

$$\bar{E}_p = \frac{\bar{q}_a}{2\pi\varepsilon_0 d_{pa}} + \frac{\bar{q}_b}{2\pi\varepsilon_0 d_{pb}} + \frac{\bar{q}_c}{2\pi\varepsilon_0 d_{pc}} = \frac{\bar{q}_a + \bar{q}_b + \bar{q}_c}{2\pi\varepsilon_0 d} = \frac{0}{2\pi\varepsilon_0 d} = 0$$

13.2 Electric Field near Objects

Structures can deform the electric field and change the field strength. This depends on the shape of the structure, its material, and connectivity to ground. An accurate computation of the electric field in this case is hard to obtain. Instead, the electric field is computed assuming the structure is not present. Then, the calculated value is multiplied by a factor.

Assume a structure is in the vicinity of a power line, as shown in Figure 13.4. To compute the electric field strength at the top right corner of the structure, we can use Equation 13.2. Since the spacing between the conductors in high-voltage towers is wide, we cannot assume that the distances between the structure and the conductors are equal.

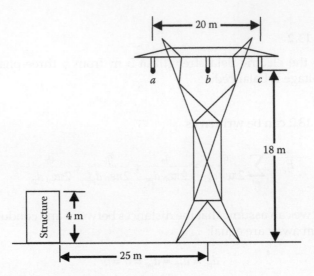

FIGURE 13.4
Electric field strength at nearby structure.

The first step is to compute the charges on the conductors, as described in Chapter 6.

$$
\begin{bmatrix} \bar{q}_a \\ \bar{q}_b \\ \bar{q}_c \end{bmatrix} = 2\pi\varepsilon_0 \begin{bmatrix} \ln\left(\dfrac{2h_a}{r_a}\right) & \ln\left(\dfrac{d'_{ab}}{d_{ab}}\right) & \ln\left(\dfrac{d'_{ac}}{d_{ac}}\right) \\ \ln\left(\dfrac{d'_{ab}}{d_{ab}}\right) & \ln\left(\dfrac{2h_b}{r_b}\right) & \ln\left(\dfrac{d'_{bc}}{d_{bc}}\right) \\ \ln\left(\dfrac{d'_{ac}}{d_{ac}}\right) & \ln\left(\dfrac{d'_{bc}}{d_{bc}}\right) & \ln\left(\dfrac{2h_c}{r_c}\right) \end{bmatrix}^{-1} \begin{bmatrix} \bar{V}_{an} \\ \bar{V}_{bn} \\ \bar{V}_{cn} \end{bmatrix} \tag{13.3}
$$

Then, we can compute the electric field strength using the dimensions in Figure 13.4.

Example 13.3

For the configuration in Figure 13.4, compute the electric field strength at the top right corner of the structure. Assume the transmission line voltage is 500 kV and the GMR of each conductor is 0.4 m.

Solution:

The first step is to compute the charges on the conductors. To do this, we need to compute the various distances between the conductors themselves, and the distances between the conductors and their images.

$$d_{ab} = \sqrt{(x_a - x_b)^2 + (y_a - y_b)^2} = \sqrt{(15 - 25)^2 + (18 - 18)^2} = 10 \text{ m}$$

$$d_{bc} = \sqrt{(x_b - x_c)^2 + (y_b - y_c)^2} = \sqrt{(25 - 35)^2 + (18 - 18)^2} = 10 \text{ m}$$

$$d_{ca} = \sqrt{(x_c - x_a)^2 + (y_c - y_a)^2} = \sqrt{(35 - 15)^2 + (18 - 18)^2} = 20 \text{ m}$$

$$d_{ab}' = \sqrt{(x_a - x_b')^2 + (y_a - y_b')^2} = \sqrt{(15 - 25)^2 + (18 + 18)^2} = 37.36 \text{ m}$$

$$d_{bc}' = \sqrt{(x_b - x_c')^2 + (y_b - y_c')^2} = \sqrt{(25 - 35)^2 + (18 + 18)^2} = 37.36 \text{ m}$$

$$d_{ca}' = \sqrt{(x_c - x_a')^2 + (y_c - y_a')^2} = \sqrt{(35 - 15)^2 + (18 + 18)^2} = 41.18 \text{ m}$$

where x_a' y_a' are the coordinates of the image of conductor a. The direct substitution in Equation 13.3 yields

$$\begin{bmatrix} \bar{q}_a \\ \bar{q}_b \\ \bar{q}_c \end{bmatrix} = 2\pi\varepsilon_0 \begin{bmatrix} \ln\left(\dfrac{36}{0.4}\right) & \ln\left(\dfrac{37.36}{10}\right) & \ln\left(\dfrac{41.18}{20}\right) \\ \ln\left(\dfrac{37.36}{10}\right) & \ln\left(\dfrac{36}{0.4}\right) & \ln\left(\dfrac{37.36}{10}\right) \\ \ln\left(\dfrac{41.18}{20}\right) & \ln\left(\dfrac{37.36}{10}\right) & \ln\left(\dfrac{36}{0.4}\right) \end{bmatrix}^{-1} \begin{bmatrix} \dfrac{500,000\angle 0°}{\sqrt{3}} \\ \dfrac{500,000\angle -120°}{\sqrt{3}} \\ \dfrac{500,000\angle 120°}{\sqrt{3}} \end{bmatrix}$$

$$= \begin{bmatrix} 4.66\angle 7.86° \\ 5.24\angle -120° \\ 4.66\angle 112.14° \end{bmatrix} \mu C$$

The second step is to compute the electric field strength at point p.

$$\bar{E}_p = \frac{\bar{q}_a}{2\pi\varepsilon_0 d_{pa}} + \frac{\bar{q}_b}{2\pi\varepsilon_0 d_{pb}} + \frac{\bar{q}_c}{2\pi\varepsilon_0 d_{pc}}$$

where

$$d_{pa} = \sqrt{(x_p - x_a)^2 + (y_p - y_a)^2} = \sqrt{(0-15)^2 + (4-18)^2} = 20.52 \text{ m}$$

$$d_{pb} = \sqrt{(x_p - x_b)^2 + (y_p - y_b)^2} = \sqrt{(0-25)^2 + (4-18)^2} = 28.65 \text{ m}$$

$$d_{pc} = \sqrt{(x_p - x_c)^2 + (y_p - y_c)^2} = \sqrt{(0-35)^2 + (4-18)^2} = 37.7 \text{ m}$$

$$\bar{E}_p = \frac{1}{2\pi\varepsilon_0} \left(\frac{4.66 \angle 7.86°}{20.52} + \frac{5.24 \angle -120°}{28.65} + \frac{4.66 \angle 112.14°}{37.7} \right) = 1.58 \text{ kV/m}$$

Example 13.4

For the system in Example 13.3, assume the structure is 15 m in height. Compute the electric field strength at the top right corner of the structure.

Solution:

The charges of the energized conductors are computed in Example 13.3 and are the same for this example. In this example, however, we need to compute the new distances considering the new height of the structure.

$$d_{pa} = \sqrt{(x_p - x_a)^2 + (y_p - y_a)^2} = \sqrt{(0-15)^2 + (15-18)^2} = 15.3 \text{ m}$$

$$d_{pb} = \sqrt{(x_p - x_b)^2 + (y_p - y_b)^2} = \sqrt{(0-25)^2 + (15-18)^2} = 25.18 \text{ m}$$

$$d_{pc} = \sqrt{(x_p - x_c)^2 + (y_p - y_c)^2} = \sqrt{(0-35)^2 + (15-18)^2} = 35.13 \text{ m}$$

Hence, the electric field at the top right point of the building is

$$\bar{E}_p = \frac{\bar{q}_a}{2\pi\varepsilon_0 d_{pa}} + \frac{\bar{q}_b}{2\pi\varepsilon_0 d_{pb}} + \frac{\bar{q}_c}{2\pi\varepsilon_0 d_{pc}}$$

$$\bar{E}_p = \frac{1}{2\pi\varepsilon_0} \left(\frac{4.66 \angle 7.86°}{15.3} + \frac{5.24 \angle -120°}{25.18} + \frac{4.66 \angle 112.14°}{35.13} \right) = 2.67 \text{ kV/m}$$

Note that the electric field intensity is expected to be higher for taller buildings.

Example 13.5

To reduce the electric field, a utility decided to arrange the conductors in inversed triangle configuration, as shown in Figure 13.5. Compute the electric field strength at the top right corner of the structure.

Solution:

We need to compute the various distances between the conductors themselves, and the distances between the conductors and their images.

$$d_{ab} = \sqrt{(x_a - x_b)^2 + (y_a - y_b)^2} = \sqrt{(18-25)^2 + (27-18)^2} = 11.4 \text{ m}$$

$$d_{bc} = \sqrt{(x_b - x_c)^2 + (y_b - y_c)^2} = \sqrt{(25-32)^2 + (18-27)^2} = 11.4 \text{ m}$$

$$d_{ca} = \sqrt{(x_c - x_a)^2 + (y_c - y_a)^2} = \sqrt{(32-18)^2 + (27-27)^2} = 14 \text{ m}$$

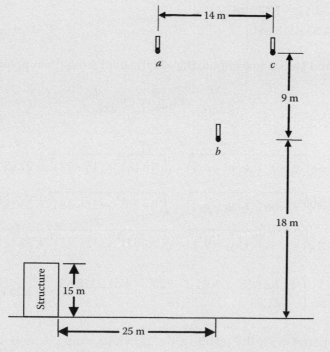

FIGURE 13.5
Conductors arranged in inverse triangle.

$$d'_{ab} = \sqrt{\left(x_a - x'_b\right)^2 + \left(y_a - y'_b\right)^2} = \sqrt{(18-25)^2 + (27+18)^2} = 45.54 \text{ m}$$

$$d'_{bc} = \sqrt{\left(x_b - x'_c\right)^2 + \left(y_b - y'_c\right)^2} = \sqrt{(25-32)^2 + (18+27)^2} = 45.54 \text{ m}$$

$$d'_{ca} = \sqrt{\left(x_c - x'_a\right)^2 + \left(y_c - y'_a\right)^2} = \sqrt{(32-18)^2 + (27+27)^2} = 55.79 \text{ m}$$

The direct substitution in Equation 13.3 yields

$$\begin{bmatrix} \overline{q}_a \\ \overline{q}_b \\ \overline{q}_c \end{bmatrix} = 2\pi\varepsilon_0 \begin{bmatrix} \ln\left(\dfrac{54}{0.4}\right) & \ln\left(\dfrac{45.54}{11.4}\right) & \ln\left(\dfrac{55.79}{14}\right) \\ \ln\left(\dfrac{45.54}{11.4}\right) & \ln\left(\dfrac{36}{0.4}\right) & \ln\left(\dfrac{45.54}{11.4}\right) \\ \ln\left(\dfrac{55.79}{14}\right) & \ln\left(\dfrac{45.54}{11.4}\right) & \ln\left(\dfrac{54}{0.4}\right) \end{bmatrix}^{-1} \begin{bmatrix} \dfrac{500,000\angle 0°}{\sqrt{3}} \\ \dfrac{500,000\angle -120°}{\sqrt{3}} \\ \dfrac{500,000\angle 120°}{\sqrt{3}} \end{bmatrix}$$

$$= \begin{bmatrix} 4.61\angle 1.15° \\ 5.04\angle -120° \\ 4.61\angle 118.85° \end{bmatrix} \mu C$$

The second step is to compute the electric field strength at point p:

$$\overline{E}_p = \frac{\overline{q}_a}{2\pi\varepsilon_0 d_{pa}} + \frac{\overline{q}_b}{2\pi\varepsilon_0 d_{pb}} + \frac{\overline{q}_c}{2\pi\varepsilon_0 d_{pc}}$$

where

$$d_{pa} = \sqrt{\left(x_p - x_a\right)^2 + \left(y_p - y_a\right)^2} = \sqrt{(0-18)^2 + (15-27)^2} = 21.63 \text{ m}$$

$$d_{pb} = \sqrt{\left(x_p - x_b\right)^2 + \left(y_p - y_b\right)^2} = \sqrt{(0-25)^2 + (15-18)^2} = 25.18 \text{ m}$$

$$d_{pc} = \sqrt{\left(x_p - x_c\right)^2 + \left(y_p - y_c\right)^2} = \sqrt{(0-32)^2 + (15-27)^2} = 34.17 \text{ m}$$

$$\overline{E}_p = \frac{1}{2\pi\varepsilon_0}\left(\frac{4.61\angle 1.15°}{21.63} + \frac{5.04\angle -120°}{25.18} + \frac{4.61\angle 118.85°}{34.17}\right) = 1.26 \text{ kV/m}$$

As compared with the result in the previous example, note that the arrangement of the conductor in triangular configuration reduces the electric field strength on nearby structures.

13.3 Electric Field Profile under Power Lines

Any transmission line occupies a corridor of land known as the *right-of-way* (ROW) (see Figure 13.6). This is the area under the power line that extends outward beyond the ends of the tower's arms. The edge of the ROW is the farthest line on the side of the corridor. Utilities use the value of the electric field strength to identify the needed width of the ROW, the height of the tower, the voltage level of the conductors, and whether structures or public access can be allowed within or near the ROW. To make such a determination, the *electric field strength profile* (EFSP) at 1 m above ground is computed using Equation 13.2, where point p is allowed to move from one edge of the ROW to the other. Such EFSP is shown at the bottom of Figure 13.6.

The electric field profile is often measured or computed at the point of maximum slack. This way, the electric field strength is the highest between two towers.

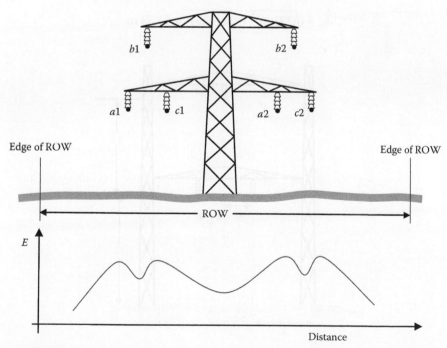

FIGURE 13.6
ROW of transmission line and EFSP.

Example 13.6

A transmission line tower is shown in Figure 13.7. The dimensions in the figure are in meters. The equivalent radius of the conductors is 12 cm. The voltage of the transmission line is 500 kV (line-to-line). Compute the electric field strength at 1 m above ground under phase *a*

Solution:

The first step is to compute the various distances between all conductors using the Pythagorean theorem.

$$d_{ab} = \sqrt{(0-6)^2 + (12-23)^2} = 12.53 \text{ m}$$

$$d_{ac} = \sqrt{(0-12)^2 + (12-12)^2} = 12 \text{ m}$$

$$d_{bc} = \sqrt{(12-6)^2 + (12-23)^2} = 12.53 \text{ m}$$

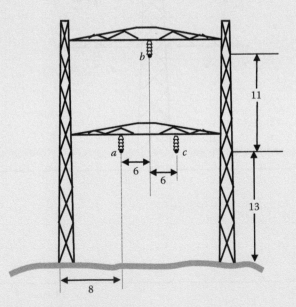

FIGURE 13.7
Configuration of a transmission line.

The next step is to compute the charges of all conductors. These charges are needed to compute the electric field strength in Equation 13.2. The charges can be computed using Equation 13.3:

$$
\begin{bmatrix} \bar{q}_a \\ \bar{q}_b \\ \bar{q}_c \end{bmatrix} = 2\pi\varepsilon_0 \begin{bmatrix} \ln\left(\dfrac{24}{0.12}\right) & \ln\left(\dfrac{23}{12.53}\right) & \ln\left(\dfrac{12}{12}\right) \\ \ln\left(\dfrac{12}{12.53}\right) & \ln\left(\dfrac{46}{0.12}\right) & \ln\left(\dfrac{12}{12.35}\right) \\ \ln\left(\dfrac{12}{12}\right) & \ln\left(\dfrac{23}{12.35}\right) & \ln\left(\dfrac{24}{0.12}\right) \end{bmatrix}^{-1} \begin{bmatrix} \dfrac{500,000\,\angle 0^\circ}{\sqrt{3}} \\ \dfrac{500,000\,\angle -120^\circ}{\sqrt{3}} \\ \dfrac{500,000\,\angle 120^\circ}{\sqrt{3}} \end{bmatrix}
$$

The solution of the aforementioned equations yields the charges of the lines.

$$
\begin{bmatrix} \bar{q}_a \\ \bar{q}_b \\ \bar{q}_c \end{bmatrix} = \begin{bmatrix} 3.62\angle 1.4^\circ \\ 3.36\angle -120^\circ \\ 3.62\angle 118.6^\circ \end{bmatrix} \mu C/m
$$

Notice that the charges are not exactly balanced even when the voltage is balanced. This is due to the unequal distances between the conductors as well as their unequal heights.

We can now use Equation 13.2 to compute the electric field strength at 1 m above ground and under phase *a*:

$$
\bar{E}_p = \frac{\bar{q}_a}{2\pi\varepsilon_0\, d_{pa}} + \frac{\bar{q}_b}{2\pi\varepsilon_0\, d_{pb}} + \frac{\bar{q}_c}{2\pi\varepsilon_0\, d_{pc}}
$$

$$
\bar{E}_p = \frac{10^{-6}}{2\pi\times 8.85\times 10^{-12}}\left(\frac{3.62\angle 1.4^\circ}{12} + \frac{3.36\angle -120^\circ}{\sqrt{6^2+23^2}} + \frac{3.62\angle 118.6^\circ}{\sqrt{12^2+12^2}}\right)
$$

$$
E_p = 2.662\,kV/m
$$

13.4 Allowable Limits for Electric Fields

There is no agreed upon standard on the limit of the electric field strength near power lines. Some countries and states have set limits to reduce the induced voltage near power lines; others have set limits to reduce the health effect of prolonged exposure to electric field. In this section, both electric safety limits and health limits are discussed.

13.4.1 Electric Safety Limits

The ranges of the known international guidelines are given in Table 13.1. The guidelines limit public access near or under power lines if the electric field strength exceeds the recommended values.

Note that for commercial parking lots, the limit is lower than that for small vehicles. This is because commercial trucks and buses tend to be longer and higher than small passengers cars. Therefore, they acquire more charge and have more induced voltage on their body. Under the midspan of a 230 and 500 kV transmission line, the expected electric field strengths are about 2 and 7 kV/m, respectively. This is more than enough to illuminate a hand-held fluorescent tube.

In the United States, a loose guideline is established where 5 kV/m is for public exposure and 10 kV/m for occupational exposure. Moreover, several states have developed their own standards (examples given in Table 13.2).

TABLE 13.1

Range for Maximum Electric Field Strength

Area	Maximum Electric Field Strength E_{max} (kV/m)
Residential areas	0.5–2.5
Within ROW	7–12
Edge of ROW	1–3
Road crossings	3–8
Shopping centers and small vehicle parking lots	3–5
Commercial parking lots	1–3

TABLE 13.2

State Standards for Electric Field Strength

State		Maximum Electric Field Strength	
		Within ROW (kV/m)	Edge of ROW (kV/m)
Florida	69–230 kV line	8.0	2.0
	500 kV line	10.0	
Minnesota		8.0	
Montana		7.0	1.0
New Jersey			3.0
New York		11.8	1.6
	Private road crossings	11.0	
	Highway crossings	7.0	
Oregon		9.0	

TABLE 13.3

Guidelines for Electric Field Strength Set by the ICNIRP for 50/60 Hz

	Exposure	Electric Field (kV/m)
Occupational	Whole working day	10
	Short term	30
General public	Up to 24 h per day	5
	Few hours per day	10

13.4.2 Health Limits

From the health point of view, several international organizations have set guidelines for electric field strength limits, including the International Commission on Non-Ionizing Radiation Protection (ICNIRP) and the American Conference of Governmental Industrial Hygienists (ACGIH). Table 13.3 summarizes the set limits by the ICNIRP for occupational exposure as well as general public exposure. Notice that the duration of the exposure is a factor in the guideline to address the health issues that might arise from exposure to electric fields.

The ACGIH has set a limit of 25 kV/m for occupational exposure. But they require workers to use protective gear such as conductive suits and gloves when the electric field strength is higher than 15 kV/m. For workers with cardiac pacemakers, the exposure must be at or below 1 kV/m.

13.5 Minimum Vertical Clearance Methods

The National Electrical Safety Code (NESC) has adopted the minimum vertical clearance (MVC) approach to calculate the safe distance from power lines. Rule 232 of the 2007 edition of the NESC is entitled *"Vertical clearances of wires, conductors, cables, and equipment above ground, roadway, rail, or water surfaces."* The NESC specifies two methods of MVC calculations in lieu of electric field calculations.

In the MVC calculations, the NESC requires that we consider the conductor temperature and loading conditions that produce the largest midspan sag. In this case, 50°C (120°F) should be used unless the maximum conductor temperature for which the line is designed to operate at is greater than 50°C (120°F). In any case, the highest temperature must be used.

13.5.1 Method 1: *MVC* for Systems with Unknown Switching Surges

NESC established the *MVC* for power lines with unknown switching transients (voltage rise due to the switching actions of the line). For phase voltages ranging from 750 V to 22 kV, the basic minimum vertical clearance MVC_0 is selected from Table 13.4.

The MVC_0 in Table 13.4 can be modified for voltages above 22 kV (phase to ground) and for heights above sea level.

1. For phase voltages between 22 and 470 kV, MVC_0 specified in Table 13.4 is increased at the rate of 10 mm/kV in excess of 22 kV.

2. For voltages exceeding 50 kV, the additional clearance specified in the previous step is increased by 3% for each 300 m in excess of 1000 m above the mean sea level.

We can generalize the MVC of method 1 as

$$MVC_1 = (MVC_0 + C_1(V-22) \times 0.01) \times \left[1 + C_2 \times 0.03 \times \text{Int}\left(\frac{h-1000}{300} \right) \right] \quad (13.4)$$

where

MVC_1 is the modified minimum vertical clearance

MVC_0 is the minimum vertical clearance obtained in Table 13.4

C_1 is equal to zero unless the phase voltage is greater than 22 kV; in this case, C_1 is equal to one

C_2 is zero unless the phase voltage is above 50 kV and the height of the tower is more than 1000 m above the average sea level; in this case, C_2 is one

Int is the integer value of the component $\left(\dfrac{h-1000}{300} \right)$

h is the height of the transmission line above sea level

TABLE 13.4

MVC_0 for Power Lines Operating between 750 V and 22 kV

Nature of Surface Underneath Power Line	MVC_0 (m)
Track rails of railroad (except electrified rails)	8.1
Roads and streets subject to truck traffic	5.6
Driveways, parking lots, and alleys	5.6
Cultivated, grazing, forest, and orchard areas	5.6
Pedestrian and restricted traffic areas	4.4
Water areas without sailboats	5.2

Source: Part of Table 232-1 of NESC, 2007.

Example 13.7

Compute the minimum vertical clearance for a transmission line whose line-to-line voltage is 500 kV and is located 1500 m above the sea level in a forested area.

Solution:

For forested area, MVC_0 is 5.6 m, as given in Table 13.4. The transmission line voltage is

$$V = \frac{500}{\sqrt{3}} = 288.68 \text{ kV}$$

By directly substituting in Equation 13.4, we obtain the MVC:

$$MVC_1 = \left(MVC_0 + C_1(V - 22) \times 0.01\right) \times \left[1 + C_2 \times 0.03 \times \text{Int}\left(\frac{h - 1000}{300}\right)\right]$$

$$MVC_1 = \left(5.6 + (288.68 - 22) \times 0.01\right) \times \left[1 + 0.03 \times \text{Int}\left(\frac{1500 - 1000}{300}\right)\right]$$

$$MVC_1 = \left(5.6 + (288.68 - 22) \times 0.01\right) \times [1 + 0.03 \times 1]$$

$$MVC_1 = 8.51 \text{ m}$$

13.5.2 Method 2: *MVC* for Systems with Voltage Exceeding 98 kV and with Known Switching Surges

Method 2 is based on selecting a reference height that is later modified according to the switching surge of the line. The reference height RH is selected from NESC Table 232-3, which is given in Table 13.5.

TABLE 13.5

RH for Power Lines Operating above 98 V

Nature of Surface Underneath Power Line	RH (m)
Track rails of railroad (except electrified rails)	6.7
Roads and streets subject to truck traffic	4.3
Driveways, parking lots, and alleys	4.3
Cultivated, grazing, forest, and orchard areas	4.3
Pedestrian and restricted traffic areas	3.0
Water areas without sailboats	3.8

Source: Part of Table 232-3 of NESC, 2007.

This reference height is modified by adding a component called "electrical component," which is an empirical formula:

$$EC = \left[\frac{V_{max} \times PU \times a}{500 \times k} \right]^{1.667} b \times c \tag{13.5}$$

where
  EC is the electrical component
  V is the maximum voltage in kV (phase-ground)
  PU is the maximum switching surge factor
  a = 1.15 to allow for 3 standard deviations
  b = 1.03 to allow for nonstandard atmospheric conditions
  c = 1.2 to provide 20% margin of safety
  k = 1.15, which is the configuration factor of conductors

Substituting all these variables in Equation 13.5 yields

$$EC = 1.236 \times \left[\frac{V_{max} \times PU}{500} \right]^{1.667} \tag{13.6}$$

The minimum vertical clearance MVC_x in this calculation is the sum of RH and EC

$$MVC_x = RH + EC \tag{13.7}$$

$$MVC_x = RH + 1.236 \times \left[\frac{V_{max} \times PU}{500} \right]^{1.667} \tag{13.8}$$

If the transmission line is 450 m above the average sea level, MVC_x is increased by 3% for each 300 m in excess of 450 m above the mean sea level. In this case, the general minimum vertical clearance by method 2 MVC_2 can be computed by

$$MVC_2 = MVC_x \times \left[1 + C_2 \times 0.03 \times \text{Int} \left(\frac{h - 450}{300} \right) \right] \tag{13.9}$$

where C_2 is zero for transmission lines at less than 450 m above the average sea level, otherwise it is one.

Example 13.8

For the system in Example 13.7, compute the minimum vertical clearance assuming that the surge factor is 2.0.

Solution:

For forest area, RH is 4.3 m, as given in Table 13.5. The maximum voltage of transmission line voltage is

$$V_{max} = \frac{\sqrt{2} \times 500}{\sqrt{3}} = 408.25 \text{ kV}$$

By directly substituting in Equation 13.8, we obtain the MVC_x

$$MVC_x = RH + 1.236 \times \left[\frac{V_{max} \times PU}{500} \right]^{1.667}$$

$$MVC_x = 4.3 + 1.236 \times \left[\frac{408.25 \times 2.0}{500} \right]^{1.667} = 7.1 \text{ m}$$

Hence

$$MVC_2 = MVC_x \times \left[1 + C_2 \times 0.03 \times \text{Int} \left(\frac{h - 450}{300} \right) \right]$$

$$MVC_2 = 7.1 \times \left[1 + 0.03 \times \text{Int} \left(\frac{1500 - 450}{300} \right) \right]$$

$$MVC_2 = 7.1 \times [1 + 0.03 \times 3] = 7.74 \text{ m}$$

Notice that MVC_2 is smaller than MVC_1 calculated in Example 13.7. The NESC states that clearances specified by Method 1 may be reduced for circuits with known switching surge factors (PU), but shall not be less than the one calculated by Method 2.

13.6 Measurement of Electric Field Strength

A quick way to measure the electric field on a surface is to use a high-sensitivity voltmeter, as shown in Figure 13.8. Keep the two probes of the voltmeter at a fixed, small distance. Place one of the probes on the surface at a point where you want to find the electric field. Rotate the other probe

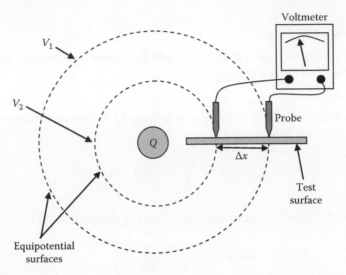

FIGURE 13.8
Simple method to measure electric field strength.

in a circular motion around the fixed probe and observe the reading of the
voltmeter. The point at which the reading is at maximum is recorded. The
direction of the electric field is the direction in which the maximum reading
is found. The magnitude of the electric field is the value of the maximum
potential difference divided by the distance between the two probes:

$$E = \frac{\Delta V}{\Delta x} \qquad (13.10)$$

where
  E is the electric field strength
  ΔV is the maximum reading of the voltmeter
  Δx is the distance between the two probes

  The direction of the electric field strength is from the point of higher potential to the one of lower potential.
  This simple method is suitable for lab experiments but is inaccurate and
cannot be used for industrial application. This is because of the following:

- The test surface must be perpendicular to the equipotential surfaces;
  otherwise, the value of Δx will be wrong.
- For more than one source of electric field, the equipotential surfaces
  are often distorted, so the measurement in one location cannot represent the electric field strength in other locations. Also, if other

objects are nearby, including humans or grounded objects, the electric field will be distorted.

• The shape of the electrodes can change the reading.

In IEEE Standard 1308, 1994, it is recommended that the electric field strength be measured by one of the following three methods:

1. Free-body meter
2. Ground-reference meter
3. Electro-optic meter

While measuring the electric field strength by these meters, the observer must be sufficiently far from the probes to avoid distorting the electric field lines. Moreover, the measurement for power lines must be made at 1 m above the ground.

13.6.1 Free-Body Meter

There are two types of free-body meters: dipole and isotropic. The dipole consists of one probe and the isotropic consists of three probes aligned along the three axes in space. In either type, the probe consists of two electrodes that are isolated from ground (thus called free-body) and are mounted on isolating rod to reduce electric field distortion.

13.6.1.1 Dipole Free-Body Meter

The probe of the free-body meter consists of two conducting electrodes separated by isolation material, as shown in Figure 13.9. The dipole probe requires

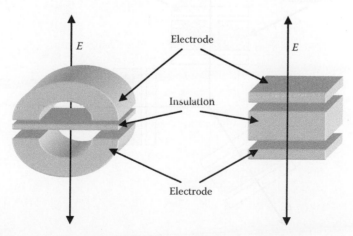

FIGURE 13.9
Dipole probes used to measure electric field strength.

the alignment of the electrodes perpendicular to the electric field lines to produce the most accurate results. The operation of the dipole free-body meter is discussed in Section 12.5. The relationship between the current of its circuit to the electric field strength is given in Equation 12.35.

13.6.1.2 Isotropic Free-Body Meter

The dipole probe requires the alignment of the electrodes perpendicular to the electric field line. This can be achieved by rotating the electrodes along the three axes (x, y, and z) to find the alignment that produces the maximum result.

An alternative method is to use the isotropic probe that consists of three dipoles placed along the three-axes, as shown in Figure 13.10. Each of the three probes measures the electric field along its own axis. The final result is the phasor sum of all probes:

$$E = \sqrt{E_x^2 + E_y^2 + E_z^2} \tag{13.11}$$

13.6.2 Ground-Reference Meter

The ground-reference meter is basically similar to the free-body meter. The main difference is that the electrodes are made of a thin, flat surface. In some designs, it uses only one electrode, and any grounded surface can be used as the second electrode. The ground-reference meter is used to measure the electric field at ground level or on flat conducting surfaces that are at ground

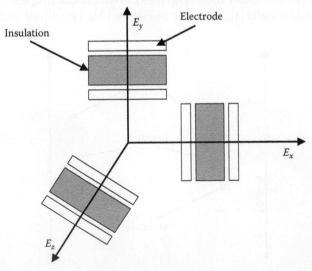

FIGURE 13.10
Isotropic probe used to measure electric field strength.

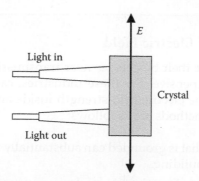

FIGURE 13.11
Electro-optic probe used to measure electric field strength.

potential. The operating principal of the ground-reference meter is similar to that of the free-body meter.

13.6.3 Electro-Optic Meter

The electro-optic meter shown in Figure 13.11 uses the Pockels effect to measure electric field strength. The Pockels effect, which is also known as the electro-optic effect, is a phenomenon that causes light passing through certain types of material to split into two rays known as *ordinary* rays (perpendicular polarizations) and *extraordinary* rays (parallel polarizations). Each of these rays has a different speed. The difference in speed depends on two factors: the type of material and the electric field surrounding the material. Materials such as calcite crystals, lithium niobate, gallium arsenide, or boron nitride can exhibit this characteristic. Since the birefringence is proportional to the electric field, it makes a good sensing device for the electric field strength.

Birefringence, or double refraction, creates two different refractive indices. The refractive index is a measure of how much the speed of light is reduced inside a given material. A refractive index of 2 means that light travels at half its air speed. The refractive index is then defined by

$$\Delta n = n_e - n_o \tag{13.12}$$

where
n_e is the speed of the extraordinary rays
n_o is the speed of the ordinary rays
Δn is the refractive index of the rays

The electro-optic probe can be used as in free-body or ground-reference meters. Lights are channeled through fiber-optic cables. Because fiber optics are electrically isolated from the ground, electro-optic meters are unaffected by the ground potential.

13.7 Mitigation of Electric Field

Industries often locate their businesses near transmission lines where land cost is relatively inexpensive. For these industries, care must be taken to reduce the effect of the electric field strength inside and outside the buildings. Some effective methods are as follows:

- Metallic siding that is grounded can substantially reduce the electric field inside the building.
- Placing a grounded conductive enclosure around the work space. The enclosure can be as simple as a grounded wire mesh screens (Faraday cage).
- Building material with reinforcement steel reduces the electric field inside buildings by creating a form of Faraday cage.
- Constructing a few grounded wires between the source of electric field and the object can cut the electric field reaching the object.
- All metallic structures must be grounded to eliminate the touch potential hazard.
- Large moving vehicles must be grounded. Grounding vehicles with insulated tires is often done by dangling chains from the vehicle to the ground surface. These chains make the potential of the vehicle body equal to the ground potential so the touch potential hazard is reduced.

Exercises

13.1 A balanced three-phase system has its conductors arranged as an equilateral triangle. The distance between any two conductors is 10 m, and the magnitude of the charge in each conductor is 3 µC/m. Compute the electric field at point p located at the orthocenter of the triangle. If one of the conductors is disconnected, compute the electric field strength at point p. Can you draw a conclusion?

13.2 For the transmission line tower in Figure 13.7, compute the electric field strength at the ground level under phase c. The radius of the conductors is 6 cm. The voltage of the transmission line is 230 kV (line-to-line).

13.3 A utility is planning to construct a 138 kV line that would pass through a pedestrian area that is 500 m above sea level. Compute the minimum vertical clearance.

13.4 A utility is planning to construct a 138 kV line that would pass through a pedestrian area that is 1500 m above sea level. The system study indicates that the switching surge factor would be 1.5. Compute the minimum vertical clearance.

13.5 State some measures to reduce the electric field near power lines.

13.6 Write a computer program to compute the profile of the electric field strength at the ground level for two adjacent towers, where each tower carries double circuits.

Test your program assuming that each tower has the configuration given in Exercise 13.5. In addition, assume that the distance between the centers of the two towers is 60 m, and each transmission line is energized at 340 kV (line-to-line). Ignore the phase shifts between the same phases in each circuit.

14

Coupling between Power Lines and Pipelines, Railroads, and Telecommunication Cables

Pipelines, railroads, and communication cables (PRC) are nonelectrical utility systems that often share the right-of-way (ROW) with power lines. This is because of several reasons such as the restraints imposed by private and governmental agencies on the routing and the environmental impact of the PRC. These restrictions are more severe near urban areas. The sharing of the ROW raises concerns regarding the induced voltage on the PRC due to mainly three variables:

1. Electric field or capacitive coupling
2. Magnetic field or inductive coupling
3. Ground current

The level of voltage induced on the PRC depends on several factors, including the following:

- Proximity of the PRC to the power line
- Voltage and current levels of the power line
- Duration for which the PRC is in parallel with the power line
- Insulation status of the PRC and its corrosion condition
- Ground resistance of the PRC
- Magnitude of the ground current

The majority of the safety problems are associated with the dielectric coating pipelines and ungrounded rail tracks. This is because the conductive parts of the PRC are isolated from the ground potential and thus retain the charges acquired from power lines. This results in elevated potential of the metallic parts of the PRC.

Besides the induced voltage, if ground current reaches the pipeline, it can corrode it. This is a major problem for underground pipelines as it reduces the withstanding pressure of the pipes and may eventually cause leaks. Although the ground current in most cases is ac and its corrosion power is less than that from dc, the power electronics and nonlinear devices that are commonly used nowadays can introduce a dc component to the ground current.

14.1 Electric Field Coupling

As covered in Chapters 6 and 7, electric and magnetic fields from overhead power lines can induce voltages on nearby conductive structures. For buried pipes or cables, the induced voltage due to electric field is very small since the electric field below the ground surface is insignificant.

For above-ground PRC, the electrical field can induce hazardous voltages that can be computed by the method described in Chapter 6. Consider the system in Figure 14.1 where a pipe above ground is in the vicinity of a three-phase power line. The induced voltage on the pipe due to the electric field is

$$\bar{V}_p = \frac{1}{2\pi\varepsilon_0} \sum_{i=1}^{n} \bar{q}_i \ln\left(\frac{d'_{pi}}{d_{pi}}\right) \tag{14.1}$$

where

V_p is the voltage induced on the pipe
q_i is the charge per unit length of energized conductor i
d_{pi} is the distance between the pipe and the energized conductor i
d'_{pi} is the distance between the pipe and the image of the energized conductor i
ε_0 is air permittivity or absolute permittivity (8.85×10^{-12} F/m)

The cases of insulated track and communication cables are shown in Figures 14.2 and 14.3, respectively. For insulated tracks, we can approximate the track by a circular solid tube whose cross section area is the same as that of the track.

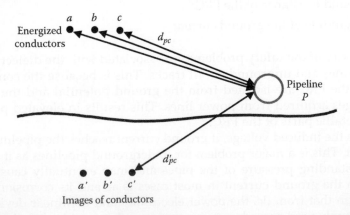

FIGURE 14.1
Insulated pipes above ground near power line.

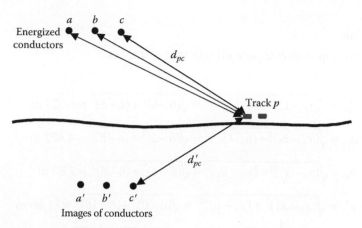

FIGURE 14.2
Insulated tracks near power line.

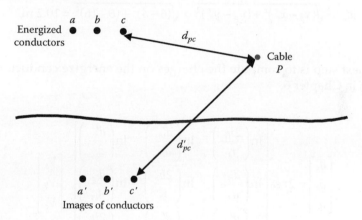

FIGURE 14.3
Elevated communication cable near power line.

Example 14.1

Estimate the induced voltage on the elevated communication cable in Figure 14.3. Assume the cable is conductive and without grounded shield. The voltage of the energized conductor is 15 kV and the conductor radius is 5 cm. The coordinates of the energized conductors and the communication cable are

Phase a = (0, 8 m)
Phase b = (2, 8 m)
Phase c = (4, 8 m)
Communication cable = (6, 6 m)

Solution:

The first step is to compute all relevant distances.

$$d_{pa} = \sqrt{(x_p - x_a)^2 + (y_p - y_a)^2} = \sqrt{(6-0)^2 + (6-8)^2} = 6.324 \text{ m}$$

$$d_{pb} = \sqrt{(x_p - x_b)^2 + (y_p - y_b)^2} = \sqrt{(6-2)^2 + (6-8)^2} = 4.472 \text{ m}$$

$$d_{pc} = \sqrt{(x_p - x_c)^2 + (y_p - y_c)^2} = \sqrt{(6-4)^2 + (6-8)^2} = 2.83 \text{ m}$$

$$d'_{pa} = \sqrt{(x_p - x'_a)^2 + (y_p - y'_a)^2} = \sqrt{(6-0)^2 + (6-16)^2} = 11.66 \text{ m}$$

$$d'_{pb} = \sqrt{(x_p - x'_b)^2 + (y_p - y'_b)^2} = \sqrt{(6-2)^2 + (6-16)^2} = 10.77 \text{ m}$$

$$d'_{pc} = \sqrt{(x_p - x'_c)^2 + (y_p - y'_c)^2} = \sqrt{(6-4)^2 + (6-16)^2} = 10.2 \text{ m}$$

The next step is to compute the charges on the energize conductors as given in Chapter 6.

$$\begin{bmatrix} \bar{q}_a \\ \bar{q}_b \\ \bar{q}_c \end{bmatrix} = 2\pi\varepsilon_0 \begin{bmatrix} \ln\left(\dfrac{2h_a}{r_a}\right) & \ln\left(\dfrac{d'_{ab}}{d_{ab}}\right) & \ln\left(\dfrac{d'_{ac}}{d_{ac}}\right) \\ \ln\left(\dfrac{d'_{ab}}{d_{ab}}\right) & \ln\left(\dfrac{2h_b}{r_b}\right) & \ln\left(\dfrac{d'_{bc}}{d_{bc}}\right) \\ \ln\left(\dfrac{d'_{ac}}{d_{ac}}\right) & \ln\left(\dfrac{d'_{bc}}{d_{bc}}\right) & \ln\left(\dfrac{2h_c}{r_c}\right) \end{bmatrix}^{-1} \begin{bmatrix} \bar{V}_a \\ \bar{V}_b \\ \bar{V}_c \end{bmatrix}$$

$$\begin{bmatrix} \bar{q}_a \\ \bar{q}_b \\ \bar{q}_c \end{bmatrix} = \begin{bmatrix} 0.1206 \angle 7.35° \\ 0.1364 \angle -120° \\ 0.1206 \angle 112.65° \end{bmatrix} \mu C/m$$

The last step is to compute the induced voltage on the communication cable using Equation 14.1.

$$\bar{V}_p = \frac{1}{2\pi\varepsilon_0} \left[\bar{q}_a \ln\left(\frac{d'_{pa}}{d_{pa}}\right) + \bar{q}_b \ln\left(\frac{d'_{pb}}{d_{pb}}\right) + \bar{q}_c \ln\left(\frac{d'_{pc}}{d_{pc}}\right) \right] = 1.314 \angle 133° \text{ kV}$$

14.2 Magnetic Field Coupling

Magnetic fields can also induce voltage on PRC. This is especially true if the power line carries heavy current and is close to the PRC.

The induced voltage on a PRC can be computed by the method explained in Chapter 7.

$$\bar{E}_p = 2 \times 10^{-7} \omega \left[\ln\left(\frac{D_a}{d_{pa}}\right) \quad \ln\left(\frac{D_b}{d_{pb}}\right) \quad \ln\left(\frac{D_c}{d_{pc}}\right) \right] \begin{bmatrix} I_a \angle(90° - \theta_a) \\ I_b \angle(90° - \theta_b) \\ I_c \angle(90° - \theta_c) \end{bmatrix} \tag{14.2}$$

where

E_p is the induced voltage on PRC per meter length. The length here is the distance by which the power line is in parallel with the PRC

D_a is the distance between the energized conductor a and a point on the ground surface under (or above) the PRC

d_{pa} is the distance between conductor a and the PRC

I_a is the current of phase a

θ_a is the phase angle of I_a

ω is the angular frequency (377 for 60 Hz system)

The magnitude of the induced voltage on a PRC depends on several factors such as the following:

- *The separation between the conductors and the PRC*: If the separation between the conductors and the PRC is wide, the denominator in the logarithms in Equation 14.2 is large. Thus, the induced voltage is reduced.

- *The length of PRC that parallels the power line*: The induced voltage increases when the length of the parallel section between the power line and the PRC increases. This is because the total induced voltage E_p in Equation 14.2 is multiplied by the length of the parallel section.

- *The magnitude of the current of the power lines*: The induced voltage is proportional to the current of the energized power line. This problem is especially hazardous when the line experiences surges due to faults, switching transients, or lighting strikes.

- *Three-phase system*: The three-phase system, especially when balanced, will induce the smallest amount of voltage on the PRC. This is because the magnetic fields of the balanced three-phase currents can cancel each other. Unbalanced, single-phase, or double-phase conductors induce higher voltage on the PRC.

Example 14.2

Estimate the induced voltage due to magnetic field between the two ends of the pipeline placed 3 m above ground. The pipe length is 10 km and has a dielectric coating. Assume the distribution line is a three-phase balanced circuit of 1000 A. The coordinates of the energized conductors and the centroid of the pipe are

Phase a = (0, 10 m)
Phase b = (4, 10 m)
Phase c = (8, 10 m)
Pipe = (10, 3 m)

Solution:

The first step is to compute all relevant distances.

$$d_{pa} = \sqrt{(x_p - x_a)^2 + (y_p - y_a)^2} = \sqrt{(10-0)^2 + (3-10)^2} = 12.2 \text{ m}$$

$$d_{pb} = \sqrt{(x_p - x_b)^2 + (y_p - y_b)^2} = \sqrt{(10-4)^2 + (3-10)^2} = 9.22 \text{ m}$$

$$d_{pc} = \sqrt{(x_p - x_c)^2 + (y_p - y_c)^2} = \sqrt{(10-8)^2 + (3-10)^2} = 7.28 \text{ m}$$

$$D_a = \sqrt{(x_p - x_a)^2 + (0 - y_a)^2} = \sqrt{(10-0)^2 + 100} = 14.14 \text{ m}$$

$$D_b = \sqrt{(x_p - x_b)^2 + (0 - y_b)^2} = \sqrt{(10-4)^2 + 100} = 11.66 \text{ m}$$

$$D_c = \sqrt{(x_p - x_c)^2 + (0 - y_c)^2} = \sqrt{(10-8)^2 + 100} = 10.2 \text{ m}$$

The next step is to compute the induced voltage on the pipe using Equation 14.2.

$$\bar{E}_p = 2 \times 10^{-7} \omega \left[\ln\left(\frac{D_a}{d_{pa}}\right) \quad \ln\left(\frac{D_b}{d_{pb}}\right) \quad \ln\left(\frac{D_c}{d_{pc}}\right) \right] \begin{bmatrix} I_a \angle(90° - \theta_a) \\ I_b \angle(90° - \theta_b) \\ I_c \angle(90° - \theta_c) \end{bmatrix}$$

$$\bar{E}_p = 2 \times 10^{-7} \times 377 \left[\ln\left(\frac{14.14}{12.2}\right) \quad \ln\left(\frac{11.66}{9.22}\right) \quad \ln\left(\frac{10.2}{7.28}\right) \right] \begin{bmatrix} 1000 \\ 1000 \angle -120° \\ 1000 \angle 120° \end{bmatrix}$$

$$E_p = 0.0124 \text{ V/m}$$

The induced voltage on the pipe is

$$V_{pipe} = E_p\, l = 0.0124 \times 10,000 = 124 \text{ V}$$

Example 14.3

Estimate the induced voltage due to magnetic field between the two ends of the pipeline placed 3 m below the ground surface. The pipe length is 10 km and has a dielectric coating. Assume the distribution line is a three-phase balanced circuit of 1000 A. The coordinates of the energized conductors and the centroid of the pipe are

Phase a = (0, 10 m)
Phase b = (4, 10 m)
Phase c = (8, 10 m)
Pipe = (10, −3 m)

Solution:

The first step is to compute all relevant distances.

$$d_{pa} = \sqrt{(x_p - x_a)^2 + (y_p - y_a)^2} = \sqrt{(10-0)^2 + (-3-10)^2} = 16.4 \text{ m}$$

$$d_{pb} = \sqrt{(x_p - x_b)^2 + (y_p - y_b)^2} = \sqrt{(10-4)^2 + (-3-10)^2} = 14.32 \text{ m}$$

$$d_{pc} = \sqrt{(x_p - x_c)^2 + (y_p - y_c)^2} = \sqrt{(10-8)^2 + (-3-10)^2} = 13.15 \text{ m}$$

$$D_a = \sqrt{(x_p - x_a)^2 + (0 - y_a)^2} = \sqrt{(10-0)^2 + 100} = 14.14 \text{ m}$$

$$D_b = \sqrt{(x_p - x_b)^2 + (0 - y_b)^2} = \sqrt{(10-4)^2 + 100} = 11.66 \text{ m}$$

$$D_c = \sqrt{(x_p - x_c)^2 + (0 - y_c)^2} = \sqrt{(10-8)^2 + 100} = 10.2 \text{ m}$$

The next step is to compute the induced voltage on the pipe using Equation 14.2.

$$\bar{E}_p = 2 \times 10^{-7} \omega \left[\ln\left(\frac{D_a}{d_{pa}}\right) \quad \ln\left(\frac{D_b}{d_{pb}}\right) \quad \ln\left(\frac{D_c}{d_{pc}}\right) \right] \begin{bmatrix} I_a \angle (90° - \theta_a) \\ I_b \angle (90° - \theta_b) \\ I_c \angle (90° - \theta_c) \end{bmatrix}$$

$$\bar{E}_p = 2\times10^{-7}\times377\left[\ln\left(\frac{14.14}{16.4}\right) \quad \ln\left(\frac{11.66}{14.32}\right) \quad \ln\left(\frac{10.2}{13.15}\right)\right]\begin{bmatrix} 1000 \\ 1000\angle-120° \\ 1000\angle120° \end{bmatrix}$$

$$E_p = 0.0069 \text{ V/m}$$

The induced voltage on the pipe is

$$V_{pipe} = E_p l = 0.0069\times10000 = 69 \text{ V}$$

14.2.1 Electrically Continuous Underground Pipeline

Pipelines are made of sections; each section is normally about 5–7 m in length. If the pipe is bare, the potential of the pipe is the same as that for the local ground. However, pipes are often coated with insulation material to reduce corrosions. The induced voltage on these coated pipes is mainly due to magnetic field as the electric field does not penetrate ground.

The joints between pipe sections are either made of electrically conductive material (electrically continuous pipes) or nonconductive material (electrically discontinuous pipes). The magnitude of the induced voltage between the two ends of a pipeline depends on the type of joint used.

Figure 14.4 shows an electrically continuous pipeline. Because the joints are electrically conductive, the total voltage across the pipeline is

$$V_{pipe} = n\times v_s \tag{14.3}$$

where
   n is the number of pipe sections in parallel with the power line
   v_s is the induced voltage in one pipe section
   V_{pipe} is the end-to-end voltage of the pipe

The induced voltage per unit length can be computed using Equation 14.2. In this case, the total induced voltage between the two ends of the parallel pipes is the induced voltage per unit length multiplied by the length of the parallel section. This case is depicted in Example 14.3. The voltage profile on the pipeline is shown in the bottom part of Figure 14.4. The voltage at any point along the pipeline with respect to the immediate ground is given as a function of distance. The difference between the positive maximum and

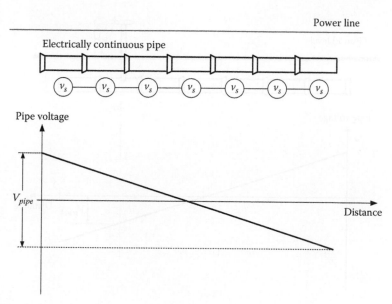

FIGURE 14.4
Electrically continuous pipe.

negative maximum points is the potential difference between the ends of the pipeline. Furthermore, if we assume symmetry, the magnitude of the maximum positive or negative is half V_{pipe}.

The voltage with respect to ground at any point on the pipeline can be represented by a straight line with negative slope:

$$V_x = \frac{-V_{pipe}}{l} x + \frac{V_{pipe}}{2}$$ (14.4)

where
V_{pipe} is the total voltage between the two ends of the pipeline
x is the distance from the beginning of the pipeline
l is the length of the pipeline
V_x is the voltage at point x with respect to the ground

For long pipelines, the induced voltage can be hazardous. Take, for example, the case in Figure 14.5 for a coated pipeline. Assume control equipment such as valve is attached to the inner part of the pipe at a given location along the pipeline. Assume the equipment is also isolated from the ground. If a person standing on the ground touches the uninsulated part of the equipment, the touch potential of the person is V_{touch} in the figure.

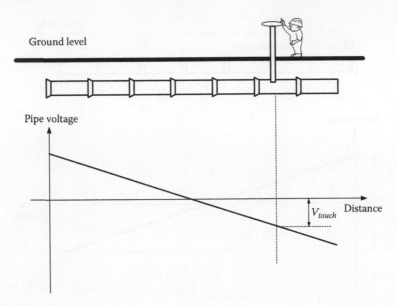

Ground level

Pipe voltage

Distance

V_{touch}

FIGURE 14.5
Hazard of touch potential in pipeline.

Example 14.4

For the system in Example 14.3, compute the touch potential at 3 km from either ends of the pipeline

Solution:

The total voltage of the pipe computed in Example 14.3 is

$$V_{pipe} = 69 \text{ V}$$

At 3 km from one end, the touch voltage is

$$V_{touch} = \frac{-V_{pipe}}{l} x + \frac{V_{pipe}}{2} = \frac{-69}{10} \times 3 + 34.5 = 13.8 \text{ V}$$

At 3 km from the other end

$$V_{touch} = -13.8 \text{ V}$$

Repeat the example and assume one of the phases is disconnected. Can you have a conclusion?

14.2.2 Electrically Discontinuous Underground Pipeline

To reduce the shock hazards, electrically discontinuous joints can be used. In this case, the pipe sections are treated as several short pipes insulated from each other. The induced voltage in this case is much smaller than

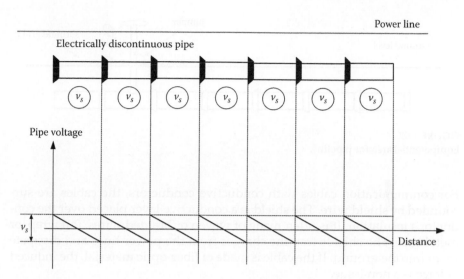

FIGURE 14.6
Electrically discontinuous pipe.

the case with electrically continuous joints, as shown in Figure 14.6. The maximum induced voltage is just the value for one section (v_s) and is often very small.

14.3 Mitigation of Electromagnetic Coupling

If the PRC is underground and insulated, the induced voltage is mainly due to magnetic field. Touching the conductive parts of the PRC often cause secondary shocks. However, in case of long parallel PRC adjacent to high-voltage power lines or adjacent to lines carrying heavy current (or during faults), the shock could be primary. To protect against this hazard, the following methods can be implemented:

- Ground all conductive components to continuously dissipate charges into the ground. The problem with this method is the continuous current that is present in the PRC, which could corrode pipes.

- Cover all PRC's conductive parts by insulation material to avert the hazard of touch potential.

- Isolate the PRC to physically prevent people from touching the conductive parts at the site

- Create equipotential area around exposed components, as shown in Figure 14.7.

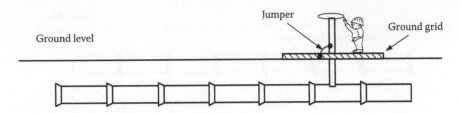

FIGURE 14.7
Equipotential area for pipeline.

For communication cables with conductive conductors, the cables are surrounded by shield wire. The shield is a conductive layer placed over the conductor's insulation. It is often made of woven or braided wires, foil wrap, or metal tube. The shields trap the charges from the power lines and dissipate them into the ground. If the cable is made of fiber-optic material, the induced voltage is a non issue.

For insulated tracks, the public is prevented from accessing the track area. Also, low-voltage surge arresters can be placed on the track to prevent charge buildup, as discussed in Chapter 12.

For coated pipelines, electrically continuous pipes, and equipotential area must be established at all control sites, as shown in Figure 14.7. For electrically discontinuous joints, the induced voltage is often not hazardous.

During the construction of a PRC, safety precautions must be established to protect construction workers from the hazard of electric shocks due to electromagnetic coupling. These are essentially the same process used for de-energized conductors such as isolating the work area, insulating the workers, establishing equipotential zone using temporary grounds, and attaching grounding chains to moving objects or vehicles.

14.4 Ground Current

Pipes fully coated with isolation material are not affected by ground current as long as the insulation is undamaged. Over time, however, the coating can be damaged due to aging, arcing when the induction voltage is high, etc. In this case, the ground current can enter and leave the pipe. Besides the corrosion problems, the current elevates the voltage of the pipe.

For coated and electrically continuous pipes shown in Figure 14.8, the total resistance of the pipeline is

$$R_{total} = n \times R \qquad (14.5)$$

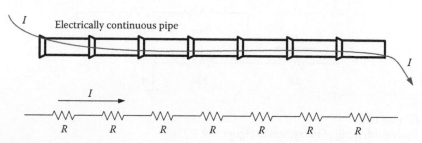

FIGURE 14.8
Electrically continuous pipe with current.

where
  n is the number of pipe sections
  R is the lateral resistance of one pipe section
  R_{total} is the total resistance of the pipeline

If we assume that the current entering the pipeline is I, we can compute the voltage across the pipeline by

$$V_{pipe} = R_{total}\, I \tag{14.6}$$

From the safety point of view, if ground current enters the pipeline, there will be a path for the current to reach a person working on metallic control equipment that is isolated from the ground, as shown in Figure 14.9. The current is branched into two paths: one exits the pipeline (I_g) and the other reaches the person (I_{man}) then to local ground. The hazard is dependent on the amount of current entering the pipe, the downstream length of the pipeline from the worker and until the current leaves the pipeline, the soil resistivity, and the resistance of the pipe sections.

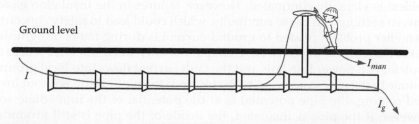

FIGURE 14.9
Hazard due to current in pipeline.

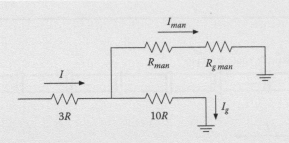

FIGURE 14.10
Equivalent circuit of the system in Figure 14.9.

Example 14.5

Consider the case in Figure 14.9. Assume the resistance of each pipe section is 0.2 Ω. If 10 A current enters the pipeline 3 sections upstream from the worker and exits the pipeline after 10 sections downstream, estimate the current passing through the person.

Solution:

The equivalent circuit of the system in Figure 14.9 is shown in Figure 14.10. Assume the body resistance of the person is 1 kΩ and his ground resistance is 30 Ω.

The current passing through the person can be computed by the current divider equation:

$$I_{man} = I \frac{10R}{10R + R_{man} + R_{g\,man}} = 10 \frac{2}{2 + 1000 + 30} = 19.4 \text{ mA}$$

This level of current is hazardous. To address this problem, the worker needs to be placed in an equipotential area, as shown in Figure 14.7.

If the pipeline is electrically discontinuous as shown in Figure 14.11, the problem is virtually eliminated. However, failures in the insulation gasket between sections can cause continuity, which could lead to safety concerns.

Another problem related to ground current is during faults or lightning strikes that elevate the GPR. An example is shown in Figure 14.12 where an insulator of a power line fails and the fault current flows into local ground. Assume that a pipe is adjacent to the tower. If the pipe is bare without insulated coating, the pipe potential is at the potential of the immediate soil. However, if the pipe is insulated, the inside of the pipe is still grounded somewhere in the system. Because of the ground potential rise, the voltage on the outer layer of the pipe is at the local GPR, which is higher than

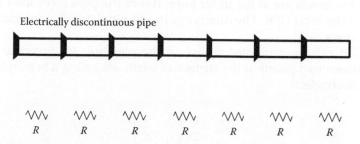

FIGURE 14.11
Electrically discontinuous pipe.

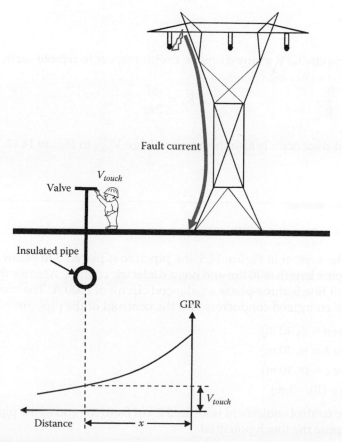

FIGURE 14.12
Ground potential rise.

the potential of the inner part of the pipe. If a person is touching the inner part of the pipe through control equipment while standing on the local ground, his hands are at the inner potential of the pipe (very low) and his feet are at the local GPR. The difference in potential (touch potential) could be hazardous.

The GPR can be computed by using the equation for potential difference between two points at the surface of earth assuming a hemisphere for ground electrode:

$$V_{ab} = \frac{\rho I}{2\pi}\left[\frac{1}{r_a} - \frac{1}{r_b}\right] \tag{14.7}$$

If we set

$$r_a = x$$

$$r_b = \infty$$

we can compute GPR at any distance x with respect to remote earth

$$\text{GPR}_x = \frac{\rho I}{2\pi x} \tag{14.8}$$

The GPR at distance x is also the touch voltage V_{touch} in Figure 14.12.

Exercises

14.1 For the system in Figure 14.5, the pipeline is placed 3 m below ground. The pipe length is 10 km and has a dielectric coating. Assume the distribution line is three-phase a balanced circuit of 1000 A. The coordinates of the energized conductors and the centroid of the pipe are

Phase a = (0, 10 m)

Phase b = (4, 10 m)

Phase c = (8, 10 m)

Pipe = (10, −3 m)

If the control equipment is located 2 km from the end of the pipeline, compute the touch potential.

14.2 In case of long parallel PRC adjacent to high-voltage power line or adjacent to lines carrying heavy current (or during faults), the shock hazard can be averted by several methods. Name them.

14.3 To avert electric shock for pipelines, we can ground all conductive components to continuously dissipate charges into the ground. What is the drawback of this method?

14.4 For the system in Figure 14.12, assume the fault current is 100 A and the pipe is 50 m away from the power line. Assume the soil is moist with 100 Ω-m resistivity. Compute the touch potential across the person.

Index

Printed and bound by CPI Group (UK) Ltd, Croydon, CR0 4YY

18/10/2024

01776263-0003